MANUAL OF METEOROLOGY

MANUAL OF METEOROLOGY

VOLUME I

METEOROLOGY IN HISTORY

BY

SIR NAPIER SHAW, LL.D., Sc.D., F.R.S.

*Late Professor of Meteorology in the Imperial College of Science and
Technology and Reader in Meteorology in the University of London;
Honorary Fellow of Emmanuel College, Cambridge; sometime
Director of the Meteorological Office, London, and President
of the International Meteorological Committee*

WITH THE ASSISTANCE OF

ELAINE AUSTIN

*of the Meteorological Office
formerly of Newnham
College, Cambridge*

CAMBRIDGE
AT THE UNIVERSITY PRESS
MCMXLII

TO

THE MEMORY

OF MY COLLEAGUES ON THE

METEOROLOGICAL COUNCIL

1897–1900

Richard Strachey

Alexander Buchan

George Howard Darwin

Francis Galton

William James Lloyd Wharton

Robert Henry Scott

CAMBRIDGE
UNIVERSITY PRESS

University Printing House, Cambridge CB2 8BS, United Kingdom

Cambridge University Press is part of the University of Cambridge.

It furthers the University's mission by disseminating knowledge in the pursuit of education, learning and research at the highest international levels of excellence.

www.cambridge.org
Information on this title: www.cambridge.org/9781107475465

© Cambridge University Press 1932

First edition 1926
First published 1932
Reprinted 1932, 1942
First paperback edition 2014

A catalogue record for this publication is available from the British Library

ISBN 978-1-107-47546-5 Paperback

PREFACE

THE four volumes which this work is intended to comprise are the expensive embodiment of a personal feeling that, for the community as a whole, there is nothing so extravagantly expensive as ignorance, however cheap it may be for any particular section of it.

The feeling developed into conviction during the war when it became my duty to supply, or alternatively to train, officers for various meteorological services. I was working in an environment which contained within its own experience or on its shelves almost all that there is to know about the weather; yet I had to send responsible officers into the services with a formula by which they could "carry on," in place of the knowledge that would enable them to become a part.

In that respect the war was like a kinema film that is run too fast: one missed what one is accustomed to see and saw things that pass unnoticed in ordinary life.

I realised that an insight into what the study of weather means was at that time and is still a privilege rather wastefully confined to a small minority of a special class, that my work for the science which, for twenty-five years, it was my duty to foster could not be regarded as finished without a definite attempt to rescue from oblivion the vast mass of information about weather that is hidden behind the backs of the books of a meteorological library.

Considering that the mere existence of the human race depends upon its capacity to adjust its behaviour to the exigencies of climate and weather, any ignorance of meteorology which is avoidable is inexcusable. Looking backward, how different the history of the world might have been, indeed must have been, if the conditions which determined the potato-famine in Ireland in 1847 had been known beforehand. Instances of the same kind are numberless and in these days there is hardly anybody so fatalistic as to contend that the anticipation of the condition for a famine is beyond the capacity

of human intelligence. But when we consider the history of the
subject it becomes evident that the real meteorology, directed to
the solution of that kind of problem in the only way in which it is
worth while to approach it, namely the study of the meteorology
of the globe as a whole, has never had a fair chance. It has
always been dependent upon the economics of other objects and
ambitions. Meteorologists have had to be content with such crops
as they could grow in other people's fields.

Even with that limitation the science has achieved enough to
justify a modest hope of ultimate success. The object of this book
therefore is to present the study of meteorology not only as making
use of nearly all the sciences and most of the arts, but also as a world-
study of a special and individual character going back inevitably to
the very dawn of history, and beyond that to the mazes of geologic
time, and still looking forward not unmindful of the advantages
which it may derive from the blazing and sometimes bewildering
light of modern science.

I have no intention of presenting any of the major problems of
meteorology as solved. My aim is as far as possible to put the actual
facts before the reader and let him draw conclusions for himself.
I know from my own experience that drawing conclusions for oneself
is an easy and pleasurable task when the facts are really known:
anybody can take that step; but to supply himself with the facts is
a labour beyond the power of the ordinary reader.

Besides expressing an effort to open up to others the material
that gradually accumulates, the book discharges, so far as they
can be discharged, two other personal obligations. The first is
contained implicitly in a note for the Royal Meteorological Society
on the educational equipment which is required for the advanced
study of meteorology. It was entitled 'Meteorology for Schools
and Colleges,' and it forms the introductory section of the collection
of lectures and essays *The Air and its Ways* which was published
in 1923 in order to " try out " the technical details of publication of
a meteorological treatise. The present volume may be regarded as a
more detailed handbook or guide to the necessary knowledge. In it
I have tried to represent the knowledge which the reader of a paper
on meteorology before a learned society of the present day will assume,
perhaps unconsciously, to be in the possession of his audience.

The other obligation is implied in a presidential address to the Royal Meteorological Society in 1919 which took up the position that, although the organisation of climatological work in this country had passed from the Societies to the Government service, the study of weather still afforded innumerable opportunities for effective scientific activity on the part of the amateur. Activity of that kind has been so characteristic of British science from the seventeenth century until recent years that even so enlightened a person as the late Lord Salisbury could not regard scientific research otherwise than as a personal hobby. Times have changed, most of the sciences have now become so specialised that the amateur is left stranded; and it seems quite likely that the study of weather, which in so far as it is specialised is devitalised, may come into its own again as an attractive subject for amateurs if they can obtain access to a point of view from which they can survey the aims and objects within reach.

These ideas were in my mind when in 1918 the Meteorological Committee concurred in a proposal that I should be relieved from administrative duty and devote my time to the translation into book form of material contained within the office shelves and store-rooms. In addition to the regular official publications, my colleagues and I had already completed *The Seaman's Handbook*, *The Weather Map*, the *Meteorological Glossary*, which ran through several issues, and *The Weather of the British Coasts* which was asked for by the Admiralty for the guidance of inexperienced skippers. An atlas of *Cloud Forms* followed, and this Manual began with the preparation of Part IV as being urgently required at that time for aeronautics. It was published by the Cambridge University Press because, among other reasons, the Manager, the late Mr J. B. Peace, was willing to interest himself in the solution of the many technical difficulties of a work that is so vitally dependent upon the effectiveness of its illustrations.

The final proof of Part IV was handed into the Press on Armistice day. Since that auspicious day many things have happened, including the attachment of the Meteorological Committee to the Air Ministry and the consequential changes, the completion of my own period of office under H.M. Treasury as Director, and Chairman of the Committee, and my appointment as professor in the Imperial

College of Science. Mr Peace's sudden death in 1923 left the work
with a number of loose ends.

The Meteorological Committee have not lost sight of the
arrangement and have done many things to assist in bringing
into book form the scientific material belonging to the Office.
Their most effective help was expressed in the work of Miss E. E.
Austin who was seconded from the Office as my official assistant
during my four years' tenure of the professorship at the Imperial
College, and, through the good offices of Dr Simpson, my successor
as director of the Meteorological Office, was allowed to continue
that assistance after the termination of my official duty.

Although when completed this Manual may have expanded to
twice the size that was originally contemplated in 1918, those readers
who are familiar with the breadth and depth of meteorological literature will probably be more impressed by its many omissions than by
what it contains. My object throughout has been to give not my
own views upon subjects of meteorological interest, though I attach
some importance to them, but to provide the reader with material
upon which he can form his own opinion, or at least indicate to him
where he can find the material upon which opinion should be based.

I have now realised that it is not so much a text-book that is wanted
as an encyclopaedia or dictionary, in ten parts, one part to be brought
up to date each year—a sort of revolving book-case or portable
substitute for a meteorological library of larger dimensions. I wish
it could have been so, but the task of providing it is beyond the
power of my personality and my purse. The book must therefore
be offered as a suggestion, I hope prophetic, of what those interested
in the study of weather may enjoy if there be the will to satisfy them.

I revert again to the difficulty in the preparation of an effective
book on meteorology arising from its dependence upon illustrations.
They must in nearly all cases be borrowed or adapted from published works or manuscript material. It was understood from the
beginning that illustrations belonging to the Office or its publications should be available, indeed one of the purposes in view was
to utilise them for the advantage of a wider public than the regular
readers of official documents. But there are many others. It will
be best to make the acknowledgments in the several volumes. The
list of illustrations in each will give the sources from which the

illustrations have been derived and acknowledgments are appended thereto.

In this volume we are particularly indebted to Mr C. J. P. Cave of Stoner Hill and Mr G. A. Clarke of the Observatory at Aberdeen for a large selection of cloud-studies. We welcome a new kind of illustration in the form of a stereo-photograph of the whole sky which we owe to the kindness of Mr Robin Hill and Mr C. S. Leaf of Cambridge. The sources of other illustrations of clouds are given in a special index of cloud-forms and cloud-groups.

The reader may be surprised to find in the chapter on the upper air a large collection of illustrations of cloud-forms and, contrary to an asserted principle, the author's views upon the subject of the classification of clouds. The reason for the apparent inconsistency is that the latter part of this volume is devoted to an exposition of the means which have been developed in the course of nearly three centuries for the collection of observations of all kinds, in order that in the succeeding volumes the author may deal with results and inferences without having to pause to discuss the methods by which the observations have been obtained. The question of observation of clouds is in the peculiar position of asking What shall be observed? and in that respect differs from the other elements which are referred to in this volume. It is a question which has to be settled before the other parts of the work can be organised.

For the conversion of our own rough sketches into book-illustrations we were dependent until 1923 upon the staff of the Cambridge Press who, under Mr Peace's direction, prepared the outlines for the maps and put into proper form the lines to be superposed thereon. These include the charts of isotherms, dew-point lines, isonephs, isohyets and isobars which illustrate comparative meteorology, the subject of the second volume. For many of the later diagrams we have been fortunate enough to obtain the practised assistance of Miss E. Humphreys who acted as artist on the staff of the Meteorological Office for press illustrations in the years before the war. Some of the illustrations are indeed reproductions of drawings made while she was at the Office, but the great majority are new, specially prepared for this work though based in many cases upon original drawings to be found in the Office collection.

In the book are a number of extracts from translations of the classics and from modern writers. For the extracts from Herodotus I have quoted Rawlinson's translation and for Aristotle I have used a free translation of the French of Barthélemy St Hilaire because, when I was writing on that part of the subject, no English translation of the *Meteorologica* was known to me. In Chapter VI however I have taken advantage of Webster's translation through the Clarendon Press. I have in this connexion to acknowledge my indebtedness to the publishers enumerated in this paragraph for permission to reproduce the extracts, and to the authors with whom I have been in correspondence, through the good offices of my friend Dr J. P. Postgate. The long transcript of Virgil's weather lore is taken from Dr J. W. Mackail's translation, published by Longmans, Green and Co. Those from Hesiod, A. W. Mair's version published by the Clarendon Press, from the *Phaenomena* of Aratus, G. R. Mair's translation in the Loeb Classical Library. I must mention also two extracts from a verse translation of the *Clouds* of Aristophanes by B. B. Rogers published by G. Bell and Sons.

I have used for the *Diosemeia* a translation made by C. Leeson Prince as a contribution to meteorology at the request of the late G. J. Symons. To a similar impulse I am indebted for the translation of the *Winds and Weather Signs* of Theophrastus by my colleague of many years at Emmanuel College, J. G. Wood, from which with his permission I have made large extracts. In the earlier chapters are a number of contemporary passages quoted from *Nature*, some of them signed by the authors, others without signature. It was to me remarkable that there should be, just at that time, so many passages which exactly met my need. And since I have put together in this volume the general ideas of the relation of weather and meteorology to humanity at large, I have been astonished at the frequency with which different aspects of the subject appear spontaneously in the public press. The sources of these and other quotations are acknowledged in the text.

In the chapter on the variability of climate I have found myself dependent upon somewhat lengthy extracts from the contributions to the subject in various books and journals: The Royal Geographical Society (Professor J. W. Gregory), the Royal Meteorological

Society (Professor Otto Pettersson), the *Geographical Review* (H. C. Butler), Huntington and Visher's *Climatic Changes, their Nature and Causes* (Yale University Press), and Sven Hedin, *Overland to India* (Macmillan and Co., Ltd.). For the concluding quotation from Jowett's translation of Plato's *Critias* I have to thank the Clarendon Press.

For the information contained in the brief biographies of meteorologists, physicists and mathematicians of the two hundred years after the invention of the barometer I am mainly indebted to the *Dictionary of National Biography*, the *Encyclopaedia Britannica* (eleventh edition) and *Chambers's Encyclopaedia* (ed. 1895).

The biographies are intended to carry the history of the development of meteorology up to the invention of the weather-map. The contributions of living meteorologists have to be considered from a different standpoint. Between these two classes are the meteorologists who gave expression to the idea of meteorology as an international science. They form the class of meteorologists who were active mainly between 1860 and the end of the nineteenth century. I have not ventured to include them in the biographies because the subject cannot be regarded as a collection of separate contributions to the same extent as it was during the two centuries covered by the biographies. Their names occur in other associations of a different kind that can be represented best by photographs of international assemblies which were a characteristic feature of the period, and the mode of representation is the more appropriate because the development of photography is practically contemporaneous with that of the weather-map.

Two groups are represented, the first that of the International Meteorological Congress in Rome in 1879, which finally established the organisation of meteorology on an international basis, and second, the Meteorological Conference at Paris in 1896 at which was initiated the International Commission for the Study of the Upper Air. They contain portraits of most of the prominent meteorologists of the period to which we have referred, but as some very notable exponents of the science do not appear in the groups, I have supplied some of the omissions by portraits from various sources that happened to be accessible. A few of the older school have been included.

The earliest photographs are those of Sir Edward Sabine and John Welsh which date from 1852 and are from a group preserved in Kew Observatory. It is a daguerreotype or collodion positive taken at Vauxhall on the occasion of one of Welsh's balloon ascents. The full picture includes Colonel Sykes, Prof. W. A. Miller and Mr J. P. Gassiot, the principal benefactor of Kew Observatory, as well as the two portraits which are here reproduced, also the table of instruments ready for the ascent.

For the reading of the proof-sheets I owe my thanks to Captain D. Brunt and Commander L. G. Garbett, R.N., of the Meteorological Office, who were at one time associated with me in the work of the School of Meteorology at the Imperial College.

The reader will share my obligation to Mr W. Lewis and his staff at the Cambridge University Press for the manner in which they have continued the work which Mr J. B. Peace began.

NAPIER SHAW

8 *July* 1926

The necessity for a second impression of this volume has been met by the reproduction of the original with only the necessary correction of the errors that have disclosed themselves in print and the rectification of statements that have become obsolete.

Some additions have been notified in square brackets in the text, and the notes at the end have been amplified to include some contributions by readers of the original volume.

N. S.

28 *November* 1931

TABLE OF CONTENTS

VOLUME I. METEOROLOGY IN HISTORY

The relation of energy to wave-length in solar and terrestrial
radiation.
   The use of colour-screens in the measurement of radiation.
   Ultra-violet radiation: actinometers.
Measurements of ozone.
The absorption by the atmosphere and the solar constant.
Instruments for the study of visibility.
Dust-counters.
Atmospherics and thunderstorm-recorders.
Lightning conductors.
Measurements of the earth's electric field and its changes. Ionisation
and air-earth current.
Magnetic forces in the atmosphere, absolute and recording
instruments.
   Record of changes in the vertical force.

LIST OF ILLUSTRATIONS

The authorities to whom acknowledgment is due for the several illustrations are noted in the respective titles in the foregoing list. Figures 2, 11, 104, 107, and 113, which are taken from official publications, are reproduced by permission of H.M. Stationery Office.

The derivation of the portrait-groups and separate portraits which face pages 156–7 and 206–7 is as follows: The International Congress in Rome, 1879, and the International Conference at Paris, 1896, are from framed copies preserved in the Director's room of the Meteorological Office at South Kensington. In the same room are the portraits of Admiral FitzRoy presented by his daughter Miss Laura FitzRoy, Sir Richard Strachey presented by Lady Strachey, Sir Francis Galton presented by his nephew Mr E. G. Wheler, and Dr Richard Assmann by his daughter Fräulein Helène Assmann. That of Alexander Buchan is from a photograph by Elliott and Fry, Ltd.

The portraits of H. W. Dove and James Glaisher are from cartes-de-visite presented by the late J. S. Harding, Admiral FitzRoy's chief clerk and secretary in the Meteorological Department of the Board of Trade.

The picture of Luke Howard is from a copy of a pencil drawing in possession of the Royal Meteorological Society; those of Sir E. Sabine and John Welsh are from a collodion positive of a portrait-group belonging to Kew Observatory, taken at Vauxhall on the occasion of one of Welsh's balloon ascents. Those of Buys Ballot and G. J. Symons are from commemorative medals in the possession of the author.

The other portraits are from published works, J. B. Biot and M. F. Maury from *Harper's Monthly*, by permission of Harper Brothers, U. J. J. Le Verrier from the volume commemorating the centenary of his birth, H. F. Blanford from the Administrative Report of the Indian Meteorological Department for 1924, S. P. Langley from the commemorative volume of the Smithsonian Institution, J. M. Pernter from the *Meteorologische Zeitschrift*, R. Abercromby and W. H. Dines from the *Quarterly Journal of the Royal Meteorological Society*.

The other picture, which is by Knudsen of Bergen and is not named, will be recognised as an effective snapshot of the meteorology of the future.

CHAPTER I

METEOROLOGY IN EUROPEAN CULTURE

Babylonian, Egyptian, Cretan civilisation	18000–1800 B.C.
Hebrew, Greco-Roman civilisation	1800 B.C.–A.D. 400
The Northmen	A.D. 400–1200
The Mediterranean revival	A.D. 1200–1600
The North-Western intrusion	A.D. 1600–

METEOROLOGY is the science of the atmosphere, or, with a certain limitation of meaning, the study of weather. The word is sometimes used as being synonymous with weather itself. It has been stated that meteorology is of great importance to armies and to navies, to ships and to farmers, although when the statement is made the seafaring man, the husbandman, and the soldier may have little interest in the science of the atmosphere. It is indeed true that weather is of urgent and vital importance for every section of the human race, it always has been so and always will be; but whether those who study the atmosphere can put their experience, and the knowledge derived from it, in such a form that practical folk like sailors or soldiers will regard it as important is quite another question; and whether it can be so represented as to challenge attention from those who are interested in science is again something quite different.

Many books have been devoted to expounding the importance of meteorology in its practical applications. This book is concerned with the possibilities of its importance as a science, as a subject of study for its own sake, interesting to those who are interested in the study of their own environment.

Hitherto the interest in meteorology has been dependent mainly upon the extent to which the community felt itself unprotected against the weather; and on that account has been subject to notable fluctuations, especially in recent times when protection has been found in buildings, clothes and new means of transport and communication. At the beginning of the nineteenth century when sea-travel was still by sailing-ships and land-travel was on horseback or by coach, attention had to be paid to the study of weather; every sailor was a practical meteorologist, all professors of natural philosophy regarded meteorology as part of their province and the subject engaged the unremitting personal attention of such influential persons as John Dalton and Luke Howard.

The interest culminated in the establishment of official meteorological departments in the early sixties and the evolution of the weather-map. But changes had already come in: sea-travel was by steamer and land-travel by rail. The community found itself more or less immune. Meteorology lost its place in the universities and was left to official organisations and special societies.

The position is most clearly expressed by noting that the Government grant for meteorology in this country which had increased fivefold between 1854 and 1882, remained quite stationary for a quarter of a century thereafter, though the operations were under the highest scientific direction. In 1899 the Royal Society when asked to obtain funds for pensions gave expression to the almost inhuman sentiment that there ought to be pensions but the funds should be provided by sacrificing activity.

Fluctuations on a smaller scale are easily remarked. When the art of flying began, all phases of weather were important; but by the time the war was over the aviator's interest was mainly confined to fog and wind above clouds.

Conditions have however changed again, the development of travel by air over long distances and the spread of wireless facilities have almost restored the enthusiasm of sixty years ago. The Government spends on meteorology eight times as much as it did before the war. The fact however remains that all along from the earliest times the importance of meteorology has been conditioned by the utility of its applications, particularly in the anticipation of future weather. The claims for interest in the subject for its own sake as giving an insight into nature have been recognised only by a comparatively few devoted observers. But these are the claims which are connoted by meteorology as a science and which are the subject of this book.

THE PRACTICAL IMPORTANCE OF WEATHER

What is it moulds the life of man?
The weather.
What makes some black and others tan?
The weather.
What makes the Zulu live in trees,
The Congo natives dress in leaves
While others go in furs and freeze?
The weather.
(W. J. Humphreys, *Weather Proverbs and Paradoxes*.)

Enthusiasm for meteorology is perhaps a peculiar experience; but it is not difficult to become enthusiastic, and, in favourable circumstances, even eloquent, about the importance to mankind of the atmosphere and its changes. If meteorologists have failed to interest their fellow-men therein it is not for lack of importance in the subject of their study, for the atmosphere is the chief element in man's physical environment. It is the breath of his life. With less than enough within him or about him he feels stifled; under the genial influence of the sun it provides his food and drink, and against its changes he is careful to provide himself with shelter; it is indispensable alike for his bodily warmth, for all his own physical energy and for that of his transport, his camels, his horses, his sailing-ships, his steamers and his motor-cars. The larger part of man's life-history consists in his endeavour to adjust himself to the ways of the atmosphere, to its habits in respect of wind and weather. His interest in the air has been vivid and unremitting. It is alto-gether insufficient to say that with some nations the study of weather has been connected with religion. The religions of mankind have been in large part formed out of the ideas which prolonged experience has engendered about the control of the atmosphere. Human lives have been sacrificed in order to propitiate or conciliate the powers that rule the air[1]. Perhaps even

[1] Mexico, *Hibbert Journal*, October 1923.

at the present day the greater part of mankind, following the examples of Greek and Hebrew poets, regard weather as under the personal control of the supreme deity into which it would be impious to inquire. In English law "the act of God" is still a legitimate plea for exoneration of damages for injuries to person or property if the circumstances are so exceptional as to be unforeseen. Small indeed is the fraction of mankind who are like-minded with the Greek philosophers and regard the vagaries of the atmosphere as subject to natural laws, and are unwilling to believe them to be past finding out.

This diversity of attitude was as conspicuous in the ancients as it is in the moderns; we shall deal with it more fully in the sequel when we glance at the references to weather and its control which are to be found in the poets and the philosophers of the ancient world. It has had some important consequences for the study because the idea of the personal control of weather by a major or minor deity has led to the confusion of those who pursue the scientific study of the atmosphere with its worst enemies—the rain-maker, the magician, and the temple-minder who accepts sacrifices in order to propitiate a deity.

CIVILISATION AND THE STUDY OF WEATHER

We may regard the religious practice of a people as the expression of their relationship to their environment, which includes inanimate nature on the one hand and living beings on the other, and therefore we might fairly expect the story of the weather to be coeval with that of civilisation, and intertwined with its records, its legends and its religions. This aspect of history is not without interest for students of weather.

According to the teaching of the new anthropology[1], human civilisation was autochthonous in ancient Egypt, and spread from there over the world with subsequent subcentres of diffusion in Babylonia and India. This view may not be accepted but it arrests attention by the circumstance that Egypt, and especially the Egypt of the early Egyptians, the Thebaid, is that part of the world which is most nearly independent of what we understand by weather. It draws its water-supplies from the river and takes nothing but dew from the sky. It has winds generally so arranged as to temper and mollify the burning effect of the sun's rays, seldom strong enough to raise a dust-storm and practically free from the terrible visitation known as simoom. At the same time it is wonderfully fertile with very little effort on the part of the husbandman.

> They take the flow o' the Nile
> By certain scales i' the pyramid; they know
> By the height, the lowness, or the mean, if dearth
> Or foison follow. The higher Nilus swells
> The more it promises; as it ebbs, the seedsman
> Upon the slime and ooze scatters his grain,
> And shortly comes to harvest[2].

(*Antony and Cleopatra*, II, vii, 20.)

---

[1] *Nature*, vol. CXII, 1923, p. 611.
[2] Measures of the height of the Nile go back to 3600 B.C. H. G. Lyons, *The Physiography of the River Nile and its Basin*, Cairo, 1906.

If then our civilisation began in Egypt, we are faced with the conclusion that primitive man found the line of least resistance to his advance towards civilisation in a country which has no weather, and yet enjoys a plentiful supply of water, with a sky so serene and genial as to make clothing a matter of little importance and the indispensable shelter beyond that of black pigment an easy artifice.

Professor Elliot Smith, in his work on the ancient Egyptians and the origin of civilisation, writes as though at the dawn of civilisation the world in every part was full of human beings, or "saturated" with humanity as well as with other forms of life, and that the spread of civilisation meant the replacement of autochthonous practices by Egyptian practices. The idea is very suggestive as expressing in the most simple terms by a physical analogy the relation of man to his environment. The possible density of population of any region has always been dependent on the capacity of production of the necessaries of life. That capacity has been increased by the arts of civilisation to many times the original productivity of the soil. It is capable of still further increase; but whenever the population increases beyond the productivity of its own area it must either restrain its numbers or extend its area. The density of population necessary to "saturate" a particular locality would vary greatly with the available means of subsistence. It is however difficult to imagine a density of population in a region of primeval forest in any way like that which would be possible in a country like irrigated Egypt. A pastoral people might be to some extent gregarious because they had to tend their flocks, but forest-dwellers must have been comparatively isolated families.

Civilisation... began when the early Egyptians invented the art of irrigation to extend artificially the area of cultivation of barley. The irrigation-engineer of early Egypt was the first man to organise the labour of his fellows. He conferred the benefits of security and prosperity upon the community and upon every individual member of it. He personified every subsequent idea of kingship. The life of the community flowed from him in a sense as real and actual as that in which the Nile was subject to his control. To identify him with these subtle forces was less an act of metaphysical ingenuity than one of unsophisticated realism. He became the incarnation of the life-giving powers which he bestowed upon his people. He became a god, assimilating to himself attributes of the shadowy Great Mother, and was apotheosised after death as Osiris. Eventually his powers were extended and transferred to his successor, Horus, himself credited with the immortalisation of the dead king.

(*Nature*, vol. CXII, 1923, p. 611.)

If this kind of experience is the condition for, or specially favourable to, the development of civilisation, the selection of Egypt is singularly apposite because no other place on the earth can be found with conditions more nearly perfect or even exactly parallel. Mesopotamia and Chaldaea give the nearest approach to the conditions, Palestine and Greece show a complete contrast. The valley of the Indus is also suitable for irrigation; but the flow of the river is irregular, and its overflow for irrigation "sporadic and fluctuating." Commencement has now been made of the construction of a dam at Sukkur in the Province of Sind which will bring $5\frac{1}{3}$ million acres under irrigation.

"The total cultivated area in Egypt is thus exceeded by half a million acres in this one scheme[1]."

The types of civilisation where there is no irrigation are quite different. People, such as those of Egypt or in a less degree of Babylon, who have no weather and grow their crops by irrigation, must naturally take a view of life entirely different from that of a people whose chance of continued existence, as in Palestine or Greece, depends upon "the former and the latter rain," perhaps better expressed as the beginning and the end of the rains. The meaning and importance of the river was well-known both to Egyptians and Greeks, as the following quotation from Herodotus will show:

> One fact which I learnt of the priests is to me a strong evidence of the origin of the country. They said that when Moeris was king, the Nile overflowed all Egypt below Memphis, as soon as it rose so little as eight cubits. Now Moeris had not been dead 900 years at the time when I heard this of the priests; yet at the present day, unless the river rise sixteen, or, at the very least, fifteen cubits, it does not overflow the lands. It seems to me, therefore, that if the land goes on rising and growing at this rate, the Egyptians who dwell below lake Moeris, in the Delta (as it is called) and elsewhere, will one day, by the stoppage of the inundations, suffer permanently the fate which they told me they expected would some time or other befall the Greeks. On hearing that the whole land of Greece is watered by rain from heaven, and not, like their own, inundated by rivers, they observed—"Some day the Greeks will be disappointed of their grand hope, and then they will be wretchedly hungry"; which was as much as to say, "If God shall some day see fit not to grant the Greeks rain, but shall afflict them with a long drought, the Greeks will be swept away by a famine, since they have nothing to rely on but rain from Jove, and have no other resource for water."

(The *History* of Herodotus, translated by Rawlinson, Book II, chap. 13.)

But the independence of weather, which has been accepted as characteristic of early Egyptian civilisation, could not have represented the experience of the Egyptians of the Delta, who according to Elliot Smith and others initiated and developed the art of navigation, by which they established communication with all parts of Europe, Asia and Africa and perhaps America. The Delta appears indeed to have evolved a goddess whose worship was carried to Greece, through Crete. "The Minoan 8-shaped shield is itself the outcome of that which formed part of the emblem of the Egypto-Libyan Delta goddess Neith. A Minoan goddess holding this shield seen at Mycenae seems to have been the prehistoric forerunner of Athena[2]."

When the Hebrews migrated from Egypt they cut themselves adrift from the security of the great river and actually entrusted their whole future to the permanence of meteorological conditions of which they could have only the vaguest assurance. Moses must indeed have been not only a great law-giver but a very competent meteorologist because the land of his choice is at best semi-arid and forms a tongue, with no adequate rivers, lying between two deserts. It depends for its seasonal rainfall upon the nearness of the Mediterranean and the suitability of the prevailing winds. It would be ruled

---

[1] *Nature*, vol. CXII, 1923, p. 699.
[2] Sir Arthur Evans, *Nature*, vol. CXII, 1923, p. 660.

out of consideration at first sight, by a meteorologist of to-day, as one of those semi-arid lands which are subject to famine.

While he accepted the change of his sources of sustenance from the river to the air, he very strongly deprecated at the same time the repetition of the Egyptian precedent of personifying and deifying the agencies of fertility or of failure. He trustfully anticipated natural conditions and had no patience with people who would placate a minor deity instead.

But the Greeks had no such scruples. Their prosperity was not based upon a single element such as the river Nile. It must have been evident to all that they were dependent upon the sky and sea as well as upon Mother Earth and, with the same idea of deifying experience, the mythological religion of Greece expresses most clearly the difference of meteorological conditions from those of Egypt. Instead of the custodian of the perennial river they imagined the control and government of the Universe entrusted in course of time to three sons of Kronos and the earth: Zeus (Jupiter) who was in charge of the sky, Poseidon (Neptune) who ruled the sea and shores, and Hades or Pluto, who ruled the under-world with its hollows and caves as well as its mines and minerals. On the human analogy the ruler of the sky, Zeus, the thunderer, the lord of rain, the cloud-gatherer, the lord of the aegis, surrounded himself on Mount Olympus with a numerous assembly of auxiliary deities, Apollo, the far-striker, and Artemis, twin-controllers of the sun, with all its life-giving power, and of the moon with her periodical changes. "Action at a distance" seems to have been included in their attributes and therewith the control of pestilences, sudden and irresistible, together with recurring ailments equally incomprehensible; it was to the benevolence of the controllers that appeal might be made for deliverance or recovery. The hierarchy of Olympus is not the subject of our study, but as our natural philosophy was born in Athens we are concerned to remark that Ruskin wrote of the patron-goddess Athene as the Queen of the Air, the personification of wisdom, who was derived directly from the brain of Zeus fully equipped with shield and spear. The mythological view is explicit testimony to the complications of the atmospheric conditions of Greece, very greatly in contrast with the simplicity of Egypt, as may be gathered from a study of the tables of chapter II. It is peculiarly significant that the patroness of Athens should have been evolved as the goddess of the atmosphere.

Besides the principal Olympian deities there were a host of mythical personifications of the controlling agencies of various aspects of weather, which are familiar enough, not only in ancient but in modern literature. Personification went to great lengths. Elsewhere I have noted that the harpies are a very apt personification of the line-squall, characterised by its propensities for snatching and fouling with dust[1]; and that Medusa, or Gorgon, is equally an apt personification of the winter cyclone of the Mediterranean. The coil of a snake, which is embodied in the very word cyclone, describes her hair; and the frigid north wind, the final expression of Mediterranean cyclonic energy, turns limpid water into stony ice.

[1] Ruskin, *Queen of the Air*.

This elaborate personification seems to have been attributable more to the poetic and aesthetic sense of the Greeks than to actual religion or to the exercise of the powers of reasoning which are their legacy in the culture of Europe. Here is a curious juxtaposition taken from Herodotus:

The storm lasted three days. At length the Magians, by offering victims to the Winds, and charming them with the help of conjurers, while at the same time they sacrificed to Thetis and the Nereids, succeeded in laying the storm four days after it first began; or perhaps it ceased of itself.

If civilisation began in Egypt, the dawn of meteorology came some time after the dawn of civilisation; and there is this further point of interest, that although the localities of genial climate, plenty of water and no weather, are the easiest for human beings to live together in, they are apparently not the best in the long run. The civilisations that spread out from the original centres, carrying with them the contrivances for their own protection and enrichment that had been invented in the favourable locality, developed faster when they faced successfully the vagaries of a weather-climate, and, coming back from the bitter North to the mild South, actually drove out the successors of their own progenitors, who seem to have become effeminate in comparison; and their own fate was to be driven out by still more hardy invaders from the North, later on. So much so that, though we may trace the dawn of civilisation to the country where cyclonic depressions are practically unknown, we must look upon the region of the maximum number of cyclonic depressions as the most favourable for the development of human energy[1].

It is useless to expect much information about weather-study from the records of the dawn of civilisation outside the range of changes of weather. We can only expect notes about weather from those countries where there is weather to be noted.

THE CONTRIBUTIONS OF METEOROLOGY TO ANCIENT CULTURE

Turning from these mythological questions to consider the real services which the study of weather had already rendered to mankind in the time of the ancients, we may call to mind all that is embodied in the kalendar, the division of the year into seasons, the selection of times for ploughing, sowing, reaping and harvesting, the arrangement of the work of husbandmen and pastoral folk, the suitable times for making voyages with ships, the contrivance of buildings for protection against the natural severity of the seasons, the drainage of lands and roads. All these imply the ability to profit by past experience, the benefit of which the great mass of mankind now take for granted; although they owe their present immunity, at least from the ordinary effects of weather and season, to the collective experience of their predecessors, who were in their own way practical students of weather. Thus, besides those who have put their ideas about the weather into words, there has been a vast body of students of weather whose only memorial is to be found in common

[1] Ellsworth Huntington, *Civilisation and Climate*, Yale University Press, 1915.

practice. The classification of the work of understanding the ways of the air from the earliest times to the present day is in itself a subject of some interest. As we survey the progress in this department of knowledge, we can discern three collateral aspects: first, the preservation of the memory of the events of past weather and their sequence, and its embodiment in common agricultural or navigational practice; second, speculations upon the relations of those events and upon their proximate and ultimate causes such as are to be found in the writings of philosophers; and, thirdly, the endeavours to use existing knowledge for the anticipation of future weather expressed from time immemorial in weather-lore. To-day we recognise the corresponding division of labour in modified forms as between the observer, who makes regular contributions to the recorded memory, the natural philosopher or knowledge-lover, who endeavours to trace the connexions of events as the relation of effects to causes, and the practical meteorologist, a government official, who draws such conclusions as he can with regard to future weather for the benefit of the general public.

To these must be added, as the result originally of the belief in personal weather-control and magic, a fourth class to which we have already referred. From very hoary antiquity there has always existed in some form or other a class of persons whose profession has been to exercise some form of control over the weather, either by appealing to the deities to whose influence the course of nature might be supposed to be entrusted, or, in more recent times, by the direct application of incantations or of physical forces to the atmosphere itself.

BEFORE AND AFTER THE WEATHER-MAP

Since the middle of the nineteenth century, the first and third classes have been co-ordinated to form a novel and very important class of persons, or groups of persons, whose business it is to organise the general public memory of the events of weather, to collect all the available information on the subject and to apply it to the needs of the community. This is the primary function of the meteorological offices of modern days which, beginning with departments of naval or military establishments, have now become a recognised part of all civilised governments; the care of these is generally entrusted to persons who are qualified also as scientific investigators. Since the middle of the nineteenth century the collection of facts has been operated not only by personal memory or by written records but also by electric telegraph and now by radio-telegraphy. With the development of this method of collecting facts there began a new era in the science of meteorology, quite different from anything with which the ancients were acquainted. The history of meteorology in its present sense begins with the weather-map.

Still, the primary classification into recording observations, discussing them from the philosophical point of view, and applying them to the use of the community, is evident throughout the ages and remains to this day.

The change in the amount of attention devoted to the different aspects of the subject is also somewhat remarkable. In the earlier days speculation and philosophical discussion were active, when the facts were wholly inadequate and records hardly existed; to-day we have a multitude of facts and records, though even now they are insufficient, but philosophers are less disposed to regard their exposition as part of their ordinary duty than they were before the introduction of the weather-map.

The place of forecasting in modern meteorology

This most recent development leads us to remark upon another misconception in the public mind with regard to meteorology, and that is that meteorology means forecasting the weather and nothing else. It is a great misfortune for the subject. A curious sense of depression for some meteorologists arises from the question which their friends, without exception, ask on the mention of any new scientific advance in meteorology: "Will it be useful in forecasting?" It is hardly an exaggeration to say that meteorologists have a natural aversion from the iteration of the duty of forecasting. No student of physical science in any of the recognised institutes, however sure he might feel about the result of a scientific calculation, with the exception of such well-tried predictions as those of the *Nautical Almanac*, would dream of announcing it in the public press without waiting for its verification, if waiting were the only avenue to verification. This feeling of premature disclosure is deep-seated in the scientific student. When Admiral FitzRoy, overpowered by the glamour of a telegraphic synoptic map, published his anticipations of coming weather as forecasts in the newspapers, his colleagues of the Royal Society were shocked and gave expression to their feelings; so much so that, after FitzRoy's death in 1865, when the Royal Society took over the administration of his office, no forecasts were allowed. In 1879 the controlling body, with full scientific authority, had arrived at the conclusion that the primary problems were solved, and forecasts have been issued day by day since that date. But the problems were not really solved with the degree of precision necessary for unfailing accuracy. It is a fair question whether forecasts would not really have been better to-day if meteorologists had not been compelled to issue a series of 50,000 sets of forecasts, only more or less correct. They might have given their attention to more purely scientific aspects, with some assurance of useful results, because the scientific aspect is the aspect from which true knowledge is derived. But any such question is *chose jugée* from the first. The universal desire for information about future weather opens the main artery of communication between the science and the public and is the chief vindication of an appeal for public funds; and when once a scientific subject enjoys public money it is difficult to persuade anybody that it is not provided for in all particulars. Public money is not always an unmixed blessing; science sometimes prospers in what appears to the outsider as poverty or adversity.

In the meantime forecasting is an invaluable touchstone, or test, of the reality of the scientific conclusions that are reached in the study of the atmosphere. If the general conclusions are correct, the inferences fairly drawn from them in particular cases must necessarily be correct too, and the comparison between inference and fact is an indispensable part of scientific reasoning which depends partly upon inductive principles and partly upon deductive calculations.

It is from that point of view that we propose to treat the subject in this book. We shall survey the methods and results of observation, we shall set out the inductive principles which have been established by their aid, and illustrate the application of physical and dynamical reasoning, with the understanding that the final test of the conclusions is their agreement with the facts of observation.

THE EARLIEST RECORDS

Reverting after this digression to the main purpose of this chapter, we note that the civilisations from which our own culture is derived include those of Egypt, Crete, Babylonia, Phoenicia, Chaldaea, Palestine and Greece; also that for satisfactory evidence of the dawn of meteorology we must not look to those civilisations which were independent of weather. We naturally look to the documents that remain of the civilisations which took their sustenance from the sky for the earliest indications of the nature and progress of meteorological study. The indications which have survived are only fragmentary until we come to the time of Aristotle, who wrote the first formal treatise with the title *Meteorologica*; but such fragments as there are, are not without interest.

There are still a considerable number of popular superstitions about the weather. In a lecture before the Royal Meteorological Society in 1908 on *The Dawn of Meteorology*[1], Dr G. Hellmann of Berlin traced some of them back to very early times. For example, the widespread belief in the first twelve days of the year as prophetic of the weather of the ensuing twelve months is traceable in the old Indian or Vedic texts, and the *signa tonitrui*, the forecasts of weather and fertility deduced from the thunder heard in each of the twelve months, contained in mediaeval almanacs, though attributed to the Persian prophet Zoroaster, are in fact of Chaldaic or Babylonian origin and date back to some thousands of years before the Christian era. From the astrological cuneiform library of Assurbanipal, now in the British Museum, we learn that predictions authorised with the forecaster's name were addressed to the king; for example: "When it thunders in the day of the moon's disappearance, the crops will prosper and the market will be steady. From Ašaridu."

The Babylonians had the wind-rose of eight rhumbs, and used already the names of the four cardinal points to denominate the intermediate directions; whereas it was till now generally supposed that we owe to Charles the Great, or perhaps to his learned monk Alcuin, who came from Yorkshire, this progress of the combination of the four principal winds to denote all others. That was indeed a great advance, for it is well-known that in the Greek and Roman periods each wind had its peculiar name, a practice still in use amongst the Italian mariners in the Mediterranean.

[1] *Q. J. Roy. Met. Soc.* vol. xxxiv, 1908, p. 221, revised and extended in *Met. Z.* 1908, Bd. xxv, p. 482.

The state of meteorology in the old Babylonian culture, namely, three to one thousand years B.C., shows quite another character than it did in those primeval times in which the weather proverbs originated.

After having been formed into the beginnings of a learned profession by the priests, the atmospheric phenomena were brought by them into connexion with the constellations of the heavenly bodies, and a complete system of consequences and combinations was established, which gave rise to the Astro-Meteorology. It even formed an integral part of the Assyric-Babylonian religion.

From the time of Meton, the fifth century B.C., the Greeks made regular meteorological observations, which were perhaps the first, as illustrated by the so-called parapegma, a kind of peg-almanac fixed on public columns for exhibiting the general data of the weather. In these parapegmata the observations of the wind prevail over all others, for they were of practical use in navigation.

THE DIVERSITY OF CLIMATES

Much more extensive information is available in the literature of Palestine and Greece, and to these we shall refer later, but here we note that the countries from which we seek the earliest information of the origins of meteorology are all related to a narrow strip of the earth's surface lying between the Atlantic Ocean and what we now call India, and including the shores of the Mediterranean Sea. That happens to be one of the most curious regions in the whole world from the point of view of weather and climate. It includes a number of arid deserts where no cultivation is possible, it includes also countries, Egypt, Babylonia, and India which, though arid, are fertile because they are irrigated by rivers, the Nile, the Euphrates, the Tigris, and the Indus. These are subject to very considerable rise and fall in the course of the year and provide in that way the seasonal variation which agriculture requires. On the other hand, the shores of the Mediterranean and the islands, although there is a very marked seasonal variation, are dependent, as we are, upon rainfall for the growth of crops.

We cannot expect to comprehend any of the references to the meteorological conditions which are found in the ancient records without bearing in mind these essential differences of climate. We therefore devote a chapter to the climates of the Indo-Mediterranean Region, which is often referred to as "the world as known to the ancients."

THE WORLD AS KNOWN TO THE ANCIENTS

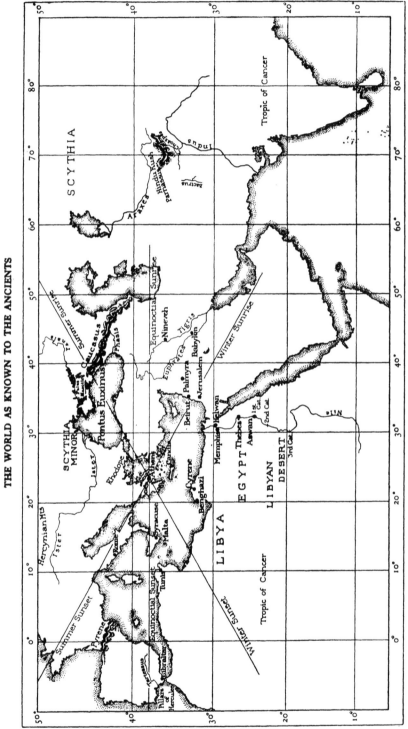

Fig. 1. A map marking the positions of stations for which climatic summaries are given on pp. 29–40, to illustrate the references to weather and climate in the works of Aristotle and other classical writers.

CHAPTER II

WEATHER AND CLIMATE IN THE "WORLD AS KNOWN TO THE ANCIENTS"

International Symbols

Symbol	Weather	Symbol	Weather	Symbol	Weather
⊕ or ●	Continuous rain	⌇	Lightning	V	Rime
●°	Drizzle	T	Thunder	~	Glazed frost
✱	Snow	℞	Thunderstorm	0	Unusual visibility of distant object
✳	Snow and rain together	≡	Fog	⋈	Mirage
● or ⊘	Passing showers	≡⦂	Wet fog	⊕	Solar halo
▲	Hail	≡°	Mist	⊥	Solar corona
△	Soft hail	≡	Ground-fog	Ⅲ	Lunar halo
←	Ice-crystals	∞	Dust haze	Ⅲ	Lunar corona
⊠	Snow on ground	⚒	Dust-storm[1]	⌢	Rainbow
↓	Snowdrift	⌒	Dew	⩊	Aurora
⊔	Gale	⊔	Hoar-frost	⋈	Zodiacal light

Intensity may be indicated by attaching "exponents" o or 2 to the symbols, thus ✱° means light snow, ✱² heavy snow. No exponents other than o or 2 should be used with the international symbols.

[1] Egyptian Weather Service [⚒ was adopted as an international symbol in 1926].

Beaufort notation

Beaufort letter		Beaufort letter	
b	Blue sky, cloudless	p	Passing showers
bc	A combination of blue sky with detached clouds	q	Squalls
		r	Continuous rain
c	Sky mainly cloudy, but with openings between the clouds	s	Snow
		sr	Snow and rain together
d	Drizzle	t	Thunder
e	Wet air, without rain falling	u	Ugly, threatening appearance
f	Fog	v'	Unusual visibility of distant objects
fe	Wet fog		
g	Gloom	w	Dew
h	Hail	x	Hoar-frost
l	Lightning	y	Dry air; relative humidity below 60 per cent.
m	Mist		
o	Completely overcast	z	Dust haze

The use of b when the sky is one quarter clouded is authorised. Detailed instructions by the Director of the Meteorological Office for the use of b and c came into operation on January 1, 1926.

The letter n is used in this book for number, to head columns which enumerate occurrences of observed phenomena in a specified period, week, month or year.

THE world "known to the ancients," fig. 1, extends, as has been said, from the valley of the Indus on the east side to the Pillars of Hercules at the Strait of Gibraltar on the west, or perhaps beyond them to the Canary Islands; and from the tropic of Cancer on the south to Scythia or Southern Russia and Central Europe on the north. The tropic of Cancer runs across Africa and Arabia from the west coast, south of the Canary Islands, to the mouth of the Indus, crossing the Nile at the first cataract, the Red Sea between Mecca and Medina, and just missing the Persian Gulf. Aristotle considered that the region to the south beyond the tropic was not habitable because it was too hot and that the north, beyond the home of the Barbarians or Scythians, was equally uninhabitable because it was too cold. The northern limit he thought might be in the countries underneath the stars that never set. Taking the latitude of Athens as 38°, the stars that never go below the horizon would be within a circle of 38° from the pole and its margin would be in the zenith of latitude 52°. That may be taken as Aristotle's idea of the northern limit of the world, and we have therefore to consider a region between the latitudes 23° N and 52° N and between the meridians of 20° W and 70° E, about one-twentieth of the area of the whole globe. We propose to give an account of the meteorology of this region because it has been the subject of observation from the dawn of history. Moreover, from the point of view of meteorology, it is perhaps the most interesting portion of the whole globe, consisting of the shores and hinterland of an inland sea thrust into a region that otherwise would certainly have been a desert, as are so many continental regions along latitude 35° in either hemisphere.

First as to its geography. In the course of his discussion of the origin of rivers, Aristotle mentions many of those which are included within the area, and also the mountains from which they take their rise, in order to controvert the idea that all rivers derive their water from a single or multiple reservoir underground, by an argument in the form of a "reductio ad absurdum." We quote his description because it gives us a rambling sketch of the greater part of the region and also because it gives a good idea of the way in which subjects were handled.

Thus in Asia most of the rivers and the largest flow down from the mountain called Parnassus which is, as everyone agrees, the highest of all the mountains which lie towards the winter sunrise; in fact when one has crossed it one sees the further ocean, the limits of which are not well-known to the inhabitants of this country. It is from this mountain also that, among other rivers, there flow the Bactrus, the Choaspes and the Araxes, of which the Tanais, that empties itself in the Palus Maeotis, is only a branch. It is from there also that the Indus flows which is the greatest of all rivers.

From the Caucasus flow also many rivers of enormous size and volume, and especially the Phasis. Of all the mountains which lie towards the summer-sunrise, the Caucasus is the most important both for its extent and its height. A good proof of its height is that it can be discerned even from "the deeps," as they are called, when one is sailing to the Palus Maeotis, and also its peaks are illuminated by the rays of the sun for a third of the night, morning and evening; and a proof of its great extent is that having on its sides many countries where many nations dwell, and having, as they say, vast lakes, one can from the highest peak see all these countries.

From Pyrene (the mountains of the Pyrenees), which is the highest mountain in Celtice towards the equinoctial sunset, flow the Ister and the Tartessus. The latter empties itself beyond the Pillars of Hercules, but the Ister, having crossed the whole of Europe, empties itself in the Pontus Euxinus. The majority of the other rivers which are in the north flow from the Hercynian mountains, which are the highest and most extensive mountains in this region. Under the constellation of the She-Bear herself, beyond furthest Scythia, are the mountains called Rhipae of which we know the size only from accounts which are too obviously fabulous. They say also that from these mountains there flow the greater number and the largest of all the rivers after the Ister.
                                    (Aristotle, *Meteorologica*, Book I, chap. 13.)[1]

In Aristotle's description some further notes are added about Libyan rivers and the rivers of Greece.

[1] The above is a free translation from *Météorologie d'Aristote, traduite en français pour la première fois*...par J. Barthélemy Saint-Hilaire, Paris, 1863.

THE CLIMATES OF THE MEDITERRANEAN REGION

We now proceed to give the climates of the regions indicated in the preceding paragraph as represented by modern statistics and in modern form. By so doing we ay ourselves open to the remark that climates change, and the weather at the period of the records which we wish to consider would be of a different character from what it is to-day. We shall devote a chapter to the consideration of that subject later on. In the meantime we may feel certain that the differences which we shall notice in the records of to-day between the climate of Egypt and Babylonia on the one hand and Palestine, Greece and the Mediterranean littoral on the other have always been operative in kind if not in the same degree. Short of an ice-age, a thunderstorm in Upper Egypt has never been anything but a miracle and never will be, and a famine in Palestine has never been very far away from the limits of the probable.

The contrast between the conditions of weather experienced in the ancient world and those of the British Isles, for example, is best expressed perhaps by stating that, in all parts of the region which we have delineated, the weather is dominated by the climate: it is quite obviously seasonal, whereas in the British Isles, although the influence of climate is discernible, experience is dominated by weather. This is clearly indicated by the fact that the Italians and the French, who derive their language from Mediterranean experience, have the same word *tempo* or *temps* as a general word for weather, and it also signifies *time*. The context has to be depended upon in order to comprehend the particular meaning in each case. And there is apparently no Greek word for weather; καιρός is said to be the word in use in modern Greek, but that again seems to have the sense of season and therefore to connote climate rather than weather. κατὰ καιρόν means unseasonable; similarly the word monsoon, one of the terms best known to us as characterising the wind and weather of the tropical east, has from its derivation the more general meaning of season. It will be at once admitted that in more Northern latitudes we could hardly get on without a special word for weather; that fact is a very clear indication of the difference to which we wish now to call attention. Our proverbially periodic phenomena are the tides and not the weather. We can understand Shakespeare writing "there is a tide in the affairs of men" when time is connoted, but we should have been quite put out if he had spoken of a weather in the affairs of men. But *temps* would fit in quite well.

Climate is the general summary of the weather for any particular locality. He who would write an effective essay on the difference between climate and weather might well choose "The Cornish Riviera" for his subject.

The differences between the various regions to which allusion is made in ancient writings can be represented most effectively by maps or climatic tables, which summarise a long series of observations of weather and now form the fundamental basis of the science of climatology. The method was not arrived at until long after the invention of instruments for measuring the physical condition of the atmosphere. With apologies for the obvious anachronism, we shall give a representation of the climates of the region under consideration by maps of the normal distribution of rainfall (fig. 3), which will make quite clear the differences of character of the climates of the different regions, and maps of the normal distribution of pressure (fig. 4), which, with the aid of Buys Ballot's law, will give sufficient indication of the prevailing winds. We shall add climatic tables for selected sites, which will indicate the range of weather which has been experienced during the periods of observation and therefore indicate also the range of variations to which the weather is liable at the present day. We do this chiefly to explain, at the outset of our work, by examples, the present method of representing climate, and at the same time to explore the real character of the climate of the Mediterranean.

THE NUMERICAL EXPRESSION OF THE METEOROLOGICAL ELEMENTS

The introduction at this stage of a representation of the climates of the Mediterranean in a modern form has the incidental advantage of enabling us to set out at the very beginning the method to be used for expressing the meteorological elements which have to be summarised in order to represent climate.

Observations without instruments

Since the introduction of meteorological instruments, a few of the elements can be measured numerically but there are still a considerable number which can only be noted and the number of occurrences counted. These are called non-instrumental observations, like fog, mist, dew, hoar-frost, thunder, lightning, and the optical phenomena—halo, rainbow, etc. With these must be associated the estimates of wind-force by noting the effects on trees, water, etc., and the amount of cloud, for which a numerical notation has been devised. These will be dealt with in due course as subjects of more detailed study; for the moment we are only concerned with the method of summarising a series of observations in a climatic table. International agreement[1] has been secured for noting an observation of one or other of these non-instrumental observations by a symbol.

The international symbols are printed at the head of this chapter. They are intended to cover all the phenomena which a meteorological observer has to note; but in the study of weather at sea the observers of the British Navy and Mercantile Marine were accustomed to add to the readings of the instruments in their log-books notes of the state of the weather, introduced or codified by Admiral Sir Francis Beaufort in the early years of the nineteenth century. They are not suitable for international use because letters connote different meanings in different languages. It is however convenient to have

---

[1] International agreement in meteorological practice is secured by periodic conferences of directors of meteorological institutes and observatories as described in chapter IX.

the Beaufort notation, as modified for modern use, set out with the international symbols.

For summarising these observations in the climatological tables the number of occurrences during the period of observation has been counted and the results are set out in columns under the appropriate symbol, the column being headed with the letter "n" to indicate that the number of occurrences per month or per year as the case may be is given in the column. Sometimes observations of a particular kind are given as a percentage of the whole number of occasions on which the phenomena might have been noted, in that case the column is headed by the symbol "%." Percentage is also required for the expression of relative humidity and possibly for other quantities in like manner. We have the intention of using the words "per cent." to head the columns in such cases and hope we may remember to adhere to that convention throughout the book.

Wind-vanes, the direction and force of the wind, the amount of cloud and visibility

The direction of the wind is shown by a wind-vane which is mounted on a vertical axis and sets itself as nearly along the lines of the air-current as movement about a vertical axis permits. The vane ought to be a body with a shape to correspond with the stream-lines or lines of flow of a current past a solid obstacle. It would appear that a wind-vane in the shape of a fish, which is endowed by nature with a "stream-line" body, is the best model. Recent vanes for recording-anemometers have been made with a long arm ending in a fish-shaped tail. The direction, when identified by the wind-vane, is expressed either by the points of the compass-card (true) along which the wind lays the vane "head on," i.e. the direction *from* which the wind comes is given. It can be still better represented by the number of degrees by which the nose of the vane deviates from true north. This can easily be indicated on a dial by mechanical contrivance but the angle is in practice often estimated to a rough approximation either in "compass points" of $11\frac{1}{4}°$ each or in degrees. The common practice of expressing directions as so many degrees West or East of North or South is very awkward when data have to be brought into computation.

Beaufort scale of wind-force[1]

A numerical estimate of the force of the wind was also introduced by Sir Francis Beaufort in 1806. He based his scale upon the behaviour of a sailing-ship of recognised type. A calm was marked 0 and a hurricane "that which no canvas could withstand" was marked 12, the intermediate eleven values were, 1 light air, 2 slight breeze, 3 gentle breeze, 4 moderate breeze, and 5 fresh breeze, the limit of agreeable wind on land, next came 6 strong breeze and 7 "moderate gale" or half a gale, for which "high wind" has

---

[1] A note on the history of the Beaufort scale and notation is given by Commander L. G. Garbett, R.N., *Q. J. Roy. Met. Soc.* vol. LII, 1926.

been substituted as being less ambiguous. Beyond these are the gale or storm-forces, viz. 8 fresh gale or gale, 9 strong gale, 10 whole gale, 11 storm.

In the beginning of the present century a new specification in terms of the velocity of the wind[1] was arrived at; it is remarkable that the values obtained as the best expression of the "forces" in terms of velocity are such as to give a straight line when plotted on double logarithmic paper with velocity as abscissa and "force" as ordinate, as represented in fig. 2.

THE VELOCITY AND THE PRESSURE OF WIND ON THE BEAUFORT SCALE

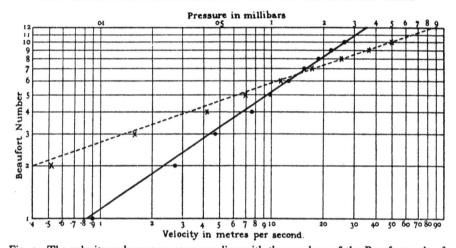

Fig. 2. The velocity and pressure corresponding with the numbers of the Beaufort scale of wind-force plotted on double logarithmic paper.
The dots set on lines drawn for successive Beaufort numbers represent the corresponding velocities on the scale at the foot, the crosses on the same lines show the pressure in millibars on the scale at the head of the diagram.

The conclusion to be drawn from the rectilinear nature of the graph in this diagram is that the numbers of the Beaufort scale, B, are very approximately related to the mean velocity of the wind by the equation $v^2 = kB^3$. Assuming that the force of the wind upon an area of about 3 metres square is represented by a pressure f in millibars and that v is the velocity in metres per second: $f = \cdot 0072v^2$. The coefficient is slightly smaller for smaller areas[2]. Thence we find $f = \cdot 005B^3$ and $v = 0 \cdot 836\sqrt{B^3}$. This would give for the mean values of the wind corresponding to the various Beaufort numbers:

Beaufort number	Velocity in metres per second	Beaufort number	Velocity in metres per second
1	0·84	7	15·5
2	2·36	8	18·9
3	4·34	9	22·6
4	6·69	10	26·4
5	9·34	11	30·5
6	12·3	12	34·7

[1] For a discussion of the tables of equivalents of the Beaufort scale used in different countries see *Report of the Tenth Meeting of the International Meteorological Committee*, Rome, 1913. M.O. publication, No. 216, London, 1914, pp. 33–40.
[2] *Notes on the Resistance of Planes in Normal and Tangential Presentation and on the Resistance of Ichthyoid Bodies*, by F. W. Lanchester, Advisory Committee for Aeronautics, Reports and Memoranda, No. 15, 1909.

In practice also the amount of cloud can be represented by a numerical estimate of the relative portions of visible sky covered by cloud and free from cloud. Thus an estimate is formed by inspection of the number of tenths of the whole sky which is covered by cloud. Various artifices may be adopted in order to obtain a satisfactory estimate and make it easy for different observers to agree in their estimates. Dividing the sky into quadrants and estimating the proportions of cloud in each quadrant is the most usual mode of procedure.

Visibility is another quantity for which in recent years numerical expression has been sought. The enumeration is arrived at by selecting a series of objects at successive distances and noting the furthest of these objects which is visible. In that way the following numerical scheme has been evolved to cover the whole range of vision from dense fog to very transparent air:

0	Dense fog, objects not visible at	50 metres	
1	Thick fog, objects not visible at	200	,,
2	Fog, very bad visibility, objects not visible at ...	500	,,
3	Bad visibility, objects not visible at	1000	,,
4	Very poor visibility, objects not visible at ...	2000	,,
5	Poor visibility, objects not visible at	4000	,,
6	Moderate visibility, objects not visible at ...	10,000	,,
7	Good visibility, objects not visible at	20,000	,,
8	Very good visibility, objects not visible at ...	50,000	,,
9	Excellent visibility, objects visible beyond ...	50,000	,,

The introduction of these numerical estimates enables the results to be summarised by numerical mean values and not exclusively by the number of occurrences, and in this way, although no instrument may be used, the representation differs from those of the typical non-instrumental observations which are represented by a simple count of the number of occurrences without regard to the intensity of any one of them.

Instrumental observations

The instruments which by international agreement form the ordinary equipment of a climatological station are the mercury-barometer, the thermometer also of mercury, the maximum thermometer, the minimum thermometer, the wet bulb or a hair hygrometer, the rain-gauge. They can be supplemented by a barograph for obtaining a continuous record of pressure, a thermograph which treats temperature in like manner, by an anemometer for recording the direction and velocity of the wind, by a sunshine-recorder for indicating the duration of sunshine, a pyrheliometer for measuring the thermal intensity of sunshine, a pyrgeometer for measuring the loss of heat by radiation from the earth, a nephoscope for finding the direction of motion of clouds, and many others; but for the present representation of Mediterranean climates we limit our attention to the more common instruments— the barometer, thermometers and rain-gauge—in addition to the non-instrumental observations.

In respect of the units employed for the representation of these quantities the world is still divided into two unequal parts. One part uses British units, the mercury inch at 32° F in latitude 45°, the Fahrenheit degree and the inch; and the other part uses the millimetre of mercury at o° C in latitude 45°, the centigrade scale and the millimetre.

The practice of giving pressure in terms of a hypothetical inch or millimetre at the freezing-point of water in latitude 45°, which was introduced into international practice only at the beginning of the current century, has seriously changed the aspect of the question of the choice of units for the expression of pressure. At a meeting of the British Association many years ago we heard the best-known British authority on units and physical constants urge that the correction of readings of the barometer for latitude was undesirable, that what was wanted in meteorology was not pressure in pressure-units but something equivalent to the height of the atmosphere at the place of observation, i.e. the "barometric height." But international agreement has decided otherwise. By that agreement the last obstacle was removed from the expression of pressure in the manner appropriate to the expression of a distributed force, namely the thrust in absolute units of force upon a unit of area.

We have accordingly selected units for the expression of these physical measurements of the atmosphere on the same ground that electricians and magneticians have chosen the units appropriate to their branch of science, namely those which are systematic and on that account the most convenient in the long run for computations involving more than one physical quantity. Those units are:

For the barometric pressure or for vapour-pressure 1000 c.g.s. units of force, called dynes, per square centimetre. This unit we call a millibar. A pressure of 1000 millibars is our standard atmospheric pressure to which other measurements are reduced when that kind of reduction is required. That is equivalent to 750·076 millimetres or 29·5306 inches of mercury at the freezing-point of water in latitude 45°. To indicate a measure of pressure in millibars in formulae or tables we use the symbol "mb."

For temperature—the tercentesimal scale of centigrade degrees measured from 273° C below the freezing-point of water. For all practical purposes the scale agrees with the absolute thermodynamic scale, but there is an actual difference which requires elaborate experiments for its determination; the value accepted at the present time for the difference between the absolute temperature and the tercentesimal temperature amounts to about a tenth of a degree. In order to indicate a measure of temperature on the tercentesimal scale in formulae or tables we use the symbol "t," or preferably "tt" to avoid confusion with the symbol for time.

For the depth of rainfall we use the millimetre with its symbol "mm."

For the velocity of the wind we use the metre per second or possibly on occasions the kilometre per hour. The symbols for these measurements are "m/s" and "km/hr."

We will not interrupt the narrative at this stage to give in detail the reasons for this selection, beyond saying that pressure and temperature as used in meteorology require much greater precision than is attached by common usage to a quotation of a reading of an ordinary domestic weather-glass and its attached thermometer. The readings used for meteorological purposes have passed through processes of correction and reduction and require some acknowledged evidence of the fact. The difference between a hall-barometer and a scientific instrument is not large but it is vital for meteorology. Our chief reason is however the necessity in meteorology for systematic units. Let anyone who wishes to convince himself of the necessity make a small collection of meteorological data and try the systematic computation of such meteorological quantities as the density of air, or of the water-vapour which it contains, the gradient-wind, the radiation from black earth, the energy of sunshine and the potential temperature. We can promise him that his only feeling after a week of computation will be a lament that the angular velocity of the earth's rotation cannot be represented by some better expression than $72 \cdot 92 \times 10^{-6}$ radians per second.

CLIMATIC SUMMARIES FOR MEDITERRANEAN STATIONS

With this preface we may pass on to consider the climatological tables in which an attempt is made to use the available material to represent the climate as concisely as possible with reference to the following elements: the direction of the wind at different hours of the day and in different seasons, and the number of gales ⟋ or the mean wind-velocity ⟋; the frequency of fog ≡ and the amount of water-vapour in the air; temperature and its ranges; the normal values and variability of rainfall; and the number of days of precipitation of any sort as well as of hail ▲, snow ✳, thunderstorms ℞ and dust-storms ⸸ . In the two cases where periodic irrigation by river forms part of the economic conditions the normal heights of the river in the several months have been given. The most conspicuous features of the common climate is the difference between summer and winter, not only in respect of solar radiation and temperature, but also in respect of rainfall and in the inverse sense the amount of water-vapour in the air.

The stations selected are:

Helwan for Cairo and the ancient Memphis, at the head of the Delta of the Nile. The level of the river has a normal range of 5·1 metres between May-June and September-October, with an almost rain-free record for the seven months May to November and really appreciable rainfall ● only in January; its yearly total is 34 mm (an inch and a quarter); the temperature is always above the freezing-point, reaching 315 t (108° F) in June and above 300 t (81° F) on every day of the four months from June to September. Winds markedly Northerly in the summer when the etesian winds blow in Greece, and tending to be Easterly in the winter, but never calm. A dry atmosphere especially in the middle of the day when humidity ranges from

20 per cent. in early summer to 44 per cent. in winter. The record shows 14 dust-storms a year, chiefly in the early part of the year.

At Aswan, not far from the site of the ancient Thebes, no rain is measured at the present day.

Babylon, now included in Iraq, the Chaldaean capital on the river Euphrates, which ranges in level through 3·5 metres between an autumn lowness in September-October and a spring flood in April[1]. It has three times as much rainfall as Helwan, and a much greater range of temperature—which reaches the freezing-point in three months of the year and 320 t (117° F) in July and August—less water-vapour, fewer dust-storms, if the standard of observation is the same, less wind, and chiefly from North and West, especially in the summer months. The rainfall is a winter-supply, amounting to 109 mm in the year (4 inches). The four months June to September are rainless.

Jerusalem lies on a high plateau 748 metres above sea-level and about 1000 metres above the gorge of the lower Jordan and Dead Sea. It has normally a good supply of rainfall during the six months November to April, 649 mm in the year (26 inches) but a rainless summer (June to September). Temperature may go as high as 312 t (102° F) and as low as 267 t (21° F). There are normally three days of snow in the year and it may be very heavy. A thickness of 29 inches was measured after a storm in February 1920 (see *Meteorological Magazine*, October 1920, p. 200). Winds are chiefly from the West in the summer and more variable in the winter. The air is not so dry as in Helwan or Babylon.

Beirut on the Syrian coast, where a rainfall which amounts to 906 mm (36 inches) in the year may occur in any month, but in the summer, June to August, falls on less than one day a year. Temperature is milder. It has reached the freezing-point in December but its normal daily minimum in that month is 12tt above the freezing-point. The winds are generally South or West in the early morning, changing to North or West in the middle of the day.

Candia on the North coast of Crete near to the site of the ancient Cnossus with 535 mm of rainfall per annum (21 inches). Being on a Mediterranean island the rainfall is more evenly distributed throughout the year, though it is still markedly characteristic of the winter season. As in the case of Beirut one-fifth of the year's supply falls in December. The temperature has ranged from just above the freezing-point to the high figure of 319 t (115° F). The winds as recorded are North or West in the summer and South in the winter but their average force is very light and there are many calms—that however may be a matter of exposure.

---

[1] It is noteworthy that, according to Sir Norman Lockyer, the Egyptian temples were oriented to the sunrise at the summer solstice which was approximately coincident with the rise of the Nile at Memphis; according to modern reckoning the flood begins nine days earlier at the site of Thebes.

The temples of Assyria on the other hand were oriented for sunrise at the equinoxes, and in spring at Babylon that would be coincident with the rise of the Euphrates. (*The Dawn of Astronomy*, pp. 230, 240.)

CLIMATE OF THE WORLD AS KNOWN TO THE ANCIENTS

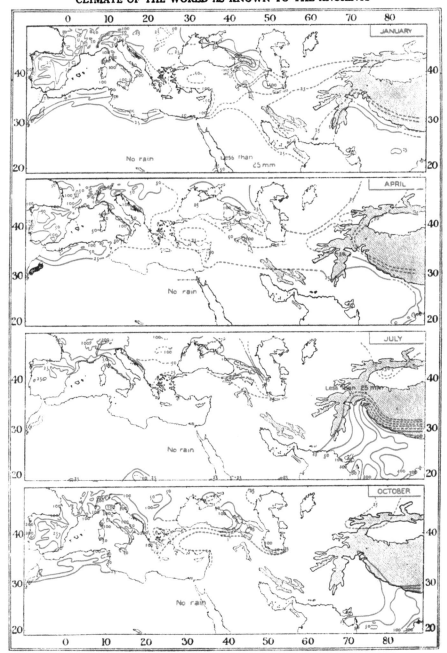

Fig. 3. Normal distribution of the rainfall of January, April, July and October over the regions of the ancient world.
The coasts are shown in black outline: the shaded areas indicate the regions above 2000 metres (6562 ft.) in height. The red lines are lines of equal rainfall (isohyets) drawn for 25 mm, 50 mm, 100 mm, 200 mm, 300 mm, 400 mm. One line of 5 mm is suggested for April.

Athens[1] has its rainfall (406 mm per annum, 16 inches) even more evenly distributed throughout the months than Candia, but the preponderance of winter rain is still marked; one-half of the whole falls in the three months November, December and January. The summer temperature is less extreme than at Candia but the winter temperatures are lower; frost may occur in any of the winter months, snow falls on six days of the year, and there may be 19 thunderstorms, which have no marked preference for the seasons. The etesian winds from the North in the summer do not come out so clearly in the tables as their reputation for persistence might lead us to expect, but the winds at a coast-station are always difficult to analyse.

Rome has 828 mm (33 inches) of rainfall with a maximum of 127 mm in October, when Northerly winds are frequent, and a minimum of 19 mm in July, when the winds are more variable. The extreme range of temperature is large, amounting to 50 t or 90° F in an 80-year period. The continentality is more marked than at Helwan but less marked than at Babylon, which has a range of 57 t in the brief period of 5½ years. There are 11 days of frost in the year and 22 thunderstorms.

Syracuse has a rainfall of 637 mm (25 inches) with a maximum of 109 mm in November and a minimum of 5 mm in June. Its temperature has been down to the freezing-point, and normally gets within 3tt of it in any year. Its maximum temperature is lower than that of Athens. Its winds are very variable, with a tendency to Westerly in the winter and Northerly in the summer. Calms are rare; nine gales a year are a normal allowance.

Gibraltar has 910 mm (36 inches) of rainfall with a maximum of 161 mm in November and a minimum of 1 mm in July, so that the seasonal character of the rainfall is preserved right up to the extreme West of the sea. But any month may apparently be rainless, or practically so, on occasions. The absolute range of temperature is 39 t (70° F); it may freeze in December but generally it does not. It has no snow and only nine thunderstorms in the year. East and West are the most notable winds; the former provides the *levanter*, an Easterly wind of gale force. Six gales a year have to be reckoned with.

Tunis for **Carthage** has the very moderate rainfall of 417 mm (16 inches), which classes the region with those known as semi-arid. There is a maximum in December and a minimum in July, but the seasonal variation is not so conspicuous as in the other stations. The range of temperature is about the same as at Helwan. It appears to be a windy site, as there is only one calm; the winds are mainly from North or West.

Benghazi for **Cyrenaica** approaches still more nearly to arid conditions with its rainfall of 276 mm (11 inches); there is hardly any between April and September. The range of temperature is however comparatively small, not reaching the freezing-point on the one side nor approaching the highest temperature of Babylon on the other. The prevailing winds are from the North in the summer, variable with a Southerly tendency in winter.

[1] Information for this section is derived from *Le Climat d'Athènes*, by D. Eginitis, a book which contains much valuable information on various aspects of the subject.

Malta, a central point of the Mediterranean, carries 516 mm of rainfall (20 inches) of which one-fifth falls in December and hardly any in June, July or August. In this respect it is extraordinarily similar to Candia and in respect also of its temperature. The winds are very variable. There are 14 thunderstorms, chiefly in the autumn months, and a curious peculiarity of this and other stations is that hail and thunderstorms do not appear to have that close association which we are accustomed to in this country.

All these figures show a cool rainy season in the winter and a hot dry season in the summer. That is the rule of the whole of the Mediterranean area. In all the countries which are represented by them the winter is the growing season and the summer the resting season for cultivation; the opposite practice prevails on the Northern side of the backbone of Europe that may be traced from the Pyrenees to the Black Sea. For the purposes of comparison we give a table in the same form for Kew Observatory, Richmond, Surrey, which shows the following results:

Richmond, Surrey. The annual rainfall is 606 mm (24 inches), which is not very different from the values observed in the Mediterranean region but is much more evenly distributed throughout the year, with a minimum of 37 mm in April and a maximum of 69 mm in October. The absolute range of temperature in the 45-year period is 48 t (86° F) but the average range of temperature within a year is only 33 t (59° F); the freezing-point is reached normally on 49 days in the year and the temperature rises above 298 t (77° F) on only 12 days. Winds are chiefly from South and West and show no very marked seasonal variation. Calms are frequent.

RAINFALL, PRESSURE AND WINDS

The positions of the historic sites are set out in the map of fig. 1. The characteristics of the region in respect of temperature and cloudiness may be inferred in sufficient detail from the maps of the distribution of temperature and cloudiness over the globe in Vol. II, and for the other meteorological elements, rainfall and pressure, the observed values at the several stations are collected in detailed maps of the distribution of rainfall (fig. 3) with its notable irregularities, and of the distribution of pressure (fig. 4) which shows by arrows along the isobaric lines the ordinary drift of air at some 500 metres above the Mediterranean level.

This is the second occasion on which we have to notice the relation between wind and the distribution of pressure, we may here therefore call to mind that the relation was expressed first by the Dutch meteorologist Buys Ballot who pointed out that in the Northern Hemisphere if you stand with your back to the wind the pressure is lower on your left hand than on your right, in other words the air always moves with lower pressure on the left hand side and higher pressure on the right. The reverse is the case in the Southern Hemisphere and the cause of the relation has been traced to the earth's rotation. It is the starting-point of modern dynamical meteorology: further

CLIMATE OF THE WORLD AS KNOWN TO THE ANCIENTS

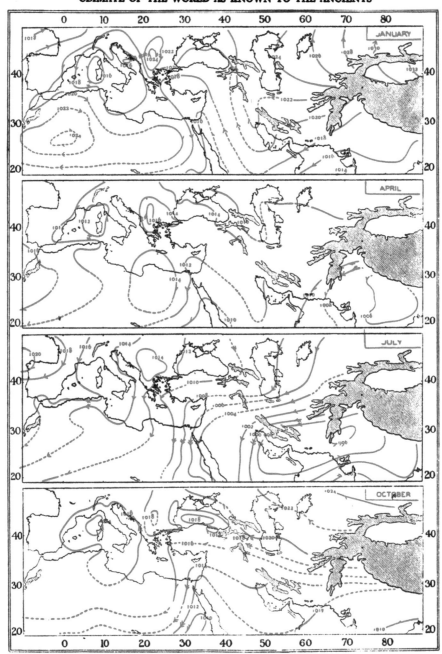

Fig. 4. Normal distribution of pressure over the regions of the ancient world in January, April, July and October.

The coasts are shown in black outline: the shaded areas indicate the regions above 2000 metres (6562 ft.) in height. The red lines are isobars drawn for intervals of 2 millibars. The geostrophic wind, computed from the separation of the isobars, is indicated by feathers on the lines, one for velocities under ten metres per second, and two for those above that limit.

investigation has shown that the relationship, or law, which Buys Ballot detected at the surface is much more complete in the upper air, so that one may with very little error regard the upper air, at some such level as 500 metres, as flowing truly along the isobaric lines, and the deviation of the surface-air from the isobars as being caused by the friction offered by the earth's surface to the flow of air in accordance with the relationship which holds above. In order to keep its path along the isobars and resist the influence of the pressure across its path a certain velocity of motion is required and if that velocity is reduced by the frictional resistance of the surface the balance fails and the air takes on a drift from high pressure to low. Hence we can always treat a map of the distribution of pressure as giving a representation also of the distribution of wind, regarding it as having a flow along isobars but at the same time deviated, at the surface, towards the lower pressure.

The summer conditions as represented by the month of July fall under a general slope of pressure from 1020 mb on the Atlantic, off the North West coast of Spain, to 1006 mb over the South East corner of the Levant, with however much irregularity and contortion of the lines as they cross the sea from the Northern to the Southern shore. There is a local area of high pressure over the Balkan peninsula. The winds form the outlying Western section of the circulation of the summer monsoon of the Indian Ocean which has a band of lowest pressure extending from the Persian Gulf to India. The monsoon circulation is dominant over the Levant. At this season of the year the general distribution of pressure is very stable and few disturbances occur, since depressions are now confined to the North Western portion of the Mediterranean. They seldom give rise to winds of any considerable strength, though thunderstorms occur frequently, and may be accompanied by violent squalls of wind and heavy rain.

The winter conditions, with their rainy season, are quite different. The sea becomes a region of extended low pressure lying between the vast anti-cyclonic system of the winter of the Eurasian continent and a large area of high pressure extending from the Nile across Africa and the Atlantic Ocean to the Western shores of North America, with minima over the Sardinian region, over the Adriatic, and over the Eastern Levant and Red Sea. High pressure is shown over the Balkan peninsula, the Alps, Spain and Africa. Rain accompanies depressions, which are numerous. Their paths are often tortuous but they proceed generally from West to East. They may traverse any part of the sea and strike any coast. They carry strong winds or gales; and when the central region of a depression is to pass any locality its advent will be announced by a strong wind from the South East, this will veer to South and South West and ultimately to North West with a fall of temperature.

In setting out a summary of the winds we have grouped the observations under the conventional headings of N, S, E and W; some attention has also been paid to the strength of winds by quoting either the mean velocity or the number of gales. That was not by any means the practice of the ancients who, as we shall see later, had separate names for separate winds. Many separate names are still used in the Mediterranean for the characteristic winds of the different localities.

The principal of these are [the *levanter* of Gibraltar] the *mistral* of Southern France [the *tramontana* of Northern Italy], the *bora* of the Adriatic, the *scirocco* of Southern Italy, the *samun* of Algeria which is also called *scirocco*;...the *bora* blows from the cold highlands to the north of the Adriatic into a cyclonic depression over the southern Adriatic or in the Mediterranean....On the north-east coast of the Black Sea (Pontus Euxinus) a similar *bora* wind rushes down from the cold mountains when a high pressure system exists to the northward and coincides with the passage of a depression across the Black Sea; it is characterised by violent gusts and an extremely low temperature....The strong north-easterly winter-winds of the Aegean Sea which occur when pressure is high over the Balkans are similar in type.

The *scirocco* as a warm damp southerly wind of southern Italy usually blows on the south-east side of a depression which is traversing the Mediterranean from west to east and thus a *scirocco* may be blowing on the south side of such a depression while a *mistral* or a *bora* is blowing on the northern side. When the *scirocco* blows from the interior of an elevated region or over the edge of a plateau it is warm or even hot, and dry.

The hot winds of the African coast are of this character and the Algerian *scirocco*, the Tunisian *chili*, the Tripolitan *gibli*, and the Egyptian *khamsin* are also dry winds of the same type blowing from the dry and elevated interior of the country towards a depression which is passing along the Mediterranean within effective distance of its southern shore. (*Monthly Meteorological Charts of the Mediterranean Basin.* M.O. publication, No. 224, London, 1919.)

The simoom is a hot suffocating local wind or whirlwind common in the deserts of Africa and Arabia. The cyclonic system with which the simoom is associated moves slowly forward, generally from South to North or from East to West. It often carries with it hugh rotating columns of sand. Spring and summer are the usual times of its appearance: it seldom lasts many minutes, not more than twenty at the outside (*Chambers's Encyclopædia*, s.v. Simoom). Its correlative in Greek literature appears to be Typhon, from which we derive the name of the storms of the China seas. The *Meteorological Glossary* suggests only a hot dry dusty wind without reference to rotation.

The name *haboob* is given to a dust-raising wind in the Sudan. "It is used generally to imply the passage of a dense mass of whirling sand usually accompanied by a strong wind." Haboobs are particularly associated with the rainy season—May to October. Thunder and lightning follow one in three during the rainy season. Half of the haboobs which occurred at Khartoum during the period May to September in the eight years 1916–23 were associated with rain, while during July-September the ratio was 70 per cent.[1].

A local name for wind of the Mesopotamian region is *shamal*. It denotes the North Westerly winds of summer which belong to the monsoon circulation.

The name *gregale*[2] is given to "a strong North East wind which occasionally blows [on the Mediterranean] in the winter months with great fury and force for two or three days together." It is said to be the modern equivalent of the Euroclydon (Euraquilo) which brought disaster to St Paul's ship.

[1] L. J. Sutton, *Q. J. Roy. Meteor. Soc.* vol. LI, 1925, pp. 25–30.
[2] *New Oxford Dictionary*, quot. from *Ency. Brit.* 1883.

EGYPT. Lat. 29° 51′ N. Long. 31° 20′ E. Alt. 115·7 m. HELWAN.

WIND-QUADRANTS AND VELOCITY																Month; and gauge-reading of Nile at Roda (Cairo)	MOISTURE						
Hours of observation																	≡	Normal vapour pressure			Normal humidity		
8 N	14 N	20 N	8 E	14 E	20 E	8 S	14 S	20 S	8 W	14 W	20 W	8 Calm	14 Calm	20 Calm	↗		8	8	14	20	8	14	20
n	n	n	n	n	n	n	n	n	n	n	n	n	n	n	m/s		n	mb	mb	mb	per cent.		
8	13	15	13	4	10	8	5	4	2	9	3	½	0	0	3·6	Jan. 14·9 m	5	8	8	8	69	42	56
4	6	12	11	2	6	9	5	5	3	15	5	1	0	½	3·5	Feb. 14·6 m	3	8	7	8	64	34	48
13	15	16	11	5	10	5	3	3	2	8	2	0	0	0	4·9	March 14·3 m	1	9	7	8	63	29	43
14	14	18	6	4	10	6	3	2	4	10	1	0	½	0	5·4	April 14·0 m	0	11	8	9	55	23	36
18	19	18	6	3	10	3	2	1	4	7	2	½	½	½	5·4	May 13·8 m	0	12	8	10	50	20	32
16	19	22	8	3	7	2	½	0	3	7	2	1	½	0	5·4	June 13·8 m	0	16	10	12	55	21	33
24	23	27	3	1	1	½	0	0	3	7	3	½	0	0	5·1	July 14·4 m	1	19	12	14	65	24	34
25	22	29	2	1	1	0	0	½	4	8	1	1	0	0	4·8	Aug. 17·4 m	0	21	13	16	69	27	40
22	24	26	5	2	4	0	0	0	2	5	0	1	0	0	5·1	Sept. 18·9 m	0	19	12	16	68	30	48
16	25	21	11	4	10	1	0	0	3	2	0	0	0	0	5·8	Oct. 18·9 m	1	19	12	14	67	33	51
11	12	16	12	4	8	6	5	4	1	9	2	0	0	0	4·1	Nov. 17·1 m	3	13	11	12	68	37	54
8	9	12	12	2	10	10	8	5	2	12	4	0	0	0	3·6	Dec. 15·6 m	4	9	9	9	70	44	58
179	201	232	100	35	87	51	32	25	33	99	25	6	2	1	4·7	Year 15·6 m	18	14	10	11	64	30	44
3 years, 1906–1908																Period 1873–1910	7 years 1904–1910	9 years, 1904–1912					

TEMPERATURE								Month	PRECIPITATION									
Extremes recorded		Mean of extremes of months		Normal daily		< 283	> 300		●			Number of days						
									Norm.	Monthly extremes		● in mm			[symbol]		[symbol]
Max.	Min.	Max.	Min.	Max.	Min.					Max.	Min.	< 1	1–10	> 10			
t	t	t	t	t	t	n	n		mm	mm	mm	n	n	n	n*		n†
302	275	299	276	291	281	27	0	Jan.	10	37	1·4	0·8	2·0	0·2	1·3		0·1
305	276	300	277	293	282	20	1	Feb.	5	25	0	0·8	1·8	0·0	1·6		0·2
312	276	304	279	297	284	14	7	March	6	25	0	0·3	0·6	0·2	1·9		0·1
317	279	311	281	301	287	3	16	April	6	50	0	0·1	0·3	0·2	2·0		0·5
319	284	314	285	305	290	0	26	May	1	10	0	0·1	0·0	0·1	1·6		0·3
319	286	315	289	308	293	0	30	June	0	0	0	0	0	0	1·4		0·1
316	289	313	291	308	294	0	31	July	0	0	0	0	0	0	0·7		0·0
315	290	312	292	308	294	0	31	Aug.	0	0	0	0	0	0	0·7		0·0
314	288	309	290	305	293	0	30	Sept.	0	0	0	0	0	0	0·4		0·0
313	283	309	287	303	291	0	24	Oct.	0	3	0	0·1	0·1	0	1·3		0·4
311	279	303	282	298	287	2	8	Nov.	2	13	0	0·3	0·5	0·1	0·7		0·1
302	274	297	279	293	283	17	0	Dec.	4	19	0	0·7	1·1	0·1	0·7		0·1
319	274	315	276	301	288	83	204	Year	34	91	5	3·2	6·4	0·9	14		2
1904–1920		1904–1912		1904–1920		1904–1918		Period	1904 to 1920	1904–1912		1905–1918			1915 to 1923		1907 to 1923

* A sandstorm is here taken to be a wind carrying dust and reaching at some time a velocity of not less than 40 km per hour.
† The data are for Cairo (Giza, Abbassia, Ezbekia).

CHALDAEA (IRAQ).   Lat. 32° 30′ N. Long. 44° 20′ E. Alt. about 30 m. BABYLON.

WIND-QUADRANTS AND VELOCITY / MOISTURE

Hours of observation (7, 14, 20·5)

Month; level of Euphrates near Babylon	N (7/14/20·5)	E	S	W	Calm	↗ m/s	≡	Normal vapour pressure (7/14/20·5) mb	Normal humidity (7/14/20·5) per cent
Jan. 1·0 m	7 / 11 / 9	5 / 4 / 7	6 / 6 / 6	13 / 10 / 7	0 / 0 / 2	3·0	2·5	7 / 7 / 7	84 / 49 / 69
Feb. 2·0 m	6 / 10 / 10	6 / 4 / 7	5 / 6 / 5	10 / 8 / 5	1 / 0 / 1	3·8	0·5	7 / 7 / 7	73 / 40 / 53
March 3·0 m	8 / 11 / 9	6 / 4 / 7	6 / 7 / 6	10 / 9 / 7	1 / 0 / 2	3·6	0·5	9 / 8 / 8	66 / 30 / 44
April 3·5 m	9 / 11 / 10	5 / 3 / 8	5 / 5 / 5	10 / 10 / 5	1 / 1 / 2	3·6	0	11 / 10 / 10	58 / 27 / 40
May 3·3 m	10 / 12 / 13	5 / 3 / 6	4 / 5 / 4	11 / 11 / 6	1 / 0 / 2	3·4	0	14 / 12 / 13	47 / 22 / 34
June 2·0 m	14 / 16 / 13	1 / 1 / 2	1 / 2 / 2	14 / 11 / 11	0 / 0 / 2	4·4	0	13 / 11 / 12	36 / 17 / 26
July 1·3 m	14 / 16 / 11	1 / 0 / 2	1 / 1 / 2	15 / 14 / 14	0 / 0 / 3	4·4	0	13 / 11 / 13	33 / 13 / 25
Aug. 1·0 m	12 / 15 / 12	1 / 1 / 2	1 / 1 / 2	17 / 14 / 12	0 / 0 / 3	3·9	0	12 / 11 / 11	32 / 13 / 22
Sept. 0·0 m	11 / 14 / 12	2 / 1 / 3	1 / 2 / 2	16 / 12 / 9	1 / 1 / 4	3·0	0	11 / 9 / 10	38 / 13 / 23
Oct. 0·0 m	10 / 13 / 11	4 / 3 / 5	4 / 5 / 4	12 / 10 / 7	1 / 0 / 4	3·0	0	10 / 10 / 10	49 / 21 / 34
Nov. 1·0 m	7 / 12 / 9	4 / 3 / 5	3 / 4 / 4	14 / 11 / 9	2 / 0 / 3	2·8	1	9 / 9 / 10	68 / 34 / 50
Dec. 0·8 m	8 / 12 / 9	3 / 3 / 5	4 / 5 / 4	15 / 11 / 10	1 / 0 / 3	2·8	2	8 / 9 / 9	84 / 48 / 68
Year 1·7 m	116 / 153 / 128	43 / 30 / 59	41 / 49 / 46	157 / 131 / 102	9 / 2 / 30	3·4	6·5	10 / 9 / 10	56 / 27 / 41

5½ years, June 1907–December 1912 | Period uncertain 1911–1912 | 5½ years, 1907–1912

TEMPERATURE / PRECIPITATION

Month	Extremes recorded Max	Min	Mean of extremes of months Max	Min	Normal daily Max	Min	Precip. Norm.	Monthly extremes Max	Min	days mm <1	days mm 1 & over & <10	days mm ≥10	⚶	❄	⚡
Jan.	297	267	294	270	287	276	24	36	2	6	2·6	1	½		1
Feb.	301	270	298	272	292	279	9	15	5	6	2·4	0	2		2
March	308	274	304	277	297	282	28	66	4	6	1·8	1	2		2
April	314	278	311	281	303	288	5	13	2	6	1·6	0	1		5
May	319	287	316	288	309	293	1	1	0	6	0·2	0	1½		5
June	322	289	319	291	314	296	0	0	0	2	0·0	0	¼	Very rare	1
July	322	289	320	294	317	297	0	0	0	0·2	0·0	0	0		0
Aug.	323	290	320	293	317	297	0	0	0	0	0·0	0	0		0
Sept.	320	287	319	289	314	294	0	42	0	0·4	0·0	0	0		0
Oct.	313	281	312	283	307	289	10	42	0	0·4	0·0	0	0		1
Nov.	306	270	303	275	297	282	11	18	0	4	1·4	1	0		2
Dec.	300	266	296	270	290	277	20	52	3	6	3·0	0·5	½		2
Year	323	266	320	270	304	287	109	156	53	46	13	3	8		21

5½ years, June 1907–December 1912 | 5½ years, 1907–1912 | 1911–1912 | 1907–1912

PALESTINE. Lat. 31° 48′ N. Long. 35° 11′ E. Alt. 748 m. JERUSALEM.

WIND-QUADRANTS† AND GALES — MOISTURE†

N 7	N 13	N 21	E 7	E 13	E 21	S 7	S 13	S 21	W 7	W 13	W 21	Calm 7	Calm 13	Calm 21	⟋	Month	≡	VP 7	VP 13	VP 21	Hum 7	Hum 13	Hum 21
n	n	n	n	n	n	n	n	n	n	n	n	n	n	n	n		n	mb	mb	mb	per cent.		
3	2	3	8	9	7	4	6	4	10	12	12	7	1	5	3·6	Jan.	2·4	7	8	8	82	66	79
2	2	2	6	7	6	4	5	3	11	14	12	5	1	5	3·8	Feb.	1·7	7	9	8	76	58	77
3	2	2	6	8	6	4	5	2	12	16	15	6	1	5	4·2	March	1·4	9	9	9	76	57	77
2	3	2	7	7	4	4	4	3	12	15	15	5	1	6	1·5	April	0·6	10	10	10	63	42	66
3	5	4	8	6	3	3	3	2	12	16	18	5	0	3	0·8	May	0·4	11	11	11	55	33	59
5	8	5	3	2	1	1	2	2	18	18	22	3	0	1	0·8	June	0·8	13	13	15	58	32	66
5	8	5	1	1	1	1	2	2	22	20	23	3	0	1	0·4	July	0·6	15	15	17	65	35	70
5	9	6	1	1	0	0	1	1	20	19	22	4	1	2	0·1	Aug.	1·4	16	15	18	71	36	74
5	9	6	2	2	1	0	1	0	15	18	22	7	1	1	0·1	Sept.	2·1	16	14	17	72	36	73
5	5	4	6	9	4	2	3	1	11	13	17	8	1	5	0·3	Oct.	0·8	13	12	14	64	36	63
3	3	3	8	12	8	2	4	1	8	10	12	8	1	6	0·8	Nov.	1·4	10	10	11	73	50	72
3	3	3	8	11	8	3	5	3	9	12	12	7	1	5	3·3	Dec	1·6	8	9	9	78	60	76
44	59	45	64	75	49	28	41	24	160	183	202	68	9	45	19·7	Year	15·2	11	11	12	70	45	71
1904–1913															About 12 years	Period	About 12 years	1904–1913			About 12 years		

TEMPERATURE — PRECIPITATION

Extremes recorded Max.	Min.	Mean of extremes of months Max.	Min.	Normal daily Max.	Min.	< 273 t	> 300 t	Month	Norm.	Extremes Max.	Min.	mm ⊥–1·1	mm 1–10	mm 10 >	❄	▲	⏚
t	t	t	t	t	t	n	n		mm	mm	mm	n	n	n	n	n	n
295	267	289	272	283	277	2·5	0	Jan.	157	324	52	2	7	5	1·4	0·3	0·4
298	271	292	274	287	278	0·3	0	Feb.	137	267	16	1·5	5	4	0·5	0·8	0·7
305	271	297	275	289	280	0	0	March	99	236	7	2	4	4	0·2	0·6	1·2
308	273	303	276	294	283	0	5	April	43	205	0	2	3	1·5	0·0	0·1	1·0
312	277	307	280	299	286	0	13	May	6	28	0	2	1	0·2	0·0	0·1	0·8
310	282	308	284	302	288	0	23	June	0	0·3	0	0·1	0	0	0·0	0·0	0·1
310	285	307	287	304	290	0	30	July	0	0·3	0	0·3	0	0	0·0	0·0	0·0
311	285	307	287	304	290	0	30	Aug.	0	0	0	0	0	0	0·0	0·0	0·0
309	281	305	285	302	289	0	23	Sept.	0·6	4	0	0·2	0·2	0	0·0	0·0	0·0
307	278	304	282	299	287	0	9	Oct.	11	68	0	0·8	2	0·5	0·0	0·0	0·8
303	273	298	277	291	283	0·1	0·2	Nov.	58	191	0	1	3·5	2	0·0	0·1	1·4
298	269	293	274	286	279	0·7	0	Dec.	138	299	29	2	5	4	0·7	0·7	1·0
312	267	309	271	295	284	3·6	134	Year	649	897	266	13	31	20	2·9	2·7	7·4
1895–1913		About 12 years				1896 to 1907	1904 to 1913	Period	1846 to 1915	1895–1913		1904–1913			About 12 years		

† The times of observation are 7 h, 13 h and 21 h up to the end of March 1908, from April 1908 the times are 7 h, 13 h and 20 h, and from the beginning of October 1908 the times are 7½ h, 13 h and 20 h.

SYRIA.   Lat. 33° 54′ N.   Long. 35° 28′ E.   Alt. 35 m.   BEIRUT.

WIND-QUADRANTS AND GALES / MOISTURE

N (8¼,14½,20½)			E (8¼,14½,20½)			S (8¼,14½,20½)			W (8¼,14½,20½)			Calm (8¼,14½,20½)			⚐	Month	≡	Normal vapour pressure (8¼,14½,20½)			Normal humidity (8¼,14½,20½)		
n	n	n	n	n	n	n	n	n	n	n	n	n	n	n	n			n	mb	mb	mb	per cent.		
2	8	3	9	3	8	12	7	12	4	10	5	3·5	3	3	0·8	Jan.	0·0	10	11	11	69	66	73
1·5	7	4	8	3	7	10	7	9	3	10	5	5	1·5	3	0·7	Feb.	0·0	11	12	12	71	67	74
4	9	7	6	3	5	9	6·5	9	7	11	6	6	2	4	0·8	March	0·0	13	14	13	70	67	74
5	8	6	3	3	4·5	8	6	8	9	11·5	7	4	1	5	0·4	April	0·2	15	16	16	71	67	78
5	10	6	2	2	4	8	5	6	12	13	7	3	1	8	0·1	May	0·3	19	19	19	71	66	79
3	6	4	1	1	2	10	6	7	14	16	9	2	0	8	0·0	June	0·0	22	22	22	69	63	79
1	3	2	0	0	1	12	8	9	17	19	12	1	0	7	0·1	July	0·0	25	26	26	67	61	78
2	5·5	4	1	0	1	11	5·5	7	16	19	10	2	0	9	0·0	Aug.	0·0	25	26	26	65	59	75
5	11	8	2	1	3	8	3	5	11	14	7	3	0	7	0·3	Sept.	0·0	22	23	23	63	59	72
6	14	11	5	3	6	7	3	5	6	9	4	7	1	5	0·2	Oct.	0·0	20	21	21	65	61	72
2	10	5	6	3	8	10	5	9	4	9	4	8	3	4	0·5	Nov.	0·0	14	16	15	66	63	70
2	8	3	9	3·5	8	13	7	11	3	9	4·5	4	3	3	0·7	Dec.	0·0	12	13	12	69	66	72
39	101	65	52	26	58	118	69	98	107	152	79	49	17	65	4·6	Year	0·8	17	18	18	68	64	75

1876–1900	Period	1876–1900	1887–1906	1876–1900

TEMPERATURE / PRECIPITATION

Extremes recorded Max	Min	Mean of extremes of months Max	Min	Normal daily Max	Min	< 278 t	< 303 t	Month	Norm.	Extremes Max	Min	mm 0·1–0·9	mm 1·0–9·9	mm ≥10	❄	▲	ℝ
t	t	t	t	t	t	n	n		mm	mm	mm	n	n	n		n	n
299	275	294	278	290	283	1·3	0	Jan.	186	380	26	1·6	7	7		1·2	1·9
301	275	295	279	290	283	1·2	0	Feb.	161	400	35	1·8	7	6		1·6	2·1
306	275	301	280	292	285	0·6	0·3	March	99	217	36	1·5	6	3		1·6	1·8
309	279	303	283	295	287	0	0·8	April	53	159	1	1·4	3	2		0·4	1·0
311	283	305	286	298	290	0	2	May	15	65	0	0·9	2	0·4		0·1	0·7
311	286	306	290	301	293	0	6	June	6	69	0	0·2	0·4	0·2		0·1	0·2
310	291	306	293	304	295	0	25	July	0·6	10	0	0·1	0·2	0·0		0	0·0
310	290	307	294	305	296	0	28	Aug.	0·6	7	0	0·1	0·1	0·0		0	0·0
310	289	306	293	303	295	0	15	Sept.	9	61	0	0·4	0·8	0·2		0	0·4
311	284	305	289	301	293	0	4	Oct.	48	187	0	0·6	2·5	1·5		0·1	1·8
305	278	301	284	296	289	0	0·2	Nov.	133	389	0	1·2	5	4		0·4	3·1
302	272	297	279	292	285	0·5	0	Dec.	194	347	6	1·1	7	6		1·2	2·5
311	272	308	277	297	290	3·7	81	Year	906	1306	591	11	41	30		6·7	15·4

(❄ column note: Does not occur in Beirut and is very rare on the mountains)

1876–1900	Period	1876–1900

CRETE. Lat. 35° 20′ N. Long. 25° 8′ E. Alt. 27·1 m. CANDIA.

WIND-QUADRANTS AND VELOCITY						Month	MOISTURE			
Hour of observation					↗ Mean of day		Normal vapour pressure		Normal humidity	
8 N	8 E	8 S	8 W	8 Calm			8	20	8	20
n	n	n	n	n	m/s		mb	mb	per cent.	
8	1	12	5	5	1·8	Jan.	9	10	69	69
5	1	11	4	7	1·6	Feb.	9	10	69	70
6	0·5	9·5	5	10	1·5	March	10	11	64	69
6	1	5	4	14	1·2	April	11·5	12	61	67
9	1	3	5	13	1·0	May	14	15	59	67
10	1	1	6·5	11·5	1·0	June	17	18	55	65
11·5	0·5	0	10	9	1·3	July	19	20	55	64
11·5	0·5	0	10·5	8·5	1·3	Aug.	20	20	57	63
9·5	0·5	2·5	7	10·5	1·3	Sept.	18	19	60	67
7	0·5	6	7	10·5	1·2	Oct.	15	16	65	69
4·5	0·5	12	7·5	5·5	1·6	Nov.	13	13	68	68
4·5	0·5	15	6·5	4·5	1·5	Dec.	11	11	73	72
93	8	77	78	109	1·3	Year	14	15	63	67
1908–1920					1911–1920	Period	1908–1918			

TEMPERATURE							Month	PRECIPITATION ●				Number of days	
Extremes recorded		Mean of extremes of months		Normal daily				Norm.	Monthly extremes		●		
Max.	Min.	Max.	Min.	Max.	Min.				Max.	Min.	mm ●·1 v	mm 1 ∧	
t	t	t	t	t	t			mm	mm	mm	n	n	
295	275	293	278	288	282		Jan.	91	197	59	2	11	
297	274	293	278	288	282		Feb.	90	176	19	2	9	
303	275	297	279	290	283		March	45	94	1	1	6	
305	280	300	282	293	285		April	26	48	8	2	3	
309	282	304	284	296	288		May	20	31	Drops	1	2	
319	284	309	288	300	292		June	2	16	0	0	1	
314	290	307	291	302	295		July	1	13	0	0	0	
313	291	306	292	302	295		Aug.	3	33	0	0	0	
311	287	305	289	300	293		Sept.	16	75	0	0	1	
308	284	303	286	297	290		Oct.	46	153	3	1	3	
304	279	300	282	293	286		Nov.	87	132	57	2	7	
298	277	295	280	290	284		Dec.	108	186	17	2	10	
319	274	311	277	295	288		Year	535	710	295	13	53	
1908–1920		1908–1918		1908–1920			Period	1908–1920	1908–1918				

● Days ≥ 0·1 mm and < 1 mm.

GREECE. Lat. 37° 58′ N. Long. 23° 44′ E. Alt. 107 m. ATHENS.

WIND-QUADRANTS AND GALES — MOISTURE

N (8 14 21)			E (8 14 21)			S (8 14 21)			W (8 14 21)			Calm (8 14 21)			⌐	Month	≡	Normal vapour pressure (8 14 21)			Normal humidity (8 14 21)		
n	n	n	n	n	n	n	n	n	n	n	n	n	n	n	n		n*	mb mb mb			per cent.		
13	11	13	7	6	6	5	7·5	6	3	6·5	4	3	0	2	2	Jan.	8	8	9	8	79	68	78
11	9	11	6	5	5	4	7	5	2	7	3	5	0	3	2	Feb.	9	9	9	9	78	65	76
8	7	8	5	4	4	7	11	10	4	9	6	7	0	3	1	March	10	10	10	10	75	61	74
6	5	6	4	4	4	10	12	10	5	9	5	5	0	5	1	April	7	12	12	12	69	54	71
7	5	6	3	3·5	3	10	13·5	10·5	6	9	5	5	0	6	1	May	8	15	15	14	63	50	67
6·5	6	5·5	4·5	4	3·5	9	11	11	6	9	5·5	4	0	4	0	June	5	17	16	17	56	43	60
11	10	11	7	7	6	6	8	7	5	6	4	3	0	3	1	July	4	18	17	18	50	38	54
12	9	10	6	7	6	5	8	6	4	6	4	4	0	5	1	Aug.	3	17	17	18	50	37	48
9		10	7	6	5·5	4	8·5	6	2	6	3·5	5	0·5	4·5	1	Sept.	6	17	16	17	60	45	61
8	7·5	7	6	4·5	5	8	10·5	9	3	8	4	6	0·5	6	0	Oct.	8	16	16	16	71	57	72
10	8	9·5	6	5	5	6	10	8	3	7	3·5	5	0	4	1	Nov.	9	12	13	12	79	67	77
12	10	12	7	5	5	5	8	6	3	7	4	4	0·5	4	2	Dec.	11	10	10	10	79	69	77
13	96	109	68	61	58	79	115	94	46	89	51	56	1	49	13	Year	88	13	13	13	68	63	68
1878–1893															1891–1911	Period	1894–1911	1885–1893			1860–1863 1885–1893		

TEMPERATURE — PRECIPITATION

Extremes recorded Max.	Min.	Mean of extremes of months Max.	Min.	Normal daily Max.	Min.	<273	Month	Norm.	Monthly extremes Max.	Min.	● ▲ or ✸ (0/∧)	✸	▲	R
t	t	t	t	t	t	n		mm	mm	mm	n	n	n	n
297	267	290	273	285	278	2	Jan.	56	166	2	13	1·8	0·1	1·0
296	267	292	274	286	279	1	Feb.	38	110	1	11	1·5	0·2	0·8
301	266	295	275	289	281	0·4	March	37	86	5	11	1·1	0·3	0·8
306	275	299	280	293	284	0	April	22	64	2	9	0·1	0·1	1·0
311	279	305	284	298	288	0	May	21	76	0	7	0·0	0·2	1·8
313	285	307	289	303	293	0	June	11	42	0	5	0·0	0·1	2·1
314	287	310	292	306	295	0	July	8	51	0	3	0·0	0·1	1·7
314	288	310	292	305	295	0	Aug.	11	63	0	3	0·0	0·1	1·7
312	282	307	288	302	292	0	Sept.	14	56	0	4	0·0	0·0	1·7
308	277	302	284	296	288	0	Oct.	45	211	0	9	0·0	0·1	2·7
301	274	296	278	291	284	0	Nov.	75	254	3	12	0·1	0·3	2·3
295	269	292	275	287	280	0·4	Dec.	63	197	2	13	0·9	0·3	2·5
314	266	311	271	295	287	4	Year	406	847	206	98	5·6	2·1	19
1840–1893		1853–1893				1847–1893	Period	1858–1894				1840–1894	1858–1894	1858–1893

* Eginitis in *Le Climat d'Athènes*, p. 192, writes: "Le brouillard est un phénomène rare à Athènes. D'après les observations faites depuis 1840 jusqu'aujourd'hui le nombre annuel de jours de brouillard est à peine 1." The figures quoted above are from the data in *Annales de l'Observatoire Nationale d'Athènes*.

ITALY.　　　　　　　Lat. 41° 54′ N.　Long. 12° 28′ E.　Alt. [50·6] m.　ROME.

WIND-QUADRANTS AND VELOCITY | MOISTURE

N (7,13,21)	E (7,13,21)	S (7,13,21)	W (7,13,21)	Calm (7,13,21)	velocity (m/s)	Month	≡ (n)	Normal vapour pressure (7,13,21) mb	Normal humidity (7,13,21) per cent
21 / 17 / 17	5 / 5 / 5	3 / 5 / 5	1 / 2 / 2	1 / 1 / 2	3·0	Jan.	1·6	7 / 8 / 8	76 / 59 / 74
18 / 10 / 10	4 / 4 / 3	4 / 8 / 9	1 / 5 / 3	1 / 1 / 4	2·8	Feb.	1·5	8 / 8 / 8	73 / 55 / 73
17 / 7 / 7	5 / 3 / 3	6 / 12 / 12	2 / 9 / 5	2 / 1 / 3	2·9	March	1·9	9 / 9 / 9	68 / 52 / 73
14 / 4 / 5	5 / 2 / 3	7 / 12 / 12	2 / 11 / 6	2 / 0 / 3	2·6	April	1·7	11 / 11 / 11	65 / 54 / 75
15 / 4 / 5	5 / 1 / 2	6 / 13 / 13	3 / 13 / 8	2 / 0 / 3	2·4	May	1·5	13 / 12 / 13	59 / 50 / 71
16 / 3 / 3	5 / 1 / 2	5 / 12 / 13	2 / 15 / ·10	2 / 0 / 2	2·2	June	1·3	15 / 15 / 16	54 / 46 / 69
17 / 3 / 3	4 / 0 / 1	5 / 11 / 14	2 / 16 / 9	3 / 0 / 3	2·2	July	1·8	16 / 15 / r8	48 / 40 / 63
18 / 3 / 4	4 / 1 / 2	4 / 12 / 13	2 / 15 / 9	2 / 0 / 4	2·1	Aug.	2·1	17 / 16 / 19	52 / 41 / 65
16 / 4 / 5	5 / 2 / 3	4 / 11 / 12	2 / 12 / 6	2 / 1 / 5	2·1	Sept.	1·7	16 / 15 / 17	62 / 49 / 71
18 / 7 / 8	5 / 3 / 4	6 / 12 / 10	1 / 7 / 4	2 / 2 / 5	2·4	Oct.	1·6	13 / 13 / 14	72 / 58 / 76
19 / 12 / 13	4 / 5 / 5	5 / 8 / 7	1 / 3 / 2	1 / 2 / 4	2·6	Nov.	1·5	10 / 11 / 11	78 / 61 / 78
20 / 16 / 17	5 / 6 / 5	4 / 6 / 6	1 / 2 / 1	1 / 2 / 2	2·8	Dec.	1·8	8 / 8 / 8	77 / 62 / 76
209 / 90 / 97	56 / 33 / 38	59 / 122 / 126	20 / 110 / 65	21 / 10 / 40	2·5	Year	20·0	12 / 12 / 13	65 / 52 / 72
1876–1905						Period	1825–1920	1879–1890	

TEMPERATURE | PRECIPITATION

Extremes recorded Max.	Min.	Mean of extremes of months Max.	Min.	Normal daily Max.	Min.	< 273 t (n)	> 303 t (n)	Month	Monthly extremes Norm. mm	Max. mm	Min. mm	0·1–5 mm (n)	5·1–10 mm (n)	>10 mm (n)	❄ (n)	▲ (n)	℞ (n)
292	265	288	271	284	277	4·5		Jan.	83	199	10	5·5	2·3	2·6	0·6	0·7	0·7
293	265	290	272	286	277	2·6		Feb.	65	186	0	4·6	2·2	2·1	0·5	0·7	0·7
297	260	293	274	288	279	0·9		March	73	190	0·1	5·9	2·2	2·2	0·3	0·8	1·0
302	271	296	277	291	282	0		April	65	229	0·1	5·6	2·0	2·3	0·0	0·6	1·4
305	276	301	281	296	285	0		May	55	148	0·7	4·3	1·6	1·8	0·0	0·4	2·0
309	282	305	285	300	289	0		June	39	134	0·2	3·2	1·1	1·3	0·0	0·2	2·5
315	284	307	288	303	291	0	18	July	19	113	0·0	1·2	0·5	0·6	0·0	0·2	2·2
310	284	307	288	303	291	0	16	Aug.	26	107	0·0	1·9	0·4	0·9	0·0	0·1	2·7
307	279	303	284	299	289	0		Sept.	64	195	0·0	3·4	1·3	2·0	0·0	0·0	3·1
305	273	299	279	294	285	0		Oct.	127	346	8	4·5	2·0	3·6	0·0	0·2	2·9
298	269	293	274	289	281	0		Nov.	114	373	16	5·5	2·1	3·8	0·0	0·3	1·7
294	267	289	271	285	278	3·2		Dec.	98	274	0·1	5·1	2·4	3·2	0·3	0·7	1·1
315	265	308	270	293	284	11		Year	828	1470	319	50·7	20·1	26·4	2	4·8	22
1831–1910								Period	1782–1910			1827–1911			1776 to 1910	1811 to 1910	1782 to 1910

SICILY. Lat. 37° 3′ N. Long. 15° 15′ E. Alt. 23·5 m. SYRACUSE.

WIND-QUADRANTS AND GALES / MOISTURE

Hours of observation					Mo..th	≡ (nebbia fitta)	Normal vapour pressure 9 15 21	Normal humidity 9 15 21
N 9 15 21	E 9 15 21	S 9 15 21	W 9 15 21	Calm 9 15 21 / 〰				
n n n	n n n	n n n	n n n	n n n / n		n	mb mb mb	per cent.
5 / 8 / 7	5 / 10 / 6	4 / 6 / 5	17 / 8 / 14	0 / 0 / 0 / 1·6	Jan.	0·0	10 / 10 / 9	73 / 64 / 73
5 / 7 / 7	6 / 10 / 7	4 / 5 / 5	13 / 6 / 9	0 / 0 / 0 / 1·4	Feb.	0·1	10 / 10 / 9	73 / 65 / 73
7·5 / 6 / 7	8 / 11·5 / 8·5	5 / 7 / 5	10 / 6 / 11	0 / 0 / 0 / 0·4	March	0·9	11 / 11 / 11	70 / 64 / 74
6 / 4 / 5	8 / 10 / 7	5 / 7 / 6	11 / 9 / 12	0 / 0 / 0 / 0	April	0·3	12 / 13 / 12	71 / 65 / 75
12 / 8 / 10	8 / 11 / 8	3 / 9 / 6	6 / 3 / 6·5	1 / 0 / 0 / 0·2	May	0·3	14 / 15 / 15	69 / 63 / 73
13 / 7 / 11	7 / 11 / 7	4 / 9 / 7	4 / 2 / 4	2 / 0 / 0 / 0	June	0·6	18 / 19 / 18	67 / 61 / 73
17 / 6 / 13	5 / 12 / 8	3 / 11 / 5	4 / 2 / 4	2 / 1 / 1 / 0	July	0·4	21 / 21 / 22	63 / 55 / 70
11·5 / 8 / 12	6 / 9 / 7	5 / 12 / 6	6 / 2 / 5	3 / 0 / 0 / 0·2	Aug.	0·0	22 / 22 / 24	65 / 59 / 73
8 / 5 / 9	6 / 10 / 5	4 / 11 / 6	10 / 4 / 10	1 / 0 / 1 / 0·4	Sept.	0·3	20 / 21 / 21	71 / 63 / 76
6 / 6 / 6	4 / 7 / 4	7 / 9 / 8	13 / 8·5 / 12	1 / 1 / 0 / 0·4	Oct.	0·4	17 / 17 / 17	72 / 66 / 76
5 / 6 / 5	4 / 10 / 5	4 / 6 / 5	17 / 8 / 15	0 / 0 / 0 / 1·8	Nov.	0·9	13 / 13 / 13	71 / 65 / 73
5 / 7 / 7	5 / 9 / 6	3 / 6 / 4	17·5 / 9 / 13	0 / 0 / 0 / 3·0	Dec.	0·1	11 / 11 / 10	73 / 65 / 75
101 / 78 / 99	72 / 121 / 79	51 / 98 / 68	129 / 67 / 115	10 / 2 / 2 / 9·4	Year	4·3	15 / 15 / 15	70 / 63 / 73·5
1879–1885					Period	1879–1883	1879–1885	1881–1890

TEMPERATURE / PRECIPITATION

Extremes recorded Max. t / Min. t	Mean of extremes of months Max. t / Min. t	Normal daily Max. t / Min. t	< 283 n	∧ 300 n	Month	Norm. mm	Monthly extremes Max. mm / Min. mm	Number of days < &1 ∧ mm	1–10 mm	∧10 mm	✳	▲ §	℞
295 / 274	290 / 277	287 / 281	24	0	Jan.	95	287 / 3	2	8	3		0·8	0·3
294 / 273	291 / 277	287 / 281	23	0	Feb.	67	239 / 3	2	6	1		0·4	0·4
298 / 276	294 / 278	289 / 282	18	0	March	41	106 / 1	2	5	1		0·7	0·4
300 / 279	296 / 281	292 / 284	6	0	April	39	113 / 1	1	5	1		0·0	1
304 / 281	300 / 283	295 / 287	2	0·7	May	21	59 / 0	0·6	4	0·1		0·1	0·4
309 / 286	305 / 288	300 / 291	0	10	June	5	26 / 0	0·3	0·9	0·1		0·1	0·7
313 / 286	309 / 291	303 / 295	0	26	July	7	109 / 0	0	0·4	0·1	Very rare	0·0	0·3
313 / 285	308 / 292	303 / 295	0	29	Aug.	7	31 / 0	0·4	1	0·1		0·1	1
312 / 286	306 / 289	301 / 293	0	15	Sept.	49	243 / 1	0·6	4	1		0·0	3
306 / 281	301 / 285	296 / 290	0·1	0·1	Oct.	93	310 / 13	2	5	2		0·0	3
301 / 279	296 / 281	292 / 286	4	0	Nov.	109	234 / 3	1	5	2		0·3	0·9
294 / 274	292 / 278	289 / 282	18	0	Dec.	105	222 / 11	2	7	2		0·3	2
313 / 273	310 / 276	295 / 287	95	81	Year	637	1107 / 251	14	52	15		2·8	14
1879–1897		1879–1894	1879–1885		Period	1880–1905		1879–1885			1879 to 1890	1879 to 1890	1879 to 1885

§ Pioggia e grandine.

WESTERN MEDITERRANEAN. Lat. 36° 6′ N. Long. 5° 21′ W. Alt. 16·2 m. GIBRALTAR.

WIND-QUADRANTS AND GALES / MOISTURE

N (7/13/21)			E (7/13/21)			S (7/13/21)			W (7/13/21)			Calm (7/13/21)			⌇	Month	≡	Normal vapour pressure mb (7/13/21)			Normal humidity per cent. (7/13/21)		
5	4	5	11	10	10	2	2	2	13	13	13	1	2	1	1	Jan.	0·2	11	11	11	83	71	80
4	3	3	11	10	11	2	3	2	11	12	12	0	1	1	1	Feb.	0·1	11	12	11	83	70	79
5	3	4	9	9	8	2	3	2	14	15	16	1	1	0	1	March	0·3	11	12	12	83	69	79
5	2	4	11	11	10	2	4	3	12	13	14	1	1	0	0·3	April	0·1	12	13	13	82	66	77
4	2	3	12	12	11	2	3	2	12	13	14	1	1	0	0·3	May	0·1	14	15	15	81	61	76
4	2	3	12	13	11	2	3	3	11	11	12	1	1	1	0·1	June	0·1	16	17	17	79	59	73
3	2	3	14	13	13	1	3	2	11	12	12	1	2	1	0·1	July	0·3	19	20	19	80	61	74
3	1	3	16	15	14	1	3	2	10	10	12	1	2	1	0·0	Aug.	0·2	20	21	21	81	61	75
3	1	2	16	15	14	1	2	2	9	10	10	1	2	2	0·1	Sept.	0·1	19	19	19	82	64	77
4	2	4	13	12	12	2	3	3	11	12	12	1	2	1	0·2	Oct.	0·2	16	16	16	83	66	79
4	3	3	12	11	11	2	3	2	12	13	13	1	1	1	0·8	Nov.	0·1	13	14	14	84	72	80
6	4	5	9	8	8	2	2	2	14	14	15	1	2	1	1	Dec.	0·1	12	13	12	84	71	81
50	28	42	145	139	133	20	34	25	140	147	155	11	16	10	6	Year	2	15	15	15	82	66	77

July 1908–Dec. 1920	33 years	Period	1881–1920	1908–1920

TEMPERATURE / PRECIPITATION

Extremes recorded Max.	Min.	Means of extremes of months Max.	Min.	Normal daily Max.	Min.	<283 t	>300 t	Month	Norm. mm	Monthly extremes Max. mm	Min. mm	mm <1	mm 1–10	mm >10	*	⚡
297	274	293	278	289	282	19	0	Jan.	130	549	0·8	1·5	5·7	4·2		0·9
300	274	294	279	290	283	13	0	Feb.	107	361	0	1·0	6·0	3·2		1·9
299	276	295	280	291	284	12	0	March	122	451	0	1·3	7·9	3·3		1·7
304	279	297	282	293	285	6	0	April	68	204	0	1·1	4·8	1·9		0·5
305	281	300	285	295	288	0·2	0·6	May	45	165	0	0·8	3·7	1·1		0·5
309	282	303	287	298	290	0	7	June	12	99	0	0·4	0·7	0·0		0·3
311	286	305	290	301	293	0	18	July	1	16	0	0·2	0·0	0·0		0·3
311	287	306	290	301	293	0	21	Aug.	4	43	0	0·2	0·2	0·0		0·1
308	283	304	288	299	292	0	8	Sept.	36	216	0	0·8	1·7	0·8		1·3
308	280	300	284	295	289	0·7	0·4	Oct.	84	247*	0·8	1·2	3·5	2·2	Not observed 1916–1920	0·5
299	275	296	281	292	285	5	0	Nov.	161	633	0	1·1	4·8	4·5		0·6
298	272	293	279	290	283	16	0	Dec.	140	599	0·8	0·8	4·5	3·6		0·7
311	272	307	277	294·5	287	72	55	Year	910	1650†	370	10·7	43·6	24·6		9·3

1852–1901 and 1911–1920	1876–1919	1912–1921	Period	1852–1910	1852–1919	July 1908–Dec. 1920	8 years

* 297 mm in 1864 with gauge 25 ft. above ground.   † Omitting the value for 1855 which is incomplete.

CARTHAGE. Lat. 36° 48′ N. Long. 10° 10′ E. Alt. 43 m. TUNIS·

WIND-QUADRANTS						Month			MOISTURE
N	E	S	W	Calm					Normal humidity
n	n	n	n	n					per cent.
11	3	3·5	12·5	1		Jan.			74
9·5	3	3·5	12	0		Feb.			73
8	4·5	7	11·5	0		March			71
8·5	4	6	11·5	0		April			68
11	4·5	5	10·5	0		May			64
10·5	6	4·5	9	0		June			59
10	8·5	3·5	9	0		July			5?
12	9·5	2	7·5	0		Aug.			59
9·5	10	4	6·5	0		Sept.			66
8·5	5	6	11·5	0		Oct.			70
10·5	4·5	5	10	0		Nov.			73
9	2·5	5	14·5	0		Dec.			75
118	65	55	126	1		Year			67
12 years between 1887 and 1905						Period			12 years

TEMPERATURE								PRECIPITATION						
Extremes recorded		Means of extremes of months		Normal daily		< 273	Month	●	Extremes recorded		Number of days			
											●			
Max.	Min.	Max.	Min.	Max.	Min.			Norm.	Max.	Min.	o ^		*	℞
t	t	t	t	t	t	n		mm	mm	mm	n			n
302	273	294	275	288	279		Jan.	49	139	1	11			0
300	274	295	275	289	279		Feb.	52	135	12	10			0
306	274	298	276	292	281		March	46	179	5	11			0·3
313	277	303	279	294	283		April	38	120	7	9			0·5
313	279	306	281	298	286		May	23	71	0	6			0·3
315	283	311	285	303	290	Less than one day per annum	June	12	113	0	3		Very rare	0
316	283	314	287	306	293		July	4	18	0	1			0·1
320	284	315	288	307	292		Aug.	6	31	0·	1			0·4
314	284	310	286	303	291		Sept.	26	108	0	4			0·7
313	278	307	281	298	287		Oct.	50	139	11	9			0·4
305	275	299	278	294	283		Nov.	48	144	3	10			0·1
297	274	293	276	289	280		Dec.	64	163	10	11			0·1
320	273	316	274	297	285		Year	417	580	279	86			2·9
12 years				14–18 years 1887–1914			Period	22–25 years			12 years			

CYRENAICA. Lat. 32° 7′ N. Long. 20° 2′ E. Alt. 9·5 m. BENGHAZI.

WIND-QUADRANTS					Month			MOISTURE
N	E	S	W	Calm				Normal humidity 8 20
n	n	n	n					per cent.
7	4	11	9		Jan.			78 / 76
7	4	9	8		Feb.			74 / 73
11	4	9	8		March			75 / 74
14	3	7	6		April			63 / 72
15	2	7	6		May			65 / 72
20	2	4	4	About 3 calms per annum	June			69 / 77
27	1	1	2		July			83 / 84
24	2	1	4		Aug.			80 / 79
18	2	5	5		Sept.			72 / 74
13	4	9	5		Oct.			72 / 74
8	4	11	7		Nov.			78 / 75
5	4	12	9		Dec.			78 / 76
169	36	87	73		Year			73 / 75
					Period			1902–1905

TEMPERATURE								Month	PRECIPITATION			
Extremes recorded		Means of extremes of months		Normal daily					Norm.	Extremes recorded		No of rain days
Max.	Min.	Max.	Min.	Max.	Min.					Max	Min.	
t	t	t	t	t	t				mm	mm	mm	n
293	281	292	281	289	284			Jan.	77	257	9	12
297	280	296	282	291	284			Feb.	37	141	0	8
304	282	303	283	293	286			March	19	59	0	6
304	284	304	287	296	289			April	4	15	0	2
311	286	309	288	299	292			May	3	14	0	1·5
313	290	308	292	300	294			June	0·5	5	0	0·7
306	292	304	294	301	296			July	0·2	4	0	0
310	294	305	296	302	297			Aug.	0	0·2	0	0
307	290	306	294	302	295			Sept.	3	14	0	2
310	289	306	290	300	293			Oct.	13	66	0	4
300	280	299	285	295	289			Nov.	54	200	0·4	7
298	281	295	283	291	286			Dec.	66	250	21	13
313	280	311	281	296	290			Year	276	617	138	55
1886–1891		1891–1905						Period	1886–1905			

MALTA.   Lat. 35° 54′ N. Long. 14° 31′ E. Alt. 56·4 m (21·5 m for wind observations).

WIND-QUADRANTS AND GALES						Month	MOISTURE		
N	E	S	W	Calm	⤢		≡	Normal vapour pressure	Relative humidity (corrected to 24 hours)
n	n	n	n	n	n†		n	mb	per cent.
9·5	5·5	5·5	9·5	1	0·8	Jan.	0·6	10·4	77
8	5·5	5	8·5	1	0·3	Feb.	0·2	10·9	77
9	6·5	4·5	10	1	0·1	March	0·4	11·6	76
9	7	5	9	0	0·0	April	0·1	13·2	75
/9·5	8·5	4·5	7·5	1	0·2	May	0·2	15·5	73
11	8	2·5	7·5	1	0·0	June	0·2	18·7	70
13·5	6·5	2	8	1	0·1	July	0·2	21·5	66
13	6·5	3	7·5	1	0·1	Aug.	0·2	22·3	68
10	8	5	7	0	0·0	Sept.	0·1	21·4	71
9	8	5·5	8·5	0	0·1	Oct.	0·6	18·5	73
9	6	5	9	1	0·0	Nov.	0·6	14·5	75
8·5	5	5·5	11	1	0·4	Dec.	0·6	11·6	78
119	81	53	103	9	2·2	Year	4·0	15·4	73
26 years					9 years	Period	1889–1923	1884–1901	62–65 years

TEMPERATURE								Month	PRECIPITATION						
Extremes recorded		Means of extremes of months		Normal daily		<273	>300		●			Number of days			
									Norm.	Monthly extremes		●	▲	*	ℝ
Max.	Min.	Max.	Min.	Max.	Min.					Max.	Min.	††			
t	t	t	t	t	t	n	n		mm	mm	mm	n		n	n
297	277	291	280	288	284			Jan.	85	199	9	13		3·4	0·8
296	274	292	279	288	283			Feb.	53	131	0	10		1·3	0·9
301	276	295	280	290	284			March	40	173	2	8		0·9	1·0
307	280	297	282	292	286			April	21	85	2	5		0·5	0·7
305	282	300	285	295	289			May	12	82	0	3		0·3	1·0
310	287	304	289	299	292	None		June	2	14	0	1		0·1	1·0
313	290	308	292	302	295			July	1	29	0	0		0·0	0·3
314	290	307	293	302	296			Aug.	3	32	0	1		0·0	0·4
311	287	305	291	300	294			Sept.	33	235	0	3		0·1	2·6
307	280	302	288	297	292			Oct.	76	224	0	7		0·3	2·3
301	279	297	284	293	288			Nov.	87	252	10	11		0·7	2·3
297	277	293	281	290	285			Dec.	104	353	6	14		1·6	1·0
314	274	309	278	295	289			Year	516	830	238	77		9·1	14·3
61–65 years between 1853 and 1923								Period	42 years			65–67 years		16 years	1877–80 1912–17

Note. It is very rare for fog to persist for any length of time, generally a few hours after sunrise the fog dissipates and it is very rare for it to keep on to 8 or 9 o'clock (MS. note by Prof. Agius).
† Beaufort force 8 and over.
†† ≥ 0·3 mm to 1906, ≥ 1 mm 1907–1911, ≥ 0·3 mm 1912–1921, ≥ 0·1 mm 1922 1923.

S.E. ENGLAND. Lat. 51° 28′ N. Long. 0° 19′ W. Alt. 10·4 m. RICHMOND.

WIND-QUADRANTS AND VELOCITY — MOISTURE

N (7 13 18)	E (7 13 18)	S (7 13 18)	W (7 13 18)	Calm (7 13 18)	↙ m/s	Month	≡ 7h	Normal vapour pressure (7 13 18) mb	Normal humidity (7 13 18) per cent.
4 5 4	3 4 5	7 7 8	9 10 9	8 5 5	3·7	Jan.	3	7 7 7	87 80 84
4 5 4	3 3 4	8 9 8	7 9 8	7 2 4	3·8	Feb.	2	7 7 7	85 75 80
5 8 6	3 3 3	8 9 9	7 10 9	8 1 4	3·9	March	1	7 7 7	86 68 74
6 7 7	3 4 5	6 8 7	7 10 9	8 1 2	3·8	April	0·5	8 8 8	84 62 66
9 10 9	4 4 5	5 8 8	6 8 6	7 1 3	3·5	May	0	10 10 10	81 61 63
7 8 7	4 4 5	6 9 9	6 8 7	7 1 2	3·1	June	0·5	12 13 13	80 60 62
5 6 6	3 3 4	5 9 8	9 12 10	9 1 3	2·9	July	0·5	14 14 14	81 59 61
4 5 5	2 3 4	8 10 11	7 11 8	10 2 4	3·0	Aug.	1	14 14 14	85 61 65
5 7 6	4 5 5	5 8 7	4 8 5	12 2 7	2·7	Sept.	3	12 13 13	89 65 73
4 6 4	4 5 5	7 10 9	4 7 5	12 3 8	3·0	Oct.	9	10 11 11	91 73 83
5 6 6	3 4 3	6 7 7	6 9 8	10 4 6	3·3	Nov.	4	8 9 9	90 79 86
3 3 3	3 4 3	11 11 12	7 9 7	7 4 5	3·6	Dec.	1	7 7 7	87 82 86
61 76 67	39 46 52	82 105 103	79 111 91	105 27 53	3·4	Year	26	9 10 10	86 69 74
1896–1915					1881–1910	Period	1911–1915	1886–1910	

TEMPERATURE — PRECIPITATION

Extremes recorded Max.	Min.	Means of extremes of months Max.	Min.	Normal daily Max.	Min.	≷273 t	≶298 t	Month	Norm. mm	Monthly extremes Max.	Min.	● in mm I∨	I–15	I5<	✸	▲	◖
286	260	285	268	279	274	11	0	Jan.	45	124	11	9	8	0	3	0·5	0·1
290	261	285	269	280	275	9	0	Feb.	39	105	2	7	8	0	2	0·5	0·1
292	265	289	270	282	275	9	0	March	43	100	6	9	9	0	3	1	0·5
300	270	292	272	286	277	2	0	April	37	101	1	6	7	0	1	1	1
302	272	297	275	289	280	0·5	0·3	May	44	104	5	5	8	0	0·1	1	1
304	276	299	279	293	283	0	2	June	55	183	6	5	7	0	0	0·3	2
305	279	301	281	295	285	0	5	July	55	124	12	4	8	0	0	0·3	3
308	278	300	281	294	285	0	3	Aug.	57	165	12	6	10	0	0	0·3	2
306	274	297	277	291	283	1	1	Sept.	48	129	11	8	6	0	0	0·2	1
297	269	292	273	286	279	2	0	Oct.	69	151	15	10	10	1	0·1	0·2	0·3
290	266	287	270	283	277	6	0	Nov.	56	101	12	9	9	0	0·5	0·2	0·1
287	261	285	269	280	275	10	0	Dec.	58	162	10	8	11	0	2	0·2	0·1
308	260	301	268	287	279	49	12	Year	606	970	423	86	101	2	12	6	11
1871–1915		1881–1915		1871–1915				Period	1881–1915	1866–1915		1891–1915			1877–1915		

Bibliography of data used in the compilation of the maps and climatic summaries of pp. 23, 26, 29–41.

Helwan.
Meteorological Reports of the Survey Department. Cairo. (Published annually.)
Climatological Normals for Egypt and the Sudan, Candia, Cyprus and Abyssinia. Government Press, Cairo, 1922.
The Weather Map. M.O. publication, No. 225 i. London.
The data for sandstorms and thunderstorms have been supplied by the kindness of the Egyptian Meteorological Service.

Babylon.
Deutsche Uberseeische Meteorologische Beobachtungen. Heft 17–Heft 22.
The Weather Map. M.O. publication, No. 225 i. London.

Jerusalem.
Jahrbücher der k. k. Zentral-Anstalt für Meteorologie und Geodynamik. Wien, 1895–1913.
Zum Klima von Palästina. By Dr F. M. Exner. Leipzig, 1911. (The period of observations on which the summary is based is not stated but it appears to be from 1895 to about 1907 or 1908.)
Über 68-jährige Niederschlagsmessungen in Jerusalem. By J. v. Hann. Met. Zeitschr. vol. xxxv, 1918, p. 310.

Beirut.
Jahrbücher der k. k. Centralanstalt für Meteorologie und Erdmagnetismus. Wien.
Untersuchungen über die klimatischen Verhältnisse von Beirut, Syrien. By Dr S. Kostlivy. Sitz. der Königl. Böhm. Gesell. der Wiss. in Prag. Prague, 1904.

Crete.
Meteorological Reports of the Survey Department. Cairo, 1908–18.
Climatological Normals for Egypt and the Sudan, Candia, Cyprus and Abyssinia. Government Press, Cairo, 1922.

Athens.
Le Climat d'Athènes. By D. Eginitis. Athènes, 1897.
The data for gales and fog are from observations published in Annales de l'Observatoire Nationale d'Athènes, 1891–1911.

Rome.
Il clima di Roma. By F. Eredia. Rome, 1910.
I venti a Roma. By Iginia Massarini. Annali dell' Ufficio Centrale di Meteorologia e Geodinamica, vol. xxvii, parte 1, 1905.
Annali dell' Ufficio Centrale di Meteorologia. Rome, 1879–90.

Syracuse.
Annali dell' Ufficio Centrale di Meteorologia. Rome, 1879–90.
Le precipitazioni atmosferiche in Italia. By F. Eredia. Annali dell' Ufficio Centrale di Meteorologia e Geodinamica, vol. xxvii, parte 1, 1905.

Gibraltar.
MS. returns at the Meteorological Office, London.

Tunis.
Annales du Bureau Central Météorologique de France. 1887–1914.

Benghazi.
Climatologia di Tripoli e Bengasi. Studio eseguito nel R. Ufficio Centrale di Meteorologia e Geodinamica per cura di F. Eredia. (Monogr. e Rapp. Col. N. 4.) Rome, 1912.
Annales du Bureau Central Météorologique de France. 1902–5.

Malta.

MS. returns at the Meteorological Office, London.

Zum Klima von Malta. Ergebnisse 43-jähriger Regenmessungen auf der Insel Malta. By J. v. Hann. Met. Zeitschr. vol. xx, 1903, pp. 71–74, and 180–181.

Richmond.

The Weather Map. M.O. publication, No. 225 i. London.

Book of Normals. M.O. publication, No. 236. London.

British Meteorological and Magnetic Year-Book, Hourly Values, etc. 1912. M.O. publication, No. 209 e. London, 1913.

In the preparation of the maps of the distribution of pressure and rainfall over the Mediterranean reproduced in figs. 3 and 4, the following works have been consulted:

Bartholomew's Physical Atlas, vol. III, Atlas of Meteorology, prepared by J. G. Bartholomew, F.R.S.E., and A. J. Herbertson, Ph.D., edited by Alexander Buchan. Archibald Constable and Co., 1899.

Atlas climatologique de l'Empire de Russie publié par l'Observatoire Physique Central Nicolas, 1849–99 à l'occasion du cinquantième anniversaire de sa fondation, 1849–99. St Pétersbourg, 1900.

Climatological Atlas of India, published by the authority of the Government of India under the direction of Sir John Eliot, K.C.I.E., F.R.S. Issued by the Indian Meteorological Department, 1906.

A Barometer Manual for the Use of Seamen. Meteorological Office publication, No. 61. 9th ed. 1919.

Monthly Meteorological Charts of the Mediterranean Basin. Meteorological Office publication, No. 224. 1917.

A. J. Herbertson. The distribution of rainfall over the land. London, 1901.

Col. H. G. Lyons, D.Sc., F.R.S. Presidential Address: The Distribution of Pressure and the Air-Circulation over Northern Africa. Q. J. Roy. Meteor. Soc. vol. XLIII, 1917, p. 116. Climatic influences in Egypt and the Eastern Sudan. Ibid. vol. XXXVI, 1910, p. 211.

C. A. Angot. Etudes sur le climat de la France. Régime des pluies. Annales du Bureau Central Météorologique, 1911, 1913. Paris, 1918, 1919.

Filippo Eredia. Le precipitazioni atmosferiche in Italia. Annali dell' Ufficio Centrale Meteorologico e Geodinamico italiano, vol. XXVII, parte 1, 1905. Rome, 1908.

Alexander Knox. The Climate of the Continent of Africa. Cambridge University Press, 1911.

Climatological Normals for Egypt and the Sudan, Candia, Cyprus and Abyssinia. Cairo, Ministry of Public Works. Cairo, 1922.

CHAPTER III

THE MEASUREMENT OF TIME: THE KALENDAR AND THE CYCLE OF THE SEASONS

Declination of a star—the angular elevation above or below the equator, the complement of the polar distance.

Right Ascension—the angular distance of the polar circle through the star measured west to east along the equator from the first point of Aries.

Angular velocity of the earth's rotation, 360° in one sidereal day, 72.92×10^{-6} radians per second.

Mean sidereal day, 23·93447 mean solar hours.

Mean solar day, 24 mean solar hours.

True solar day varies from maxima of 24·0036 hours and 24·0083 hours on June 19–20 and December 22–23 to minima of 23·9949 hours and 23·9940 hours on March 26–27 and September 18–19.

One mean solar hour corresponds with rotation through 15° of longitude.

The length of "day" for a sun or star in hours from sunrise to sunset, allowing for refraction, is $n/15$, where

$$\sin \frac{n}{4} = \sqrt{\frac{\sin \{45° + \frac{1}{2}(\phi - \delta + r)\} \cdot \sin \{45° - \frac{1}{2}(\phi - \delta - r)\}}{\cos \phi \cdot \cos \delta}},$$

ϕ is the latitude, δ the northerly declination of the sun or star, r the apparent elevation due to refraction, about 34′

Mean obliquity of the ecliptic, 23° 26′ 56″·55 (1925).

The variation in the length of daylight:

Thebes	(lat. 25° 41′)	from	10·5 hours	to	13·7 hours
Athens	(,, 37° 58′)	,,	9·5	,,	14·8 ,,
Rome	(,, 41° 54′)	,,	9·1	,,	15·2 ,,
London	(,, 51° 28′)	,,	7·7	,,	16·6 ,,
Shetland	(,, 60° 9′)	,,	5·7	,,	18·8 ,,
N. Iceland	(,, 67°)	,,	0·0	,,	24·0 ,,

Mean lunar day, 24·9 hours.

Mean synodic lunar month, 29·53059 days, varying to the extent of 13 hours in the course of the year.

Mean solar or tropical year, 365·2422166 mean solar days.

Perihelion (1924), January 2, 2 h. Sun's distance, 91,341,000 miles.

Aphelion (1924), July 3, afternoon. Sun's distance, 94,495,000 miles.

Precession of the equinox, 50·2564″ + 0·02220″ t per annum, t being the time reckoned from 1900 in centuries.

Annual increase of declination of a star in consequence of precession, 20″ cos a, where a is the Right Ascension.

Vernal equinox, March 20 or 21; Summer solstice, June 21 or 22.

Autumnal equinox, September 23 or 24; Winter solstice, December 22 or 23.

Commencement of quarters of the May Year: May 6, August 8, November 8, February 4.

1st Olympiad of 4 years commences 776 B.C. 293rd Olympiad of 4 years commences A.D. 393.

Lustrum of 5 solar years = 6 Romulian years.

IN the summaries of observations which represent the climates of the Mediterranean the data have been grouped according to what is conventionally called a year, and further according to "kalendar months." The year therein indicated is in reality a group of 365 days, or 366 days for leap year, taken as an approximation to the mean solar or tropical year of 365·2422166 days. The kalendar months are twelve in number, of unequal lengths adjusted arbitrarily to make up 365 or 366 days. That is the ordinary vogue in dealing with the representations of climate and it has thousands of years of history behind it; but it was not the common practice of the ancients. The kalendar as a practical system of recording the time of events was only brought into something like order by Julius Caesar and the months, as we know them, date only from an edict of Augustus. The latter, in making the final adjustments, introduced an unnecessary complication by claiming the maximum allowance of thirty-one days for the month of his own name as well as that of his great-uncle Julius, which used to have the more appropriate names of Quinctilis and Sextilis.

SPECIFICATION OF SEASONS BY THE STARS

There was always the day with its alternation of light and darkness, and the year with its regular variation of the sun's times and position of rising and setting and the consequent cycle of the seasons, but there was no acceptable method of marking the position of an event within the year by counting the days. The gradual march of the sun during the year through the twelve constellations of the Zodiac was made out by the Babylonians[1], and the moon with its regularly recurring phases must always have been within the cognisance of the dwellers in those countries where the sky is proverbially clear, and where night is the most suitable time for making land-journeys. But to identify the times or seasons for making voyages by sea, or for the operations of agriculture on land, it was apparently not the number of moons from the commencement of the year that was remembered but the face of the sky about sunrise or sunset. The conspicuous features of the sky, the prominent stars and constellations were recognised and the seasons were identified by the stars that rose just before, or set just after, the sun: the so-called "heliacal rising and setting" of the stars just visible when the sun itself was perhaps 10° below the horizon.

DAY AND NIGHT. THE HOUR

Before we enter into further details of the use of the stars for the identification of the seasons we must devote some attention to the general question of the rising and setting of the sun, moon or stars, the length of the day and of the night and its variations with the seasons.

The length of the true solar day is the period of rotation of the earth with respect to the sun. It is the actual interval between two consecutive passages of the actual sun, the true sun as the astronomers call it, over any meridian of the earth, as established by setting up a true North and South line and watching "apparent" successive passages of the sun across the vertical plane of the meridian. The actual interval between two consecutive transits varies at different times of the year from 24·0083 hours on December 22–23 to 23·9940 hours on September 18–19.

But the transits of a star are subject to no such variation within the year; apart from the variation due to very slow alterations in the position of the earth's polar axis among the stars, which causes a gradual change in the star's declination, any star rises always at the same point on the horizon, sets at the same point and remains above the horizon always for the same length of time. The duration of its day at a selected point on the earth's surface is given by the equation which is set out at the head of this chapter: when the latitude of the place and the declination of the star are duly inserted.

The length of time that the sun is above the horizon is no fixed quantity as

---

[1] From 2330 B.C. the Babylonians used a regular kalendar with a week of seven days and a year of twelve months, named after the zodiacal signs. Their year was of 360 days which probably suggested the division of the circle into degrees which we have derived from them. *Chambers's Encyclopædia*, s.v. Chronology.

in the case of a star. The fact that the earth moves in its orbit so as to complete the circumnavigation of the sun in a year adds one complete sidereal revolution to the year and gives 366 complete sidereal days within the year; and the elliptical shape of the orbit causes the variation in the length of the solar day that has been already referred to. Further than that, the inclination of the polar axis or of the plane of the earth's equator to the plane of its orbit, the ecliptic, causes the sun to vary its declination during the year from its maximum of $23\frac{1}{2}°$ above the equatorial plane at the summer solstice (June 21 or 22) to $23\frac{1}{2}°$ below the equatorial plane at the winter solstice (December 22 or 23), passing through the equinoctial points in the upward direction on March 20–21, and in the downward direction on September 23–24, when the sun is in the plane of the equator

THE TOWER OF THE WINDS

All these variations can be traced upon sun-dials, which have been used from time immemorial for that purpose. In meteorological work they are represented in a very interesting manner by the instrument for recording sunshine devised originally by John F. Campbell of Islay in 1853 to show the scorching effect of the sun upon a wooden bowl, and transformed into an effective meteorological instrument by Sir George Stokes in 1879. The instrument makes use of the image of the sun formed by a spherical glass lens. The ordinary sundial uses the shadow thrown by a gnomon or straight edge set parallel to the earth's polar axis, so that, as the earth rotates, the sun, apart from any alteration of its declination during the day, keeps at a fixed angle with

Fig. 5. The tower of Andronikos Kyrrhestes at Athens bearing sculptures of winds on the frieze (see fig. 9) and sun-dials with appropriate graduation on the entablatures beneath the sculptures.

the axis of the instrument, and the dial can be graduated in a systematic manner.

But a graduation-line, such as we are accustomed to, is not really necessary for an effective dial, the shadow of a small object, a knob at the end of a rod, will serve. If we suppose the shadow thrown upon a vertical wall by a knob at the extremity of a rod projecting from the wall, the shadow will pass across the wall during the day as the sun travels between the point of sunrise and that of sunset, and so a line of track can be drawn on the wall. The track will be lower down on the wall next day if the sun is higher in declination and higher if the change is the other way. A series of lines drawn down the wall across the tracks can mark equal stages from sunrise to sunset.

Each of the eight sides of the Tower of the Winds at Athens, which dates back to the first or second century before Christ, is used for a sun-dial of this character; the lines on the walls divide the period between sunrise and sunset into twelve hours. The tower is represented in fig. 5; a shadow of one of the gnomons can be seen under the frieze on the front face. The sides are quite well oriented according to the true points of the compass; no single one of the faces can show all the hours, all the year round, but each face takes its part, and even the North side is duly graduated, though the shadow of its gnomon can only fall on it in the earliest and latest stages of the day in the summer half year.

The division of the day and night each into twelve hours of length dependent upon the season has now been replaced by the division of the mean solar day into twenty-four equal hours carefully guarded by watches and by clocks.

HELIACAL RISING

For the purpose of studying the relation of the stars to the sun in the face of the sky, as a guide to the seasons, we may consider the sun as a star which changes its declination backwards and forwards through 47° in a year. Consequently the length of day is different from day to day, whereas the declination of a fixed star swings backwards and forwards through the 47° only in about 26,000 years. For a single year or even for a short period of years it may be regarded as constant and the star will be above or below the horizon for the same period throughout the year, though the solar time of rising and setting will vary.

The result of this periodical change in the polar axis is that the sun's position among the stars in the Zodiac as seen from the earth at an equinox or solstice will be altered by a degree in about 70 years on the average or about 30° in 2200 years. It will complete its cycle in about 26,000 years.

As the most notable example let us consider the Dog-star, "Sirius," the Egyptian "Sothis," which now has a declination of 16° 36′ 44″ S, and a consequent period above the horizon of 10¾ hours in latitude 30°. The sun's declination will vary from 23½° N to 23½° S; it passes through the position 16° 36′ S on November 8 on the way down, and on February 3 on the way up. In ancient times Sirius would be actually rising simultaneously with the sun in May and consequently invisible, after coincidence it would rise before the sun by a gradually increasing interval until its rising would take place just after sunset when it would be visible the whole night through. The separation would go on until the star would rise in the daytime and be visible setting, just after the sun. All the time since its heliacal rising it has been visible some time of the night but, later on, its setting is earlier than sunset, it will be 10¾ hours before it will rise again and, if the sun is up before the hours have elapsed, the star will not be seen at all. There is therefore a certain period in the year when the star is not visible day or night. Hence rising and setting have a new kind of significance for a brilliant celestial

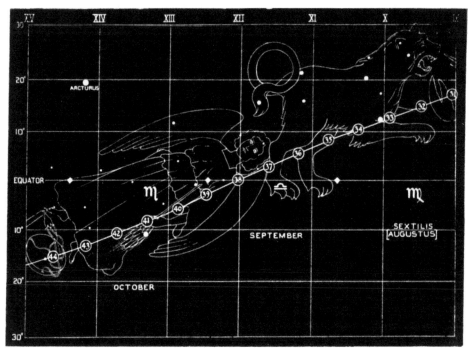

Fig. 6. The positions of the twelve constellations, the principal stars and the Milky Way in 1900.

object like Sirius; after it rises with the sun it is going to be visible at some time in the night, until it sets with the sun, when it will be invisible until the next heliacal rising.

The calculation of the period when the star will not be visible is a complicated astronomical problem. A star that is within $66\frac{1}{2}°$ of the North Pole, and has always therefore higher declination than the sun, can have no period of disappearance from the Northern Hemisphere; part of its journey must be in the dark because it is always above the horizon longer than the sun. For stars below that declination a period of invisibility depends upon latitude and seasons; and the fact that the ancients were so familiar with the face of the sky as to be able to recognise these features and their relation to the seasons is very striking evidence of the thoroughness of their scrutiny, which may be recognised also in the amazing ingenuity of the Zodiac.

The stars most frequently used to indicate the seasons were Sirius, Arcturus, the Pleiades, Procyon and Castor and Pollux, and the period of the first appearance of the star for the year when it was a morning star was the most important sign of its influence.

In the early historic times the heliacal rising of Sirius ushered in the rising of the Nile in Egypt, the etesian winds of Athens, the hot summer, the resting period for crops of the Mediterranean countries, "the weary

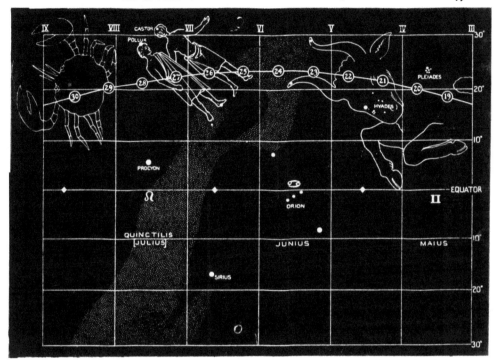

Fig. 6 (*cont.*). The position of the sun in the ecliptic counting in weeks numbered from
January 1.

season." Arcturus "rising at eventide" marked the beginning of spring.
The heliacal rising of the Pleiades marked the vernal equinox, the sign for
harvest, and their heliacal setting the autumnal equinox, the time for ploughing.
Both the Pleiades and the Hyades marked the autumnal rains. Procyon is
not very different in its date and consequent influence from Sirius.

A collection of notes on the stellar influence by Cassianus Bassus, generally
ascribed to a date A.D. 900 to 1000, is quoted in A. W. Mair's *Hesiod*. This
gives the rising of Sirius as on July 24 and the etesian winds as beginning
on July 26. In the same work we find cited a rustic kalendar of 1669 which
gives Sirius as rising on July 31 (O.S.) and the head of Castor on June 10.
The correction of the dates thus assigned for the variation of the position of
the sun in the Zodiac, or the corresponding wandering of the Julian kalendar
through the seasons, is an intricate problem which may be noted here but
need not be pursued.

The "heliacal rising" of a known star was indeed an index of the position
of the sun in the Zodiac with which the months came to be associated. From
the time of Hipparchus the first point of Aries has been the conventional
description of the position of the sun when it crosses the plane of the equator
in the spring, but the sun no longer crosses the equator when the sun "enters
Aries" but about 30 days earlier when the sun is far away in Pisces. Ancient
records require correcting for this variation of the equinoctial point.

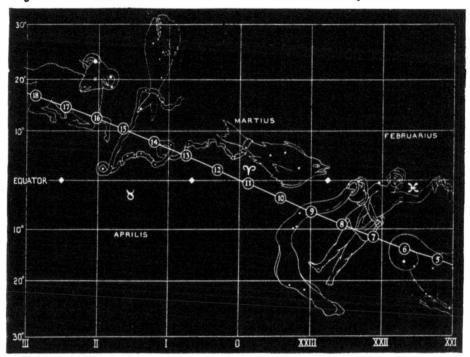

Fig. 6 (*cont.*). The projections of the monthly travel of the sun shown by diamond marks on the equator, with the sign and the name of the month above or below.

Even so late as the nineteenth century the astrologers' symbols for the signs of the Zodiac were used to represent the months:

♈	Aries	March	} Spring signs	♎	Libra	September	} Autumn signs
♉	Taurus	April		♏	Scorpio	October	
♊	Gemini	May		♐	Sagittarius	November	
♋	Cancer	June	} Summer signs	♑	Capricornus	December	} Winter signs
♌	Leo	July		♒	Aquarius	January	
♍	Virgo	August		♓	Pisces	February	

The months indicated in fig. 6 by names and signs as portions of the sun's path in the Zodiac are now a long way from the constellations.

From various sources we have noted the following indications of the cycle of the seasons in classical times.

Hesiod.

Commencement of spring	The evening rising of Arcturus 60 days after the winter solstice
Commencement of summer or reaping time	Heliacal rising of the Pleiades after they have remained concealed for 40 days and 40 nights
Threshing time	Heliacal rising of first star of Orion
Period of most oppressive heat	Heliacal rising of Sirius (about July 12 in 800 B.C. in latitude 38°)
End of summer	Fifty days after the solstice
Period of vintage	Heliacal rising of Arcturus. Culmination of Sirius and Orion (September)
Commencement of winter (ploughing time and close of navigation)	The morning setting of the Pleiades, of the Hyades and of Orion (end of October)

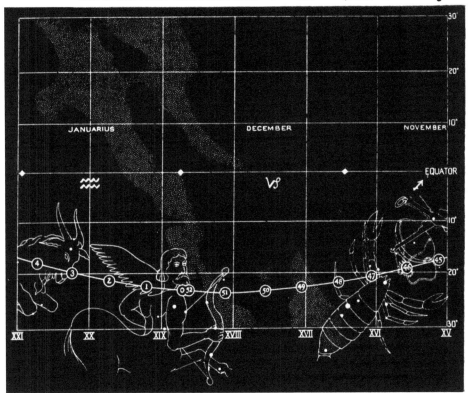

Fig. 6 (*cont.*). The astrological signs for the months showing by their distance from their respective constellations the secular displacement along the Zodiac.

Hippocrates.

Commencement of spring			Vernal equinox
,,	,,	early summer	Heliacal rising of the Pleiades
,,	,,	late summer	Heliacal rising of Sirius
,,	,,	autumn	Heliacal rising of Arcturus
,,	,,	ploughing and sowing season	Morning setting of the Pleiades
,,	,,	winter proper	Winter solstice
,,	,,	planting season	Evening rising of Arcturus

Julius Caesar.

Commencement of spring	Breezes of Favonius begin to blow	February 7
Vernal equinox		March 25
Commencement of summer	Heliacal rising of the Pleiades	May 9
Summer solstice		June 24
Commencement of autumn	Morning setting of Fidicula	August 11
Autumnal equinox		September 24
Commencement of winter	Morning setting of the Pleiades	November 11
Winter solstice		December 25

December 25 now misses the winter solstice by three days on account of the imperfection of the Julian system, which has been corrected for the accumulation of error between the Council of Nice A.D. 325 and the present day but not for the period between its institution in 46 B.C. and A.D. 325.

After the lapse of a year the sun crosses the equator again before it has reached its previous position in the Zodiac to the extent of one day in about seventy years, so in two thousand years, the lapse of time since Hipparchus fixed the position of the sun at the equinox as the first point of Aries, the sun has slipped back through the Zodiac by nearly 30° or a whole constellation The slipping back is going on still and has been going on from before the dawn of history. Hence if we know the relation of the "heliacal rising" of a star to the cycle of the seasons, which is controlled by the equinoxes and solstices, we can by noting the change in the position of the equinoctial line in the Zodiac determine the interval that has elapsed.

Thus Mr Knobel[1] has computed that the Sothic or "heliacal rising" of the star Sirius, the Dog-star, in the latitude of Memphis (30° N), expressed in terms of our present kalendar would have been approximately as follows:

A.D. −7171 May 21	A.D. +139 July 17
−5900 May 30	[+1922 August 3]
−4400 June 10	
−2900 June 22	
−1400 July 4	

Hence what was the sign of the commencement of the rise of the Nile in Upper Egypt and of the summer solstice was the sign of the very hot days that we are accustomed to think of as "dog days" at the beginning of our era.

This wandering along the Zodiac of the position of the sun at solstice or equinox has been turned to account by Sir Norman Lockyer with the assistance of Mr F. C. Penrose and others to form an estimate of the age of the great stone monuments of Egypt, Greece and Britain. They are regarded as temples arranged to celebrate the rising of the sun on some notable day or days of the cycle of the seasons. The great day of the celebration may be either the summer solstice for the solstitial year, a system of reckoning which still finds expression in our festivities of midsummer and Christmas, or May day, for what is known as the "May year," which is marked by the sun attaining the declination 16° 20′ N or S of the equator[2]. That is regarded as still surviving in the customary celebrations of May day in England, of Lammas in Scotland, and Roman veneration of ancestors—the festivals of All Saints or St Martin or November 5—and Candlemas in February. If we take the dates as May 6, August 8, November 8 and February 4, we have a very notable division of the year by weeks into four parts:

May 6 to August 8	High sun, long days
November 8 to February 4	Low sun, short days
February 4 to May 6	Changing sun, lengthening days
August 8 to November 8	Changing sun, shortening days

Sir Norman Lockyer sought to identify by the alignment of the stones, first the kind of year which the different monuments were intended to commemorate, secondly the stars which were used to announce the coming sun by their heliacal rising, or the "clock stars" which belonged to those which

---

[1] E. B. Knobel, *The Heliacal Rising of Sirius.*
[2] Sir Norman Lockyer, 'On the Observations of Stars made in some British Stone Circles: Second Note,' *Proc. Roy. Soc.* A, vol. LXXVII, 1906, p. 468.

never set and could be used to anticipate the rising of the true heliacal stars or the sun on the festal day. These being identified the age of the monument is determined by the displacement of the solstice-sunrise or the May-sunrise along the Zodiac or the change in the declination of the clock star.

Whatever may be the opinion upon the final calculation of age it will be conceded at any rate that the face of the sky before sunrise is the oldest form of kalendar, and that conclusion is confirmed as we shall see by its general use in ancient literature.

THE MOON AS TIME-KEEPER

For a permanent kalendar the position of the sun in the Zodiac is clearly not constant enough. The concurrent or alternative endeavour to use the moon for that purpose has not met with greater success. The moon is very attractive as a means of measuring intervals less than a year because its age can be read on its face—first quarter, full moon and last quarter are easily identifiable. They are separated by seven full solar days and a little more, less than a half day. Hence we come by the week of seven days, which had been arrived at by the Babylonians with the Zodiac as a kalendar. But in the long run the little more has given rise to a good deal of trouble. The period from new moon to new moon is the most clearly defined interval of time longer than the solar day, much more so than the customary epochs in the cycle of the seasons such as the rains, or seed-time or harvest. It requires an astronomer to say when the solstice or the equinox or "May day" has arrived, but anybody can identify the new moon. Moreover there are quite a number of periodic events which keep time with the moon. The tides are among them and so regular is their succession that "time and tide" form one idea in Northern minds as time and weather do in the South. Hence the counting of time by moons, or months, is as ancient as any other form of counting. It was used by both the Hebrews and the Greeks. But they had to make it fit in somehow with the inexorable cycle of the seasons; the fact that the average lunar month is about $29\frac{1}{2}$ days, so that twelve new moons make up only 354 days, is very awkward; there are $11\frac{1}{4}$ days to spare to make up the cycle of the seasons which must be adhered to. In three years the reckoning by twelve moons to the year would be more than a month out of the cycle. Accordingly in the lunar kalendars a month was intercalated before the new year was allowed to start—seven times in a cycle of nineteen years in the Jewish reckoning, and three times in a cycle of eight years by the Greeks. That meant an extra month every alternate year with one omission in eight years.

Lunar reckoning is still in vogue in India. The dates of our Easter and other variable feasts are still determined by the moon, and, though reckoning long periods in moons has not survived, payments by the month are almost universal.

> *Strepsiades.* I've found a notion how to shirk my debts.
> *Socrates.* Well then, propound it.

Strepsiades.	What do you think of this?
	Suppose I hire some grand Thessalian witch
	To conjure down the Moon, and then I take it
	And clap it into some round helmet-box,
	And keep it fast there, like a looking-glass,—
Socrates.	But what's the use of that?
Strepsiades.	The use, quotha:
	Why if the Moon should never rise again,
	I'd never pay one farthing.
Socrates.	No! why not?
Strepsiades.	Why, don't we pay our interest by the month?

(*The Clouds of Aristophanes*, translated by Benjamin Bickley Rogers. London, G. Bell and Sons, Ltd., 1919, p. 54, ll. 747–756.)

THE SUN AND MOON AS JOINT TIME-KEEPERS

The kalendar as we have it now is the result of a series of efforts to bring the sun and moon into accord with one another as time-keepers by adjusting the length of the month. Two years out of three there are twelve new moons in the year, rather oftener than once in three years there are thirteen, an unlucky number; twelve months would fill out the year if only they were long enough. The Romans elaborated a kalendar with an eight-day week in which the so-called months had days added to them to make up the eleven days by which the twelve real months were short. But the original project was not very successful; to make good, an intercalary month of 22 or 23 days had to be added from time to time at the discretion of the Pontifical College which regulated such matters in Rome. The reckoning had got into such disorder in the century before the Christian era that Julius Caesar, in his capacity as Pontifex Maximus, or more probably as Dictator[1], intercalated no less than three months, aggregating 90 days, in one year, and introduced what is known as the Julian kalendar. The plan was devised by the astronomers of Alexandria, it took the year at 365·25 days and it was arranged that, of every four consecutive civil years, three should have 365 days each, and the fourth, with a numeric divisible by four, should have 366 days. The cycle of the months began with a March of 31 days, it was followed by months of 30 and 31 days alternately, the final month of February having 29 days, except in leap year when it counted 30. The calculation was misapplied by the pontiffs who introduced the intercalary day every third year. This was corrected by Augustus in 8 B.C., who also arranged the nomenclature of the months in the order which has remained until the present time. When this was done, as we have already mentioned, the name Augustus was given to the month Sextilis, as previously the name of Julius had been given to Quinctilis, and at the same time August was enriched by a day taken from February. This inconvenient result may perhaps be explicable if we are correct in understanding from *The Legacy of Rome*[2] that the deification

[1] J. S. Reid, 'Chronology' in *A Companion to Latin Studies*, Cambridge University Press, 1910, p. 99.
[2] Clarendon Press, Oxford, 1923.

of the emperor, though it may strike a modern mind as a merely complimentary formula, did in effect result in a vitality of personal adoration which overshadowed and replaced the lifeless formulas of the traditional Roman pantheon. Perhaps the fifth and sixth months were chosen because they happened to contain the birthdays of Julius and Augustus.

The associations of months of the Roman year seem therefore to have been as follows:

Martius	the month of Mars, the patron of spring-time and youth.
Aprilis	the month of opening (*aperire*).
Maius	possibly related to Maia (or to *major*, the month of increase).
Junius	related to *juvenis*, the month of youth.
Julius ⎫ Augustus ⎭	in commemoration of the emperors.
September	the seventh month.
October	the eighth month.
November	the ninth month.
December	the tenth month.
Januarius	the month of Janus (the beginning).
Februarius	the month of purification.

The Julian method of reckoning, which assumed the year to be 365·25 days instead of 365·242 days, nearly kept pace with the sun but got behind it by three days in four hundred years: thirteen days had actually been lost when Pope Gregory XIII ordained the system, by which we now reckon years, by a Bull, dated March 1, 1582, prescribing the skipping of ten days in order to bring the kalendar back to the state in which it was at the Council of Nice in A.D. 325, and to move the beginning of the year to the first day of January; hence in the countries which obeyed the Bull the day after October 4, 1582, was counted October 15, and the corresponding change was made in this country by Act of Parliament of 1751 to take effect on September 2, 1752; the day following was called the 14th just as on April 1 of this year or other suitable date we were enjoined to call the hour following 1 a.m., 3 a.m., instead of 2 a.m. The year of a century not divisible by four is now no leap year.

Such changes in the plan of reckoning time make it difficult to interpret accurately any ancient records, and as a matter of coincidence it is remarkable that the error of the Julian reckoning nearly compensated, so far as the kalendar in use at the time is concerned, for the change of date of the heliacal rising of Sirius. So difficult is it to frame a satisfactory system that for long intervals of time astronomers use mean solar days in preference to years.

These considerations are of much more importance to astronomers, who in dealing with ancient records have to concern themselves with such well-defined events as eclipses, than they are to meteorologists; but they fall to be considered whenever questions of change of climate or change of season are discussed, and generally in the interpretation of the ancient records and the philosophical treatises concerning them.

With the revision of the kalendar the months, in spite of their arbitrary and unequal lengths, have been used alternatively with the week and the quarter for periods of time less than a year.

THE RELATION OF THE KALENDAR TO THE CYCLE OF THE SEASONS

Having now a scheme for measuring time that identifies the different parts of the year, we can give greater precision to the cycle of the seasons by relating thereto the times of sowing and harvesting which occur in different countries in different astronomical seasons according to the special character of the climate.

In Egypt the round of the year begins with the rising of the Nile at the end of June. "There are three agricultural seasons: (1) summer, April 1 to July 31, when crops are grown only on land under perennial irrigation, (2) flood, August 1 to November 30, and (3) winter, December 1 to March 31. Cotton, sugar and rice are the chief summer crops; wheat, barley, flax and vegetables are chiefly winter crops; maize, millet and "flood" rice are flood crops. Millet and vegetables are also, but in a less degree, summer crops[1]."

Concerning Egypt Sir Norman Lockyer writes: "From the earliest times the year was divided into twelve months of 30 days each, the leading month being dedicated to the God of Wisdom, Thoth (Tehuti)." He gathers the twelve months beginning with the summer solstice into groups of four each, the first four (end of June to the end of October) correspond with inundation, the second four (the end of October to the end of February) are allotted to seed-time, and the third four (the end of February to the end of June) are assigned to harvest. Each year five epacts, or intercalated days, were required to make up the cycle of the seasons.

In Palestine the solar year was divided into six seasons, seed-time, winter, cold weather, harvest, summer, hot weather, which may be divided between the months as follows:

Seed-time	September October November	Harvest	March April May
Winter	December January	Summer	June July
Cold weather	February March	Hot weather	August September

So, in India, at the present day, the year is divided into North East monsoon, cold weather, hot weather, and the monsoon rains.

In Greece the harvest of wheat and barley was before the summer in April or May, a month later in mountainous districts. The vintage was after the summer in the beginning of September in the plains, or in the end of September in the hills. Olives were gathered in the late autumn[2].

---

[1] *Ency. Brit.* 1910, s.v. Egypt, Agriculture and Land-Tenure.

[2] Leonard Whibley, *A Companion to Greek Studies*, Cambridge University Press, 1905, p. 540. A conspectus of the cycle of the seasons, or farmers' year, in classical times, is given by A. W. Mair, *Hesiod, The Poems and Fragments, done into English prose with introduction and appendices*, Oxford, Clarendon Press, 1908.

I owe the following information about crops in the Mediterranean countries to the kindness of the International Institute for Agriculture in Rome. It was sent to me in reply to a request for the dates of sowing, flowering, harvest and resting in the various countries.

Egypt.

Crop	Month of sowing	Month of flowering	Time of harvest	Time of resting
Wheat	November	February	May	—
Barley	November	February	April	—
Cotton	March	June	September to October	—
Flax	November	March	April	—
Grapes	October to February	March to April	June to August	October to January
Figs	February	March	August to September	October to January
Dates	August to September	March	August to November	Depends on the variety, generally in October
Olives	February to August	April	September	October to January
Bersim	September to October	January	June (seed)	—

Cyprus.

Crop	Month of sowing	Month of flowering	Time of harvest
Wheat	October to January	March to April	May to June
Barley	September to November	February to March	March to April
Oats	October to January	March to May	May to June
Cotton	April to June	July to August	August to October
Flax	October to January	February to March	May to June
Grapes	February to May	April to May	August to October
Figs	November to February	April to May	July to September
Dates	November to February	March to April	October to November
Olives	November to February	March to May	September to October
Grass	No grass is cultivated		

"Non-irrigable land lies fallow in summer except where sown with native dry-staple cotton. Where irrigation is possible summer crops are sown, such as sesame, linseed, Indian corn, potatoes, tobacco, cowpeas and various vegetables. Throughout the corn-growing area in the plains, which is mostly non-irrigable, the land lies fallow, between harvest and sowing, and occasionally from harvest till the sowing-season of the year after."

Greece.

Crop	Sowing	Flowering	Harvest
Wheat	*Peloponnesus: mountainous regions* September or October	April	Mid-June to mid-July
	plains End of September to mid-November	April	June to beginning of July
	Macedonia: A little later than in the Peloponnesus	May	—
Barley	Beginning or end of the sowing of wheat	April	May to June
Oats	Beginning or end of the sowing of wheat	May	June to July
Spring oats	February	—	—
Vines: (de Corinthe et des Sultanines)	—	April to May	July or August
(other varieties)	—	First fortnight in May	First fortnight of September
Figs: (figues fleurs)	—	—	End of May to end of June. Spring
(figues d'été)	—	—	August
(figues de séchage)	—	—	August 15 to September 15
Olives	—	—	October to March

Iraq[1].

Crop	Month of sowing	Month of flowering	Time of harvest	Time of resting
Wheat	November to January	February	May to June	—
Barley	November to January	February	May	—
Cotton	March to April	May to November	July to November	—
Flax	November	February and March	April to May	—
Grapes	November *or* February	March and April	July to September	December to January
Figs	February	(a) White: March (b) Black: April	July, August August, September	December to January
Dates	October *or* March	March	August to November	—
Olives	February and March	April	September	—
Bersim	October and November	February	June (seed)	—
Melons	March to June	April to July	June to October	—

[1] Forwarded by Flight Lieutenant R. C. Bryant, R.A.F., Baghdad, from information supplied by the Director of Agriculture.

Sicily.

In Sicily the time of sowing of wheat is very variable depending on the autumn rains, but sowing is possible from October to January, the time of earing varies from April 12 to May 5. Flowering takes place five to ten days after earing. The average date of harvest is about the first week in June but may vary from May 26 to July 5 according to the height above sea-level.

More Northern Countries.

In the South and East of England our own arrangement of the seasons would be approximately:

November December	winter sowing	May June	flowering
January February	resting	July August	grass-harvest grain-harvest
March April	spring sowing	September October	root-harvest and tree-fruit harvest clearing and cleaning

The seasons are still later further North; for the North of Scotland we might find an almost continuous crop of grass, and a harvest of oats, distributed over the months from September to November.

The intermediate position between the Northern experience with regard to crops and seasons as compared with the Southern which we have been considering, is well expressed by the new kalendar devised by the National Convention of the first French Republic in 1793. Seasonal names were given to the 12 groups of 30 days each between September 22 and the next following September 16, as follows:

Vendémiaire	Vintage	Sept. 22—Oct. 21.
Brumaire	Foggy	Oct. 22—Nov. 20.
Frimaire	Sleety (or white frosty)	Nov. 21—Dec. 20.
Nivôse	Snowy	Dec. 21—Jan. 19.
Pluviôse	Rainy	Jan. 20—Feb. 18.
Ventôse	Windy	Feb. 19—Mar. 20.
Germinal	Budding	Mar. 21—Apr. 19.
Floréal	Flowering	Apr. 20—May 19.
Prairial	Grass crop	May 20—June 18.
Messidor	Grain crop	June 19—July 18.
Thermidor	Heat	July 19—Aug. 17.
Fructidor	Tree-fruits	Aug. 18—Sept. 16.

The remaining five or six days were devoted to holiday celebrations.

So in different parts of the world different climates have different associations of seasons, the details of which might form an introductory section of any comprehensive treatise on the application of meteorology to agriculture. Here we are only concerned with the subject as helping us to make out the times and seasons of ancient climates.

THE PRESENT POSITION OF THE KALENDAR

As time passes and statistics are compiled in a systematic manner the defects of our kalendar become more and more apparent, and this is the case especially with the compilations of meteorological data, which are now available for 50 years or more at many stations and for as long as 200 years in some. We are accustomed to group the data according to the unequal kalendar months. For their "Dekadenbericht" the Deutsche Seewarte divided the months into 10 days, 10 days and 8 days, 9 days, 10 days or 11 days. For a number of years the British Meteorological Council published the hourly values of the records at their observatories in the form of five-day means but their most effective contribution to meteorological statistics, the Weekly Weather Report, which runs from 1878 to the present time, is based on the year and the week. All the weeks are of equal length and the year consists occasionally of 53 weeks and its beginning ranges from December 29 to January 3.

So far as meteorology is concerned the problem seems to compel us to make up our minds whether it is reasonable to continue to work with 12 months which are of unequal lengths in any case, even if the worship of the Emperor Augustus be now disregarded and all months made of 30 or 31 days. Five of the months are indivisible into two or three equal parts. They give us no practical clue to the dates of new and full moons, which are really pertinent as regards tides, so that it is really seeking trouble to call them months. On the other hand a year could be made up of 52 weeks with an additional or fifty-third spell of a single day or of two days in leap year.

No doubt there are some discomforts or awkwardnesses; quarters of the year would be 13 weeks, not three months, and if groups of four weeks were used, the two solstices and two equinoxes would be unsymmetrically placed with regard to them. Thirteen weeks are awkward in harmonic analysis.

The rational conclusion seems to be that the grouping of meteorological data for statistical purposes to form the basis of comparison with the statistical data of agriculture, trade, commerce and public health for each year should be arranged in the form most convenient for the effective representation of the material irrespective of the minor details of the kalendar. From this point of view the week is quite definitely the most suitable unit intermediate between the day and the year, both of which are fixed by causes beyond human control. Meteorological statistics grouped in months are of little use in reference to agriculture, trade, commerce or public health, even if the months be of equal length, for the simple reason that a month is too long a period; it is not generally possible to characterise effectively the weather of 28 days but a period of seven days is more naturally susceptible of definition and is indicated for the statistics of trade and commerce which in the present organisation of society and religion have inevitably a weekly period. Medicine seems to have special associations with the month, the use of the words lunacy and lunatic implies as much, yet the statistics of public health are already based generally on the week.

The Meteorological Section of the International Union of Geodesy and Geophysics have given expression to these ideas in a resolution of a meeting at Madrid in 1924, in favour of the mean solar day (with its subdivisions), the week and the kalendar year as suitable units of time for meteorological statistics[1].

[1] *Journal of the Royal Statistical Society*, 1925. [See also vol. II, pp. 49 and 304.]

THE QUARTERS OF THE MAY YEAR

The question may perhaps be raised whether the division into as many as twelve intervals is really necessary for a general climatic summary and whether a group into four quarters with the possibility of more detailed expression for weeks or fortnights would not do instead. By way of examining this question we may note that with our present practice of using a solstitial year, that is to say a year beginning with the winter solstice, or nearly so, each group of three months, January to March, April to June, July to September and October to December, covers the whole interval between a solstice and an equinox; at those two extremes we have the most stationary condition and the most variable condition respectively, so each quarter includes half a stationary period and half a period of change; if, however, we could revert to the prehistoric practice of the May year, and group our figures for each of the 13 weeks, May 6 to August 8, August 8 to November 8, November 8 to February 4, and February 4 to May 6, we should get the data for the steady solstitial periods separated from those of the variable equinoctial periods. Counting in weeks the periods would be:

Spring equinoctial period	13 weeks	(6th to 18th)	Feb. 4—May 5.
Summer solstitial period	,,	(19th to 31st)	May 6—Aug. 7.
Autumn equinoctial period	,,	(32nd to 44th)	Aug. 8—Nov. 7.
Winter solstitial period	,,	(45th to 5th)	Nov. 8—Feb. 3.

The grouping of the days of the year in the manner thus indicated isolates the periods of thirteen weeks of greatest and least income of solar energy between two equal periods of rapid transition. Regarded from the point of view of solar influence the difference between the two periods thus isolated is remarkable. In illustration we give the results of observation of solar radiation at South Kensington, a metropolitan station, and at Rothamsted, a rural station, in the year 1924. The figures represent the average daily income of thermal energy from sun and sky expressed in kilowatt-hours received on a square dekametre of horizontal surface, about 1000 square feet. Since the electric energy which we buy is commonly counted in kilowatt-hours and would be cheap at 1d. per unit, we can assign a cash figure to the income of solar energy at that price.

Daily average income of solar energy per square dekametre during the quarters of the May Year, in kilowatt-hours and pounds sterling.

	South Kensington		Rothamsted	
	No. of units fs	Value at 1d. per unit	No. of units fs	Value at 1d. per unit
Summer quarter	39284	£164	38916	£162
Autumn ,,	19019	79	19110	80
Winter ,,	3312	14	5244	22
Spring ,,	13741	57	19292	80

We have already noticed in the Scottish quarter days and certain English customs the survival of this interesting division of the year. It seems to be deeply ingrained in English rural life, for in a volume of addresses by the late Dr James Gow, Headmaster of Westminster School, we read that "on most of the commons of England the Lord of the Manor has the right of pasture and the right of the hay from Candlemas (Feb. 2) to Lammas (Aug. 1) and the tenants have the right of common from August to Candlemas."

We give a tabular representation of the climate of Babylon on this plan. It contains only one-third of the number of data included in the monthly table of p. 30 but it gives a fair idea of the nature of the climate.

CHALDAEA (IRAQ).　　Lat. 32° 30′ N.　Long. 44° 20′ E.　Alt. about 30 m.　BABYLON.

WIND-QUADRANTS							Season	MOISTURE		
Hours of observation								≡	Normal vapour pressure	Normal humidity
7 14 20½ N	7 14 20½ E	7 14 20½ S	7 14 20½ W	7 14 20½ Calm		↗			7 14 20½	7 14 20½
n 38 44 37	n n n 7 4 10	n n n 6 8 8	n n n 40 36 31	n n n 1 0 6	m/s 4·1		May–July	n 0	mb mb mb 13 11 13	per cent. 39 17 28
33 42 35	7 5 10	6 8 8	45 36 28	2 1 11	3·3		Aug.–Oct.	0	11 10 10	40 16 26
22 33 27	12 10 17	13 15 14	42 32 26	3 0 8	2·9		Nov.–Jan.	5½	8 8 9	79 44 62
23 32 29	17 11 22	16 18 16	30 27 17	3 1 5	3·7		Feb.–April	1	9 8 8	66 32 46
116 153 128	43 30 59	41 49 46	157 131 102	9 2 30	3·4		Year	6½	10 9 10	56 27 41
5½ years, June 1907–December 1912							Period	1911–1912	5½ years, 1907–1912	

TEMPERATURE								PRECIPITATION							
Extremes recorded		Means of extremes of quarters		Normal daily			Season	● Norm.	Seasonal extremes		● Number of days			☷	✳
Max.	Min.	Max.	Min.	Max.	Min.				Max.	Min.	mm I V	mm I 10 ∧ V	mm 10 ∧			⏅
t 322	t 287	t 321	t 288	t 313	t 295		May–July	mm I	mm I	mm 0	n 8	n 0	n 0	n 2	n	n 6
323	281	321	283	313	293		Aug.–Oct.	10	42	0	4	0	0	0	Very rare	1
306	266	303	269	291	278		Nov.–Jan.	55	89	4	16	7	2½	1		5
314	270	311	272	297	283		Feb.–April	42	74	15	18	6	1	5		9
323	266	320	270	304	287		Year	109	156	53	46	13	3	8		21
5½ years, 1907–1912							Period	1907–1912			1911–1912				1907–1912	

CHAPTER IV

POETS AND HISTORIANS: THE APPLICATIONS OF METEOROLOGY TO AGRICULTURE AND NAVIGATION. HERODOTUS

		B.C.
Homer (*Iliad* and *Odyssey*)		960–915
Hesiod (*Works and Days*)	before	700
Hebrew Captivity		588–538
The book of the Psalms of David		700–100
The book of Job		600–300
The book of the Preacher		200
Defeat of the Persians under Darius at Marathon		490
Defeat of the Persians under Xerxes at Salamis		480
Herodotus (*History*)	about	440
Aristophanes (*Clouds*)		423
Greek conquest of Asia by Alexander		330
Aratus (*Book of Signs*)	about	278

THUS, having summarised the experiences which are characteristic of the Mediterranean climates at the present time and having indicated the natural associations of the cycle of the seasons with the information to be gathered from the face of the sky, we may now revert to the co-ordination of the experiences of weather to which reference has been made in the first chapter and the different ways in which the co-ordination was expressed by the poets on the one hand and by the philosophers on the other.

It has been suggested that the experiences of weather led the ancients to regard the forces of nature as being under personal control and that they were in that way the origin of a religion, mythological or theological, which might regard an endeavour to trace the sequence of weather to natural causes, the scientific inquiry of the philosopher, as being an improper or even impious curiosity about an almighty power.

But due submission to the divine will was not regarded as inconsistent with such anticipation of the future providence as a careful collation of past experience would indicate; so we find, together with exhortations to an implicit trust in providence, the beginnings of a weather-lore, or signs of coming weather and seasons, many of which are observed even to this day, and in some cases the form which the ancients gave still survives.

This attitude is expressed by the Greek and Hebrew poets. We shall quote a number of extracts to illustrate that statement. As already indicated in this connexion we have nothing to say about Egypt because the Egyptians were dependent only on the rising of the Nile, and raised the structure of their mythology upon the sun and his risings and settings with which the changes in the Nile were so closely associated as to need no assistance from the signs of rain or storm[1].

The moon was probably worshipped before the sun, being connected with the animal gods, the baboon and the ibis of Tehuti, the god of lunar measure, of wise

[1] Sir Norman Lockyer, *The Dawn of Astronomy*, Cassell and Co., Ltd., London, Paris and Melbourne, 1894, chap. XXIII.

reflection, and of research. In a country where the need of the sun for growth was not obvious, and its heat was not sought, and where the night was favourable for travel, the moon was looked on as a helper; the month was more obviously important to man than the year, as in Arabia. After this, the moon was linked with Isis, especially in late times; and the great goddess of the dynastic people, Hat-her, was grafted upon both ideas: the moon became her head-dress, and she was identified with Isis.

(W. M. Flinders Petrie, *Religious Life in Ancient Egypt*,
Constable and Co., 1924, p. 8.)

THE GREEK AND HEBREW POETS

We have made no exhaustive examination of ancient literature, the quotations which we give are such as we have met with in a very imperfect survey of a very limited range; but they are sufficient to enable the reader to trace the development of the idea that the control of weather is a personal control, and that the intentions of the controller are sufficiently manifest by signs of the coming seasons and coming weather for it to be worth while for the farmer and sailor to pay careful attention to the inferences which have been drawn.

In the *Iliad*, Homer almost invariably attributes the changes in wind or weather to the direct interposition of one or other of the gods. The most impressive events are thunder-storms which are in the personal charge of Zeus. When the Greeks press the Trojans to the gates of Troy itself Zeus from Mount Ida disperses them completely by a lightning-flash that ploughs up the sand at their feet. In the *Odyssey* Poseidon seems to be in command of the weather at sea. When Odysseus, at large on a raft after seventeen days "keeping the Bear, which they likewise call the Wain, on the left" had come within sight of the land of the Phaeacians, Poseidon espies him and

With that he gathered the clouds and troubled the waters of the deep, grasping his trident in his hands; and he roused all storms of all manner of winds, and shrouded in clouds the land and sea: and down sped night from heaven. The East Wind and the South Wind clashed, and the stormy West, and the North, that is born in the bright air, rolling onward a great wave. Then were the knees of Odysseus loosened and his heart melted, and heavily he spake to his own great spirit.

(S. H. Butcher, M.A., and A. Lang, M.A. *The Odyssey of Homer done into English prose*, Macmillan and Co., London, 1890, Book v, p. 85.)

Later on in the *Odyssey* we find Zeus in charge of the rain:

Now it was so that night came on foul with a blind moon, and Zeus rained the whole night through, and still the great West Wind, the rainy wind, was blowing.

(Book xiv, p. 236.)

Compare these passages with that from Psalm cvii, Prayer-book, 1559:

> They that go down to the sea in ships:
> and occupy their business in great waters;
> These men see the works of the Lord:
> and his wonders in the deep.
> For at his word the stormy wind ariseth:
> which lifteth up the waves thereof.

They are carried up to the heaven, and down again to the deep:
    their soul melteth away because of the trouble.
They reel to and fro, and stagger like a drunken man:
    and are at their wits' end.
So when they cry unto the Lord in their trouble:
    he delivereth them out of their distress.
For he maketh the storm to cease:
    so that the waves thereof are still.
Then are they glad, because they are at rest:
    and so he bringeth them unto the haven where they would be.
O that men would therefore praise the Lord for his goodness:
    and declare the wonders that he doeth for the children of men!

Here are the dedicatory lines of Hesiod's poem entitled *Works and Days*:

Muses of Pieria, who glorify with song, come sing of Zeus, your Father, and declare His praise, through whom are men famous and unfamed, sung or unsung, as Zeus Almighty will. Lightly He giveth strength and lightly He afflicteth the strong: lightly He bringeth low the mighty and lifteth up the humble: lightly He maketh the crooked to be straight and withereth the proud as chaff: Zeus, who thundereth in Heaven, who dwelleth in the height.

(A. W. Mair. Hesiod, the poems and fragments done
into English prose. Oxford, Clarendon Press, 1908.)

And here are some examples of Hesiod's weather-lore, the earliest in classical writings:

But when Zeus hath completed sixty days after the turning of the sun[1], then the star Arkturos, leaving the sacred stream of Ocean, first riseth in his radiance at eventide. After him the twittering swallow, daughter of Pandion, cometh into the sight of men when spring is just beginning. Ere her coming prune the vines: for it is better so. (p. 21.)

Take heed what time thou hearest the voice of the crane from the high clouds uttering her yearly cry, which bringeth the sign for plowing and showeth forth the season of rainy weather, and biteth the heart of him that hath no oxen. (p. 16.)

For fifty days after the turning of the sun, when harvest hath come to an end, the weary season[2], sailing is seasonable for men. Thou shalt not break thy ship, nor shall the sea destroy thy crew, save only if Poseidon the Shaker of the Earth or Zeus the King of the Immortals be wholly minded to destroy. For with them is the issue alike of good and evil. Then are the breezes easy to judge and the sea is harmless. Then trust thou in the winds, and with soul untroubled launch the swift ship in the sea, and well bestow therein all thy cargo. And haste with all speed to return home again; neither await the new wine and autumn rain, and winter's onset and the dread blasts of the South Wind, which, coming with the heavy autumn rain of Zeus, stirreth the sea, and maketh the deep perilous. Also in the spring may men sail; when first on the topmost spray of the fig-tree leaves appear as the foot-print of a crow for size, then is the sea navigable. This is the spring sailing, which I commend not, for it is not pleasing to my mind, snatched sailing that it is. (p. 24.)

Some centuries later we have in Hebrew Scriptures:

Then the Lord answered Job out of the whirlwind, and said,
    Who is this that darkeneth counsel
    By words without knowledge?
    ....
Hast thou entered the treasuries of the snow,
    Or hast thou seen the treasuries of the hail,

---

[1] End of February.   [2] July-August.

> By what way is the light parted,
> 　　Or the east wind scattered upon the earth?
> Who hath cleft a channel for the waterflood,
> 　　Or a way for the lightning of the thunder;
> To cause it to rain on a land where no man is;
> 　　On the wilderness, wherein there is no man?
>
> 　　　　　　　　(The Book of Job, xxxviii, 1–2, 22–26.

He that observeth the wind shall not sow; and he that regardeth the clouds shall not reap. As thou knowest not what is the way of the wind, ...even so thou knowest not the work of God who doeth all. In the morning sow thy seed, and in the evening withhold not thine hand: for thou knowest not which shall prosper, whether this or that, or whether they both shall be alike good. 　(The Book of the Preacher, xi, 4.)

We come finally to a complete exposition of the view which we seek to present by a Greek religious poet Aratus who was born some little while after the time of Aristotle. He was a physician and astronomer, attached to the court of Antigonus Gonatas, King of Macedonia, about 278 B.C., and he is known as the "one of your own poets" referred to by St Paul in his speech at Athens on the fatherhood of God.

We may preface a number of quotations from the *Diosemeia* or *Book of Signs* by the introduction to his other work on *Phaenomena* or the *Astronomical constellations*.

From Zeus let us begin; him do we mortals never leave unnamed; full of Zeus are all the streets and all the market-places of men; full is the sea and the havens thereof; always we all have need of Zeus. For we are also his offspring; and he in his kindness unto men giveth favourable signs and wakeneth the people to work, reminding them of livelihood. He tells what time the soil is best for the labour of the ox and for the mattock, and what time the seasons are favourable both for the planting of trees and for the casting of all manner of seeds. For himself it was who set the signs in heaven, and marked out the constellations, and for the year devised what stars chiefly should give to men right signs of the seasons, to the end that all things might grow unfailingly. Wherefore, him do men ever worship first and last. Hail, O Father, mighty marvel, mighty blessing unto men. Hail to thee and to the Elder Race. Hail, ye Muses, right kindly every one! But for me, too, in answer to my prayer direct all my lay, even as is meet, to tell the stars.

　　　　　　(Aratus, *Phaenomena*, ed. G. R. Mair, Loeb Classical
　　　　　　Library, London, W. Heinemann, 1921.)

The following extracts are from the *Book of Signs*:

Do you not see when the Moon appears on the west side with thin horns that it marks the commencement of the month? and when she first becomes strong enough to cast a shadow she is at her fourth day? At the eighth day she is half illuminated, but afterwards she shows her whole face, and the phases always declining in an inverse order, she tells at each dawn what part of the month is at hand. The twelve signs of the Zodiac suffice to show the termination of the nights, and during all the year the proper seasons for tillage, seed-time and planting. All these announce that we are everywhere watched over by Providence. Whoever has experienced storm and rain upon a vessel recollects the violent Arcturus, also certain other signs which rise from the horizon in the morning, and those which set at the commencement of the night. The Sun, indeed, travels over all of them during the year, and making a long circuit, approaches sometimes one and sometimes another, sometimes when he rises, sometimes when he sets, according as one star or another star greets the dawn.

　　　　　　(Aratus, *Diosemeia*, translated by C. Leeson Prince for G. J. Symons;
　　　　　　Lewes, Farncombe and Co., Printers, 1895, ll. 1–19.)

Be discreet concerning these things likewise, if you have a ship confided to your care, to learn the precursory signs of the winter wind and tempestuous sea. It is of little trouble and soon becomes of great use to a diligent observer. (ll. 26–30.)

When in Autumn Wasps collect in many groups before the return of the Pleiades in the evening, one may be certain that the following winter will be cold in proportion to the extent and closeness of the groups. (ll. 333–6.)

The husbandman also likes to see flocks of Cranes come in season, for when they show themselves out of season and irregularly, the winter comes just as irregularly, for the earlier and the more crowded they appear, the earlier will the winter be; but when you see them late, not in flocks, but flying for a long time in small numbers, the delay of the winter will enable you to finish your last work. (ll. 343–9.)

The Dogs, too, when they scratch the earth with their two paws, scent the coming rain. The Crab is accustomed to go out of the water quickly upon the land as the tempest bursts. (ll. 403–6.)

THE CONFLICT BETWEEN RELIGION AND SCIENCE

We are thus introduced to a view of the study of weather which has been in vogue for more than two thousand years and is still quite a customary view. In the meantime the more strictly scientific method of inquiry, the co-ordination of the facts of weather into climate and the search for an explanation of the sequence of weather on a physical basis, had dawned among the Greeks. The great physician Hippocrates, whose writings belong to the end of the fifth century B.C., explains in his treatise on *Airs, Waters and Places* how the knowledge of climate ought to be used and he traces to the differences of climate the differences of habit of life and character between the East and the West. He is the first to express the idea that it is the weather that moulds the life of man.

There was even a conflict between the two schools of thought, which is represented by Aristophanes in the dialogue between Strepsiades, a country-man living in Athens, and Socrates, a rationalist philosopher, in the comedy of *The Clouds*:

Strepsiades.   No Zeus up aloft in the sky!
    Then you first must explain, who it is sends the rain; or I really must think you are wrong.
Socrates. Well then, be it known, these[1] send it alone: I can prove it by arguments strong.
    Was there ever a shower seen to fall in an hour when the sky was all cloudless and blue?
    Yet on a fine day, when the Clouds are away, He might send one, according to you.

He is pressed by Strepsiades as to the relation of the clouds to rain and thunder regarded by Socrates as a necessary physical consequence of their motion:

Strepsiades. But is it not He who compels this to be? does not Zeus this Necessity send?
Socrates. No Zeus have we there, but a Vortex of air.

[1] The clouds.

Strepsiades.   What! Vortex? that's something, I own.
  I knew not before, that Zeus was no more, but Vortex was placed on his throne!
  But I have not yet heard to what cause you referred the thunder's majestical roar.
Socrates. Yes, 'tis they, when on high full of water they fly, and then, as I told you before,
  By Compression impelled, as they clash, are compelled a terrible clatter to make.

> (*The Clouds of Aristophanes*, translated by B. B. Rogers. London, G. Bell and Sons, Ltd., 1919, ll. 367–384.)

In this controversy between the religion and the science of past ages the contemporary historian did not find it necessary to intervene, as we may gather from the quotations from Herodotus which we have cited in the first chapter. His duty was limited to relating what he had seen or heard. He was quite prepared to believe that sacrifices were offered for days together to induce the Nereids to put an end to a storm at sea, but he did not feel himself bound to encourage the practice, "perhaps it ceased of itself." The uncertainties of that "perhaps" have had a long history in the study of weather. Even the worst weather comes to an end but it needs must be treated with the utmost respect while it lasts and no effort can be spared.

The attitude of the great Historian deserves to be more fully expressed. We will accordingly present a survey of his ideas about weather derived from the reading of his *History*.

HISTORICAL NOTES OF WEATHER

Thunder and lightning, the direct charge of Zeus himself, seem to have produced the most profound impression and indeed to have been invariably a cause of terror, but storms of wind and sand were not allowed to pass unnoticed.

As thus with tears he besought the god, suddenly, though up to that time the sky had been clear and the day without a breath of wind, dark clouds gathered, and the storm burst over their heads with rain of such violence, that the flames were speedily extinguished. (The *History* of Herodotus, translated by Rawlinson, Book i, chap. 87.)

In the midst of their hauling suddenly there was a thunderclap, and with the thunderclap an earthquake; and the crew of the trireme were forthwith seized with madness, and, like enemies, began to kill one another. (Book v, chap. 85.)

Seest thou how God with his lightning smites always the bigger animals, and will not suffer them to wax insolent, while those of a lesser bulk chafe him not? How likewise his bolts fall ever on the highest houses and the tallest trees? So plainly does he love to bring down everything that exalts itself. Thus ofttimes a mighty host is discomfited by a few men, when God in his jealousy sends fear or storm from heaven, and they perish in a way unworthy of them. For God allows no one to have high thoughts but himself. (Book vii, chap. 10.)

Quitting this, the troops advanced across the plain of Thebé, passing Adramyttium, and Antandrus, the Pelasgic city; then, holding Mount Ida upon the left hand, they entered the Trojan territory. On this march the Persians suffered some loss; for as they bivouacked during the night at the foot of Ida, a storm of thunder and lightning [and whirlwind] burst upon them, and killed no small number. (Book vii, chap. 42.)

The barbarians had just reached in their advance the chapel of Minerva Pronaia, when a storm of thunder burst suddenly over their heads—at the same time two crags split off from Mount Parnassus, and rolled down upon them with a loud noise, crushing vast numbers beneath their weight—while from the temple of Minerva there went up the war-cry and the shout of victory. (Book VIII, chap. 37.)

The most effective account of a storm of wind is in the description of the wreck of Xerxes' fleet:

The fleet then, as I said, on leaving Therma, sailed to the Magnesian territory, and there occupied the strip of coast between the city of Casthanaea and Cape Sepias. The ships of the first row were moored to the land, while the remainder swung at anchor further off. The beach extended but a very little way, so that they had to anchor off the shore, row upon row, eight deep. In this manner they passed the night. But at dawn of day calm and stillness gave place to a raging sea, and a violent storm, which fell upon them with a strong gale from the east—a wind which the people in those parts call Hellespontias. Such of them as perceived the wind rising, and were so moored as to allow of it, forestalled the tempest by dragging their ships up on the

THE DESERT CLIMATES OF ANTIQUITY

Fig. 7. A modern sandstorm of the Sudan, from a photograph by G. N. Morhig,
the English Pharmacy, Khartoum.

beach, and in this way saved both themselves and their vessels. But the ships which the storm caught out at sea were driven ashore, some of them near the place called Ipni, or "The Ovens," at the foot of Pelion; others on the strand itself; others again about Cape Sepias; while a portion were dashed to pieces near the cities of Meliboea and Casthanaea. There was no resisting the tempest.

It is said that the Athenians had called upon Boreas to aid the Greeks, on account of a fresh oracle which had reached them, commanding them to "seek help from their son-in-law!" (Book VII, chaps. 188–189.)

This must have been a summer gale, or at most in early autumn, for the fighting men apparently spent the summer, the weary or resting season of the year when they could be spared from agricultural pursuits, for military and naval adventures.

The barbarian is clean gone—we have driven him off—let each now repair his own house, and sow his land diligently. In the spring we will take ship and sail to the Hellespont and to Ionia! (Book VIII, chap. 109.)

When it was now late in the autumn, and the siege still continued, the Athenians began to murmur that they were kept abroad so long; and, seeing that they were not able to take the place, besought their captains to lead them back to their own country. But the captains refused to move, till either the city had fallen, or the Athenian people ordered them to return home. So the soldiers patiently bore up against their sufferings. (Book IX, chap. 117.)

Few words are required to give a vivid picture of a characteristic feature of desert weather known in the Sudan as a *haboob* and in Lower Egypt as a *simoom* (fig. 7).

On the country of the Nasamonians borders that of the Psylli, who were swept away under the following circumstances. The south-wind had blown for a long time and dried up all the tanks in which their water was stored. Now the whole region within the Syrtis is utterly devoid of springs. Accordingly the Psylli took counsel among themselves, and by common consent made war upon the south-wind[1]—so at least the Libyans say, I do but repeat their words—they went forth and reached the desert; but there the south-wind rose and buried them under heaps of sand: whereupon, the Psylli being destroyed, their lands passed to the Nasamonians. (Book IV, chap. 173.)

Further than this, the Ammonians relate as follows:—That the Persians set forth from Oasis across the sand, and had reached about half way between that place and themselves, when, as they were at their midday meal, a wind arose from the south, strong and deadly, bringing with it vast columns of whirling sand, which entirely covered up the troops, and caused them wholly to disappear. (Book III, chap. 26.)

THE RELATION OF CLIMATE TO THE STUDY OF WEATHER

The first result to be drawn from any collection of observations of weather is a perception of climate which has been defined already as a summary of the sequence and recurrence of weather. In modern meteorology its expression takes many forms of one of which we have given some examples. But it is not necessary to have records and forms of expression; experience is unconsciously woven into the practice of ordinary life, and people adjust themselves automatically to the ordinary vicissitudes of weather and climate. In Plutarch's account of Alexander's expedition to Asia is a description of a discussion about climate between the philosophers Callisthenes and Anaxarchus, attached to the expedition.

One day a dispute had arisen at table about the seasons and the temperature of the climate. Callisthenes held with those who asserted, that the country they were then in was much colder, and the winters more severe, than in Greece. Anaxarchus maintained the contrary with great obstinacy. Upon which Callisthenes said, "You must needs acknowledge, my friend, that this is much the colder; for there you went in winter in one cloak, and here you cannot sit at table without three housing coverlets one over another." This stroke went to the heart of Anaxarchus.

(Plutarch's *Lives, Alexander*, translated by John and William Langhorne.)

Herodotus also clearly recognises what we understand by differences of climate.

The whole district whereof we have here discoursed has winters of exceeding rigour. During eight months the frost is so intense that water poured upon the ground does not form mud, but if a fire be lighted on it mud is produced. The sea

---

[1] This passage probably means that they "marched against the South wind."

freezes, and the Cimmerian Bosphorus is frozen over. At that season the Scythians who dwell inside the trench make warlike expeditions upon the ice, and even drive their waggons across to the country of the Sindians. Such is the intensity of the cold during eight months out of the twelve; and even in the remaining four the climate is still cool. The character of the winter likewise is unlike that of the same season in any other country; for at that time, when the rains ought to fall in Scythia, there is scarcely any rain worth mentioning, while in summer it never gives over raining; and thunder, which elsewhere is frequent then [in winter], in Scythia is unknown in that part of the year, coming only in summer, when it is very heavy. Thunder in the winter-time is there accounted a prodigy; as also are earthquakes, whether they happen in winter or summer. (Book IV, chap. 28.)

These statements must be regarded as an exaggeration although the coldness of the climate of the Crimea and the Northern shore of the Black Sea produced a profound and painful impression upon troops during the Crimean war. A reference to the *Atlas climatologique de l'Empire de Russie* published in 1900 shows that the neighbourhood of the Palus Maeotis is under snow for from 20 to 80 days according to locality, there is ice for from 80 to 100 days and mean temperature below the freezing-point for from 90 to 110 days. Snow falls on 20 to 30 days in winter and rain on from 10 to 20 days in summer. There are from 10 to 20 thunder-storms in the year of which as an average of four stations in the Crimea nine-tenths are recorded from May to September and only one-thirtieth, or one every two years, from November to March[1]. Herodotus's warning in 440 B.C. was not altogether inapplicable in A.D. 1854 and it must have been quite well-known in the class-rooms of Eton.

But little rain falls in Assyria, enough, however, to make the corn begin to sprout, after which the plant is nourished and the ears formed by means of irrigation from the river. For the river does not, as in Egypt, overflow the cornlands of its own accord, but is spread over them by the hand, or by the help of engines. (Book I, chap. 193.)

The country of the Cyrenaeans, which is the highest tract within the part of Libya inhabited by the wandering tribes, has three seasons that deserve remark. First the crops along the sea-coast begin to ripen, and are ready for the harvest and the vintage; after they have been gathered in, the crops of the middle tract above the coast-region (the hill-country, as they call it) need harvesting; while about the time when this middle crop is housed, the fruits ripen and are fit for cutting in the highest tract of all. So that the produce of the first tract has been all eaten and drunk by the time that the last harvest comes in. And the harvest-time of the Cyrenaeans continues thus for eight full months. (Book IV, chap. 199.)

A discussion of the periodic change of level of the Nile shows that the actual conditions were well understood.

Some of the Greeks, however, wishing to get a reputation for cleverness, have offered explanations of the phenomena of the river [Nile], for which they have accounted in three different ways. Two of these I do not think it worth while to speak of, further than simply to mention what they are. One pretends that the etesian winds cause the rise of the river by preventing the Nile-water from running off into the sea. But in the first place it has often happened, when the etesian winds did not blow, that the Nile has risen according to its usual wont; and further, if the etesian winds produced the effect, the other rivers which flow in a direction opposite to those winds ought to present the same phenomena as the Nile, and the more so as they are all smaller streams, and have a weaker current. But these rivers, of which

[1] Klossovsky, *Les orages en Russie*.

there are many both in Syria and Libya, are entirely unlike the Nile in this respect. (Book II, chap. 20.)

The reference to the etesian or annual winds, from ἔτος, year, requires a note because these winds were, and are still, the most persistent and the steadiest winds of the Mediterranean region. They are in fact notably summer-winds and a part of the circulation which forms the South West monsoon over India and the Arabian Sea and are often referred to as the monsoon, but over the Eastern Mediterranean and Egypt the direction is from the North. The regime of winds is most clearly indicated by the isobars on the maps of fig. 4.

The climate and weather of Egypt were very carefully borne in mind in ancient times, as the following extracts show:

On the difference between the Danube and the Nile:

The Ister remains at the same level both summer and winter—owing to the following reasons, as I believe. During the winter it runs at its natural height, or a very little higher, because in those countries there is scarcely any rain in winter, but constant snow. When summer comes, this snow, which is of great depth, begins to melt, and flows into the Ister, which is swelled at that season, not only by this cause but also by the rains, which are heavy and frequent at that part of the year. Thus the various streams which go to form the Ister are higher in summer than in winter, and just so much higher as the sun's power and attraction are greater; so that these two causes counteract each other, and the effect is to produce a balance, whereby the Ister remains always at the same level. (Book IV, chap. 50.)

On desert climates where the sun is frequently regarded as an enemy:

...rain and frost are unknown there [in Egypt]. Now whenever snow falls, it must of necessity rain within five days; so that, if there were snow, there must be rain also in those parts. (Book II, chap. 22.)

The Atarantians, when the sun rises high in the heaven, curse him, and load him with reproaches, because (they say) he burns and wastes both their country and themselves. (Book IV, chap. 184.)

We close the quotations with an account of the circumnavigation of Africa which carries us beyond the regions which Aristotle thought to be habitable.

The Phoenicians took their departure from Egypt by way of the Erythraean Sea, and so sailed into [over] the southern ocean. When autumn came, they went ashore, wherever they might happen to be, and having sown a tract of land with corn, waited until the grain was fit to cut. Having reaped it, they again set sail; and thus it came to pass that two whole years went by, and it was not till the third year that they doubled the Pillars of Hercules, and made good their voyage home. On their return, they declared—I for my part do not believe them, but perhaps others may—that in sailing round Libya they had the sun upon their right hand. In this way was the extent of Libya first discovered. (Book IV, chap. 42.)

Quotations of this kind, which give an insight into the common knowledge of the day about the climate and weather of the regions which came within the knowledge of the ancients, could be multiplied manifold by extracts from ancient authors, and a rough idea of the various climates of the Mediterranean can be formed which is, in its general features, quite consistent with the experiences of the same neighbourhoods at the present day. These experiences differ from those of more Northern latitudes by being dominated by a seasonal incidence of rainfall and winds as well as temperature.

CHAPTER V

METEOROLOGY IN THE TIME OF ARISTOTLE

		B.C.
Pythagoras	about	530
Anaxagoras		500–428
Socrates		469–399
Hippocrates	about	461–357
Democritus	about	460–361
Plato		427–347
Aristotle		384–322
Theophrastus	about	373–286

In the previous chapter we have noted that the writings of Hippocrates, somewhere about the year 400 B.C., give evidence of the treatment of the experiences of weather according to the method of scientific inquiry which is the guiding principle of natural philosophy at the present day and which forms so important a part of the legacy of Greece. Hippocrates discloses a definite conception of climate and its possible influences. The Greek philosophers had already begun to speculate upon the relationships and causes of the various natural phenomena. The views of Anaxagoras, Democritus, Hippocrates, Aeschylus (a pupil of Hippocrates), Anaximenes, Empedocles, Plato, Cleidemus are referred to by Aristotle in the discussion of those phenomena generally with a view to the exposure of their defects.

We do not propose to enter into any general discussion of meteorology in Greek literature, it has been done already in ample detail by Otto Gilbert in a well-known work[1]. We intend only to give some account of Aristotle's *Meteorologica*, the earliest treatise on the subject, and of the contribution of his pupil and successor Theophrastus, who wrote treatises on the winds and on weather-signs. These treatises disclose the high level of attainment of the Greek philosophers in the study of weather.

The system established by Aristotle remained for nearly 2000 years the standard text-book of our science....All text-books of meteorology issued on the continent till the end of the seventeenth century are exclusively based on Aristotle, whereas, curiously enough, in England his influence was much less. If I except Duns Scotus, I do not know any British scholar who has written a commentary on the meteorology of Aristotle, and even this one has quite recently been disputed.

(G. Hellmann, *Q. J. Roy. Met. Soc.* 1908, p. 228.)

In the Bakerian Lecture of 1892[2] Professor James Thomson cited, as an example of the crude state of knowledge at the time, a paper by Dr Martin Lister before the Royal Society in 1684 which regarded the trade-winds as the "constant breath" of the sargasso weed "because the matter of that *Wind*, coming (as we suppose) from the breath of only one *Plant* it must needs make it constant and uniform: Whereas the great variety of *Plants* and

---

[1] Otto Gilbert, *Die Meteorologischen Theorien des Griechischen Altertums*, Leipzig, 1907.
[2] *Phil. Trans.* A, vol. CLXXXIII, 1892, p. 654.

Trees at land must needs furnish a confused matter of *Winds*." Dr Lister's suggestion is based upon Aristotle's theory of exhalation or emanation as the cause of winds and, regarding the matter from a very modern point of view, if Dr Lister had described the trade-winds as the surface-exhalation or emanation from an anticyclone instead of from the plants that grow therein he would have made an effective contribution to the subject.

ARISTOTLE'S Meteorologica

It is from Aristotle's treatise that the name meteorology is derived, although Aristotle cites the name as already used by his predecessors. In modern form the treatise would be represented by an octavo volume of some hundred pages. It is divided into four books which cover a much wider range than is covered now by the name meteorology. Everything of a physical nature in sky or air or earth or sea is included. The information which is given here is taken from a French translation of the *Meteorologica* by J. Barthélemy Saint-Hilaire[1]. Our version is a free translation of the French. An English translation[2] has been published since the text of this chapter was written so that the details are now within reach of the English reader.

To begin with we give the table of contents of the first book.

1. Recapitulation of previous works on natural science. Object and extent of meteorology. Indication of further works on animals and plants.

2. General principles and elements of the terrestrial world, its relations with the rest of the universe.

3. Arrangement and nature of the four elements [earth, air, fire, water]. Anaxagoras's opinion about the aether. On the special nature of the bodies which fill the space between the earth and the stars. Double nature of exhalation. On the formation of clouds and their height. On the upper regions of the air.

4. Glowing flames (bolides), shooting stars and their cause.

5. Other celestial phenomena and their causes.

6. The comets. Opinions and explanations of Anaxagoras, Democritus, Hippocrates of Ceos, of Aeschylus. Refutation of these erroneous opinions.

7. On the nature and cause of comets.

8. On the milky way. Opinions of the Pythagoreans, of Anaxagoras, and of Democritus; another explanation. Refutation of these opinions. New theory of the milky way.

9. On the formation of clouds and of mist.

10. On dew and hoar-frost.

11. On rain, snow, hail and their relation to hoar-frost.

12. On hail: remarkable phenomena which accompany it; the opinion of Anaxagoras refuted by the facts.

13. On winds. Erroneous opinions on this subject. The formation of rivers: refutation of false theories. New theory of the origin of rivers, geographical details[3].

14. Permanent changes and reciprocal revolutions of seas and continents by the erosion of rivers. Various examples especially in Egypt. General remarks on the migration of races. Periodic cataclysms.

---

[1] *Météorologie d'Aristote traduite en français*, par J. Barthélemy Saint-Hilaire, Paris, 1863.

[2] *The Works of Aristotle, Meteorologica*, translated into English by E. W. Webster, Oxford, Clarendon Press, 1923.

[3] Already quoted p. 14.

The second book treats in like manner of the sea and its saltness, the general theory of winds, the two exhalations dry and moist, of which the dry forms the winds. Relation of winds to rain and dryness. Variation of weather. Classification of winds into North and South, their sequence. Influence of the sun and stars on the wind, the etesian winds of the North and South. Relation of wind to the configuration of the earth. Arctic and Antarctic Poles. Extent of the North and South winds, the monsoons. General scheme of winds, their number and names. Reduction of all winds to two principal kinds and their different actions. Earthquakes, lightning and thunder, with the author's theory.

In the third book the theory of thunder and lightning is pursued in relation to whirlwinds or typhoons and thunderbolts—the air is always the cause of them. Haloes and rainbows, mock suns and sun-pillars, personal observations by the author, the theory of refraction with graphic demonstrations, notes on the three colours of the rainbow and their reversal in the second bow.

Book IV is devoted to the theory of the four elements, two active, heat and cold, and two passive, dryness and humidity. The application of the theory to the changes in matter as in cooking and to the physical properties of matter in general.

In the whole work fifteen chapters out of forty-two are devoted to the science of the atmosphere. In form and manner there is an extraordinary resemblance to a treatise on meteorological theory of the present day, the same dependence for its facts on common knowledge or the recorded experience of others, the same recognition of the use of personal experiment and observation, the same habit of quoting other people's theories for the purpose of showing them to be erroneous, the same kind of argument by analogy where the analogy is assumed rather than demonstrated and may not really exist, the same impulse to provide an explanation of the facts (always said to be easy and complete) by some form of words which appeals to those who are interested in the subject as a branch of learning for learning's sake, but which carries nothing to the husbandman or sailor except the résumé of the facts which is the preliminary of all scientific argument.

There is the same appreciation of the liberty which poets and fabulists are allowed to take with regard to scientific facts.

In poetry explanations of this kind may indeed appear satisfactory for metaphor is eminently poetic; but they are evidently inadequate for the comprehension of nature. (Book II, chap. 3, par. 13.)

But to imagine, as Democritus does, that the amount of the sea is diminishing unceasingly and that at the end it will disappear altogether, such an opinion seems quite equal to Aesop's fables. Thus Aesop tells us that Charybdis having twice swallowed up the waters, first made the mountains appear, then next the islands, and at the third gulp will finally dry up the whole earth. (Book II, chap. 3, par. 3.)

What could be more reminiscent of modern treatises on meteorology than the remark "everything that is warm tends naturally to rise" (Book I, chap. 4, par. 10), or more suggestive of the most modern meteorological theory than "for those things which escape the direct appreciation of our

senses we consider we have demonstrated them in a manner satisfactory to
our reason when we have succeeded in making it clear that they are possible"
(Book I, chap. 7, par. 1).

We append a passage recalling descriptions of experiments much nearer
to our own time than the writings of Aristotle.

That which really proves that the saltness of the sea is due to the admixture of
some substance is, besides all that we have said above, the following experiment.
If you place in the sea a vase of wax, shaped for the purpose, the opening being
closed by a stopper impermeable to sea-water, that which gets through the divisions
of the wax is drinkable water. The earthy part is retained as by a sieve in the same
way as that which produces saltness by its admixture. It is this part also which causes
the heaviness and thickness of sea-water which is heavier than fresh water. (Book II,
chap. 3, par. 35.)

One wonders what actual permeable membrane is connoted by the word
which is translated "wax."

The chief basis of his meteorological argument is that the atmosphere is
made up of dry air, regarded as a sort of smoke, and water; he ridicules his
predecessors who regard wind as air in motion and who consider that ordinary
air and not exclusively water-vapour is condensed to form rain.

He explains wind on the analogy of rivers which represent the gradually
accumulated flow of water from the mountains downwards; so wind is due
to the gradual accumulation of emanation from the earth just as rain is the
gradual accumulation of water due to the condensation through cold of the
emanation of water-vapour.

The winds come mainly from the North and from the South; those from
the North emanate from the cold regions under the Great Bear, the Northern
limit of the habitable world, and are therefore cold. Those from the South
come, not from the South Pole, but from the tropic of Cancer, which is the
Southern limit of the habitable world of the Northern Hemisphere, because
beyond it the heat is too great for life: the South winds are hot winds on
account of the heat of the region from which they emanate.

Since he bases his arguments upon the facts which had been accumulated
in accessible form we may regard the *Meteorologica* as representing the state
of knowledge in his time. The facts are collected from historians and epic
poets and from common experience. We translate freely from Saint-Hilaire
an account of the formation of dew and hoar-frost. There is an equally
excellent account of clouds, rain, snow and hail.

On Dew and Hoar-Frost

(1) That part of the vapour which forms in the day, but which is not carried to the
upper regions because there is too little fire for lifting it compared with the mass of
water which it lifts, after being cooled falls to the ground afresh during the night,
and it is this which is called dew and hoar-frost. (2) It is hoar-frost when the vapour
freezes before being changed into water, and it is produced more often in winter and
in cold regions. It is dew, when the vapour changes into water, and when it is not
warm enough for it to be dried up during its ascent, nor cold enough for the vapour
to freeze because either the place or the season is warmer. (3) Dew is formed in

clear weather and in calm localities; hoar-frost, on the contrary, as I have just pointed out; for it is obvious that vapour is warmer than water because it still contains the fire which raises it, and so more cold is needed to freeze it. (4) Both dew and hoar-frost are formed in fine weather and when there is no wind; for if the weather were not clear they would not be able to rise in the air, and they would not be able to form if the wind were blowing; and a good proof that they are produced because the vapour has not risen very high, is that hoar-frost is never seen on mountains. (5) The primary cause of this phenomenon is that the vapour rises from low and damp regions, in such a way that the warmth which carries it, as if it were bearing a burden beyond its strength, cannot lift it to a great height but soon lets it fall. A second cause is that the air which flows away and which destroys this sort of combination, flows away chiefly in high regions. (6) Dew is formed everywhere with south winds and never with north winds, except in *Pontus* where the phenomenon is reversed: always with north winds and never with south winds. It is for exactly the same reason that it forms in clear weather and not in bad weather; for the south wind brings clear weather and the north wind bad weather, and the latter wind is cold enough to destroy by the bad weather the warmth of the emanation. (7) In *Pontus* on the contrary the south wind does not bring weather sufficiently good for the vapour to form, and the north wind by its coldness accumulates the warmth which surrounds it, so as to give much more opportunity of evaporation. (8) This is sometimes seen also in countries outside *Pontus*. The wells give out more vapour during north winds than during south winds. But the north winds destroy the warmth before it has accumulated in large quantities, the south winds on the contrary allow the evaporation to accumulate as much as it likes. (Book 1, chap. 10.)

We have not thought it necessary to deal in detail with the chapters upon the rainbow, the halo and other optical phenomena. The use of geometrical diagrams for the purpose of illustrating the argument is a reminder of the antiquity of that aid to reasoning. A curious fact about atmospheric optics is that the solution of its most obvious problems, the explanation of the rainbow, the halo and other associated phenomena, has hitherto made very little contribution to the more general problem of the sequence of weather. The problems are, therefore, more interesting as a challenge to the ingenuity of the physicist than as a stage in the progress of the meteorologist.

ON THE CLASSIFICATION OF WINDS

A rose of the names of the winds is here reproduced (fig. 8) and in juxtaposition therewith a card of degrees and compass-points with the deviation of the magnetic needle from true North in London.

We start with A the wind from the equinoctial sunset Zephyros and give the wind B from the equinoctial sunrise, Apeliotes, as its opposite. Then H the North wind, Boreas, coming from the Great Bear, and its opposite Notos, Θ. We then seek for the names of opposing winds and give Z Kaikias a North East wind from the summer sunrise and opposite to it T the South West wind Lips (coming from Libya). From the South East, from the winter sunrise, comes Euros Δ and opposite to it the wind E from the summer sunset called sometimes Argestes, sometimes Olympias, and sometimes Skiron. These are the eight winds which are diametrically opposed the one to the other.

There are other winds of which the directions are not exactly opposite. Thus from I blows the wind called Thraskias which is between Argestes and the North wind, and from K blows a wind called Meses or mean between Kaikias (the North East) and the North; the line IK very nearly touches the circle of the stars which are always visible, but not quite. There are no opposites for these winds; neither for Thraskias nor for Meses; for if there

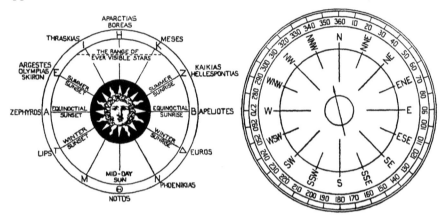

Fig. 8. A Greek head of Apollo within the wind-rose of Aristotle and Theophrastus and the deviation of the compass in London (1925) within a modern wind-dial.

were an opposite for Thraskias it should blow from N. There is however a wind which blows from a point very near to that: the country people call it Phoenikias.

Such are the principal winds which have been determined and such is their general disposition. If there are more winds coming from the northern regions than there are coming from the regions of midday sun it is because the inhabited earth is situated in the former regions, and also because much more rain and snow is repelled from the latter regions because they lie under the sun, and under his path. The water and the snow melt and filter into the earth and are warmed by the sun and by the earth; it follows necessarily from this cause that evaporation is much more considerable and is produced over a much greater range.

Of all the winds which have just been named the most distinct is Boreas called also the wind of the Bear. Thraskias partakes of Argestes and of Meses; Kaikias of Apeliotes and of Boreas. One calls a south wind both that which comes directly from the south and also Lips, one calls Apeliotes both that which comes from the equinoctial sunrise and also Euros. The name Phoenikias is common to several winds; and one calls Zephyros that which is truly Zephyros and at the same time that which is called Argestes. In a general way the winds may be divided into those from Boreas and those from Notos; Zephyros winds may be grouped with Boreas winds for they are cold because they blow from the sunset and we may put with Notos all those which come from Apeliotes because they are warmer considering they blow from the sunrise. It is by recognising the winds according to cold or heat and by the mildness of temperature that the winds are named as we have just seen. Those are warmer than these because those which blow from the east are longer under the sun. As for those which come from the west the sun departs more quickly and is later in approaching the place from which they blow.

The winds being thus classified it is evident that the opposite winds cannot blow at the same time. In effect since they are diametrically opposed one of the two must perforce cease to blow, but those which are not arranged in that way the one in relation to the other may quite well blow at the same time: for example Z and Δ (Kaikias and Euros). There we see that sometimes the two favourable winds may blow together to drive a ship towards the same port, and they do not come from the same point of the horizon and are not confused in a single wind. It is in opposite seasons that the opposite winds blow most. So at the spring-equinox, it is Kaikias, and generally all the winds from beyond the summer turning, and at the autumn-equinox Lips; at the summer-solstice Zephyros and at the winter-solstice Euros[1].

As a general rule it is the winds from Aparktias, Thraskias and Argestes which continue after the other winds and make them cease, for that they are so frequent and that they blow so violently; it is because their point of origin is very near, also they give the clearest weather of all the winds; blowing from close by they have correspondingly more force and they suppress the other winds and dispersing the congested clouds they bring fine weather, at least when they are not at the same time very cold; then in effect there is not fine weather for if they are colder than they are strong they cause condensation before they have chased away the clouds. Kaikias is not a fine weather wind because it draws clouds on itself, whence the popular proverb "He draws everything to him as Kaikias attracts the clouds." When the winds come to an end the changes in those which follow them take place according to the veering with the sun, because it is the sun which initiates the strongest movement and the initiative of the winds is set afoot exactly like the sun himself. Opposite winds produce either the same effect as their opponents or a contrary effect. Thus Lips and Kaikias which is called also Hellespontias are moist...........................
.... Argestes and Euros are dry, the latter is dry at the beginning and wet at the end, Meses and above all Aparktias are snowy for they are colder than any. Aparktias brings hail, so do Thraskias and Argestes. Notos, Zephyros and Euros are hot. Kaikias covers the sky with thick clouds, with Lips the clouds are less congested; as regards Kaikias it is because he makes them return on himself and that he partakes of Boreas and of Euros in such a way by its cold condensing the air which evaporates so that it forms in clouds, and as from its position it approaches the east wind it brings much stuff and vapour which it chases in front of it. Aparktias, Thraskias and Argestes are clear-weather winds and we have already given the cause. It is those last and Meses which most frequently bring lightning; they are cold because they blow from near by and it is by the cold that lightning is made; for it is driven out of the clouds when they combine [meet]. That is why also some of the winds bring hail because they produce a rapid congelation. They become stormy cloud-bursts especially in [late] autumn and then in spring, particularly Aparktias, Thraskias and Argestes; and what makes them stormy is especially when the winds come to the middle of other winds which are blowing, and it is especially the winds which I have just named which behave in this manner. We have already given the cause[2].

For those who live towards the sunset the etesian winds vary from Aparktias to Thraskias, Argestes and Zephyros, for Zephyros has to do also with Aparktias and the etesian winds commence from the great bear and finish with winds wide of that point. For those who live towards the dawn the etesian winds oscillate and range as far as Apeliotes[3].

That is all that I have to say about the winds, about their production starting from their origin, on their nature, their general character and the particular character of each one of them. (Book II, chap. 6, pars. 10–24.)

Further details about the winds of Athens are given in an extract from the peripatetic treatise on signs entitled *Ventorum Situs et Appellationes*[4]. It adds

---

[1] This summary should be compared with the climatic table for Athens in chapter II.

[2] The characteristics of the winds may be compared with the inferences drawn from the sculptures on the Tower of the Winds, p. 83.

[3] Compare the general flow of winds indicated by the maps of pressure distribution, fig. 4.

[4] *The Works of Aristotle*, translated into English under the editorship of W. D. Ross, M.A., vol. VI, *Opuscula*, Clarendon Press, 1913.

another name Iapyx for the North West wind Argestes and supplies names for the winds on either side of Notos, the wet South wind, namely Euronotos, for the South South East and Leuconotos (the "white" or clearing South wind) for the South South West. But for the most part it is devoted to giving the local names for the winds which have already been identified.

This extract is attributed to Theophrastus, a pupil of Plato, and of Aristotle after Plato's death in 347 B.C. He was Aristotle's literary executor and his successor in the direction of the Lyceum. He wrote a treatise *On Winds*, in which he amplifies the work of Aristotle, and a further treatise *On the signs of rain, winds, storms and fair weather*[1].

He introduces his treatise very philosophically:

We must now endeavour to show that each wind is accompanied by forces and other conditions in due and fixed relation to itself; and that such conditions in fact differentiate the winds one from another.

(Theophrastus of Eresus, *On Winds and on Weather Signs*, translated by Jas. G. Wood and edited by G. J. Symons, F.R.S. London, Edward Stanford, 1894, § 1.)

And we cull from his remarks some reminder of the time-honoured question of the monsoon or seasonal wind, which appears in Greece as the etesian wind, and of the diurnal variation of wind:

We will now consider the nature of the monsoon [etesian wind or the north wind of summer]. Why it blows at this particular season and for a particular number of days and why it ceases as the day closes in and almost universally does not blow at night are to be explained on the following principles. The movement of the air is caused by the melting of the snow.... (§ 11.)

If, then, it is true (as some and particularly the dwellers in Crete say) that the winters are more severe and more snow forms than formerly [as proved by more restricted cultivation] it follows that the monsoon also has greater duration. (§ 13.)

We add a characteristic attempt at physical explanation.

Why can it be that it is said:

Fear not as much a cloud from the land as from ocean in winter;
But in the summer a cloud from a darkling coast is a warning?

Can it be because in winter the sea is warmer than the land, so that if a cloud is formed over it, its formation is obviously due to a powerful active principle? For otherwise it would have been dissolved by the air by reason of the warmth of its situation; while in summer the sea is cold and so are the winds from the sea; and the land is warm; so that if a cloud is borne from the land seawards its formation must be due to some active principle more powerful than usual; for the cloud would have been dissolved, if the active principles had been weak. (§ 60.)

The second treatise of Theophrastus gives an excellent summary of astronomical signs which are of two kinds, derived from the risings and settings of stars, and then the indications from the general appearance and colour of the sky or of the sun or moon, the signs from the behaviour of animals, and from electrical and optical phenomena. Of necessity it covers part of the ground proper to the treatise on winds. The two can therefore be regarded as more or less continuous. Here we draw a line of separation between them because

[1] A new translation of the Weather Signs by Sir Arthur Hort has recently been published in the Loeb Classical Library.

the second treatise affords the best starting-point for the account of *Weather-lore* which is the aim of chapter VII.

The representations of the winds by Aristotle and Theophrastus lack the precision which a reader likes to see in a scientific treatise; they have given rise to some discussion[1] as to whether the Greeks used a twelve-part wind-rose in place of the eight-part wind-rose of the Babylonians. The opportunity for discussion seems to rest largely on the fact that the Greeks of the time of Aristotle had very limited means of expressing directions. Apparently the mid-day sun, winter sunrise, summer sunrise, equinoctial sunrise, and the corresponding directions at sunset, North under the Great Bear, and the range of the stars that never set, were the limit of the representation of direction. The summer and winter sunrise being about 30° North and South of the equinoctial sunrise, the choice of that method of indicating direction suggests a division of the circle into twelve parts. Aristotle's writing does not suggest the idea of a compass-card divided into equal sectors with names for the several radii but merely the sorting of winds known by their names into opposites or neighbours.

The question seems to be settled in favour of an eight-part wind-rose by the Tower of the Winds at Athens, which is octagonal and carries on its faces the names of the winds and sculptured figures intended presumably to represent the general character of the eight selected winds[2]. It is a very celebrated monument called the Horologium or Time-indicator of Andronikos Kyrrhestes by whom it was erected in the second century B.C.

Apparently it was a practical demonstration of the use of scientific knowledge such as we see to-day in the public parks of this or other countries. Each side carried a sun-dial giving the hours of the day as explained on p. 47, a clepsydra or water-clock was housed within and would count the hours of the night. In so far as the sculptures represent the general character of the weather which the several winds bring it may be regarded as a very early form of weather forecast. Drawings of the sculptures are reproduced overleaf as fig. 9.

With these we take leave of the first meteorological treatise and its supplement. We note that although the poets and the philosophers differ very widely in their attitude towards the study of weather they are on the common ground of weather-lore when they wish to find a practical use for their knowledge. We owe to the Greek philosophers not only the beginnings of the scientific inquiry into the phenomena of weather and their relation to each other but also the most compendious exposition of the practical application of weather-knowledge.

---

[1] D'Arcy Thompson, 'The Greek Winds,' *Classical Review*, vol. XXXII, 1918, p. 49.
[2] A full account of the "Tower of the Winds" with scale drawings is given in the *Antiquities of Athens*, by Stuart and Revett, London, 1762.

THE WINDS OF HELLAS

Fig. 9. Drawings of sculptured figures on the eight sides of the Tower of Andronikos Kyrrhestes.

(From Stuart and Revett, *Antiquities of Athens*, London, 1762.)

THE SCULPTURES OF THE EIGHT WINDS

NOTE: The names with their alternatives are followed by descriptions of the figures by J. G. Wood. Our estimate of the corresponding weather is given in italics and is followed by Aristotle's account of the wind.

Boreas, Aparktias with **Meses.** North or North North East. An old man very warmly clothed holding a conch-shell. *Cold roaring wind*—Final, dominant, bringing fine clear weather when not too cold; colder than any, with snow, hail, lightning, cloud-bursts especially in autumn; prevalence the same as Kaikias. A primary wind.

Kaikias. North East. An old man with severe countenance, holding a shield with hailstones (or perhaps olives) in it. *Blustering weather*—Aparktic in character, heavy clouds travelling windward, wet; prevalent at spring equinox.

Apeliotes. East. A young man with flowing drapery, bearing fruits, ears of corn and a honey-comb. *Second-harvest weather*—"Notic" in character, warm because it comes from sunrise.

Euros. South East. An old man with morose countenance, well-wrapped up, his mantle concealing his right hand, and held up by the left to protect his face. *Ugly, threatening weather, the precursor of a Mediterranean cyclonic storm*—"Notic," dry at the beginning and wet at the end. Prevalent at the winter solstice.

Notos. South. A young man emptying a jar of water. *Rainy weather*—A primary wind, warm.

Lips. South West. A man of middle age bearing an "aplustre," i.e. the ornamental finial of the stern of a Greek ship under which the helmsman stood. *Good sailing weather for a homeward voyage*—"Notic," cloudy, wet; prevalent at the autumn equinox.

Zephyros. West. A fair almost effeminate youth, nude except for a loose mantle, the folds of which are filled with flowers. *Pleasant warm weather*—Aparktic yet warm; prevalent at the summer solstice.

Skiron, Thraskias and **Argestes.** North West. An old man holding a large inverted jar, which may be a brazen fire-pot, different from the water-jar of Notos. *The cold wind with squalls at the end of a depression*—Aparktic, final, dominant, bringing fine clear weather when not very cold. Cold, dry, with hail, lightning, cloud-bursts especially in autumn; prevalent at the spring equinox.

The terms Aparktic and Notic are used to indicate Aristotle's grouping of the winds under the two primaries, Aparktias (Polar) and Notos (Equatorial). We may infer that the idea of a polar front need not be regarded as entirely modern.

The etesian winds, the steady winds of summer, beginning with Aparktias may range from Zephyros for dwellers in the West to Apeliotes for dwellers in the East.

CHAPTER VI

VARIABILITY OF MEDITERRANEAN CLIMATES
IN HISTORICAL TIMES

B.C.		
—	First glacial period (Gunz)	
430,000–370,000	Second glacial period (Mindel)	Pleistocene glaciation
130,000–100,000	Third glacial period (Riss)	
40,000–18,000	Last glacial period (Wurm)	
18,000–12,000	Beginnings of neolithic civilisation (Susa in Persia and Crete)	
8000	Mud deposit of the Nile began	
5000	Early forest period. The ice-age ended: a halt of the retreating Scandinavian ice at Ragunda	
3000	Forest period in Europe between two peat-bog periods	
2000	Horizon of tree-records	
1800	Beginning of the classical period of rainy seasons and of the bronze-age in Europe	
400	Maximum of the rainy period for temperate regions	
A.D.		
400	End of the classical period of great rainfall, beginning of the dry warm period of the middle ages	
530	Minimum of storminess (O. Pettersson)	
1300–1400	Century of violent storms and extremes	
1434	Last maximum of storminess (O. Pettersson)	

THE question of the variation of climate within the period of historical records has been a subject of prolonged discussion. Aristotle may indeed be said to have opened the discussion by his note on the changes of climate in Greece.

The same parts of the earth are not always moist or dry, but they change according as rivers come into existence and dry up. And so the relation of land to sea changes too and a place does not always remain land or sea throughout all time, but where there was dry land there comes to be sea, and where there is now sea there one day comes to be dry land. But we must suppose these changes to follow some order and cycle.

. . . .

This has been the case with Egypt. Here it is obvious that the land is continually getting drier and that the whole country is a deposit of the river Nile. But because the neighbouring peoples settled in the land gradually as the marshes dried, the lapse of time has hidden the beginning of the process....

For the parts that lie nearer to the place where the river is depositing the silt are necessarily marshy for a longer time since the water always lies most in the newly formed land. But in time this land changes its character, and in its turn enjoys a period of prosperity. For these places dry up and come to be in good condition while the places that were formerly well-tempered some day grow excessively dry and deteriorate. This happened to the land of Argos and Mycenae in Greece. In the time of the Trojan wars the Argive land was marshy and could only support a small population, whereas the land of Mycenae was in good condition (and for this reason Mycenae was the superior). But now the opposite is the case, for the reason we have mentioned: the land of Mycenae has become completely dry and barren, while the Argive land that was formerly barren owing to the water has now become fruitful. Now the same process that has taken place in this small district must be supposed to be going on over whole countries and on a large scale.

(The Works of Aristotle, Meteorologica, translated into English by E. W. Webster, Oxford, Clarendon Press, 1923, Book 1, chap. 14.)

The subject is also very appropriate in this place because it is a good example of a certain type of meteorological problem wherein the difficulty of forming a final conclusion arises from the magnitude of the time-scale and an unavoidable lack of precision in the available data. The Mediterranean countries and Central Asia are specially involved because they have been continuously inhabited since the beginning of civilisation and historic records or marks of ancient habitation are, comparatively speaking, abundant. Moreover, it is apparent that places which in the past were famous, such as Carthage and Cyrene on the North African coast, Babylon and Nineveh in Mesopotamia, or Palmyra and the surrounding district east of Palestine, are now practically desert. Cyprus and Palestine which formerly were regarded as highly prosperous are now poor. People of experience are convinced by what they see to-day that regions which were formerly productive have gone out of cultivation for lack of the rainfall which they once enjoyed. Others are equally certain from the records of antiquity that winters used to be more severe than they are now. Others again see no necessity for assuming a progressive change of climate.

A SURVEY OF THE POSITION

It is not our purpose to pronounce judgment on these questions: we shall confine ourselves to a few extracts which exhibit the nature of the problem, leaving the reader to pursue the subject to its logical conclusion.

That the Earth is now becoming drier is consistent with many impressive facts. The post-glacial lakes have grown smaller; they have been replaced by marshes, or even by meadows. The great inland sea of south-western Asia has shrunk until it is now represented only by the Caspian and the Sea of Aral. Asiatic travellers bear unanimous testimony that in large areas of Central Asia, lakes and swamps are still disappearing, fertile plains are still being converted into deserts, rivers are ceasing to run, and the deserted ruins of once busy cities are being buried in sand. All these changes are interpreted to mean that the rainfall over Central Asia has decreased and is still decreasing.

Similar evidence is reported from many widely separated parts of the world. The desolation of Palestine is attributed to a decrease in its water-supply. In tropical Africa, many lakes have vanished, and Lake Chad has become so much shallower and smaller that it has been described as "a disappearing lake." Livingstone's first contribution to South African geography, which was unluckily lost, described the increasing desiccation of the country beside the Kalahari desert. In North America, great freshwater lakes have dwindled to salt pools; Indian cities, once numerous in certain districts, have been deserted, and these districts are said to be incapable of now maintaining a population corresponding to the number of ruined settlements. In South America, Dr Moreno[1] and Sir Martin Conway[2] have referred to evidence of cultivation in land now abandoned owing to the failure of the water-supply; and Bowman[3] has described recent desiccation in southern Bolivia. The maps of the interior of Australia mark very extensive lakes which are to-day mere clay pans or salt swamps, while the once fertile plains around them have withered into desert.

Many of the geographical changes have been proved by unmistakable evidence, and

[1] F. P. Moreno, 'Notes on the Anthropogeography of Argentina,' *Geogr. Journ.* vol. XVIII, 1901, pp. 574–89.

[2] Sir Martin Conway, in discussion, *Geogr. Journ.* vol. XXIII, 1904, pp. 736–7.

[3] I. Bowman, 'Man and Climatic Change in South America,' *Geogr. Journ.* vol. XXXIII, 1909, pp. 267–78, 602.

according to Prince Kropotkin they occasioned some of the greatest historic movements among men. Thus he attributes the overthrow of the Roman Empire to the dwindling rainfall of Central Asia, which turned whole tribes of agriculturists into nomads by the repeated failure of their crops, and finally drove the drought-stricken barbarians into Europe, owing to their own lands having become uninhabitable. Similarly, Prof. J. L. Myres[1] has attributed the various invasions of Egypt to droughts in Arabia and Libya. If this desiccation of the earth is still in progress it must lead to further great political changes; for in time the populations of the world will be forced out of the centres of the continents and crowded into the continental margins.

(Prof. J. W. Gregory, 'Is the Earth Drying Up?'
Geogr. Journ. vol. XLIII, 1914, p. 149.)

Syria

The ruins of Palmyra, in the Syrian Desert, show that it must once have been a city like modern Damascus, with one or two hundred thousand inhabitants, but its water-supply now suffices for only one or two thousand. All attempts to increase the water-supply have had only a slight effect and the water is notoriously sulphurous, whereas in the former days, when it was abundant, it was renowned for its excellence. Hundreds of pages might be devoted to describing similar ruins. Some of them are even more remarkable for their dryness than is Niya, a site in the Tarim Desert of Chinese Turkestan.

(Ellsworth Huntington and S. S. Visher, *Climatic Changes, their nature and causes*, p. 66.)

Palmyra, the city of palm trees, was situated in the desert between the coast range of Northern Syria and the Euphrates. It was a station of great importance on the caravan route between the Mediterranean and the Persian Gulf. Its importance was prominent from the first century A.D. to the fourteenth. The celebrated Zenobia, queen of Palmyra when it was taken by the Emperor Aurelian, belonged to the end of the third century.

The results of the explorations of the last twenty years have been most astonishing in this regard. It has been found that practically all of the wide area lying between the coast range of the eastern Mediterranean and the Euphrates, appearing upon the maps as the Syrian Desert, an area embracing somewhat more than 20,000 square miles, was more thickly populated than any area of similar dimensions in England or in the United States is to-day if one excludes the immediate vicinity of the large modern cities. It has also been discovered that an enormous desert tract lying to the east of Palestine, stretching eastward and southward into the country which we know as Arabia, was also a densely populated country.

.   .   .   .

From a distance it is often difficult to believe that these are not inhabited places; but closer inspection reveals that the gentle hand of time or the rude touch of earthquake has been laid upon every building. Some of the towns are better preserved than others; some buildings are quite perfect but for their wooden roofs which time has removed, others stand in picturesque ruins, while others still are level with the ground. On a far-off hilltop stands the ruin of a pagan temple, and crowning some lofty range lie the ruins of a great Christian monastery. Mile after mile of this barren gray country may be traversed without encountering a single human being. Day after day may be spent in travelling from one ruined town to another without seeing any green thing save a terebinth tree or two standing among the ruins, which have sent their roots down into earth still preserved in the foundations of some ancient building. No soil is visible anywhere except in a few pockets in the rock from which it could

[1] J. L. Myres, in discussion of 'The Burial of Olympia,' *Geogr. Journ.* vol. XXXVI, 1910, pp. 679–80.

not be washed by the torrential rains of the wet season; yet every ruin is surrounded with the remains of presses for the making of oil and wine. Only one oasis has been discovered in these high plateaus.

.   .   .   .

There was soil upon the northern hills where none now exists, for the buildings now show unfinished foundation courses which were not intended to be seen; the soil in depressions without outlets is deeper than it formerly was; there are hundreds of olive and wine presses in localities where no tree or vine could now find footing; and there are hill-sides with ruined terrace-walls rising one above the other with no sign of earth near them. There was also a large natural water-supply. In the north as well as in the south we find the dry beds of rivers, streams and brooks with sand and pebbles and well-worn rocks but no water in them from one year's end to the other. We find bridges over these dry streams and crudely made washing boards along their banks directly below deserted towns. [Many of the bridges span the beds of streams that seldom or never have water in them and give clear evidence of the great climatic changes that have taken place.] There are well heads and well houses, and inscriptions referring to springs; but neither wells nor springs exist to-day except in the rarest instances. Many of the houses had their own rock-hewn cisterns, never large enough to have supplied water for more than a brief period, and corresponding to the cisterns which most of our recent forefathers had which were for convenience rather than for dependence. Some of the towns in southern Syria were provided with large public reservoirs but these are not large enough to have supplied water to their original populations. The high plateaus were of course without irrigation; but there are no signs, even in the lower flatter country, that irrigation was ever practised; and canals for this purpose could not have completely disappeared.

(H. C. Butler, 'Desert Syria, the Land of a Lost Civilisation,' *Geographical Review*, Feb. 1920, pp. 77–108, quoted by Huntington and Visher in *Climatic Changes*.)

Central Asia: Sin Kiang

The importance of the Tarim Basin lies in the fact that it was in the direct line of communication which once formed the great trade-route between China and the West. Two thousand years ago it seems to have been the only means of interchange of commerce between the Far East and the West....

Sir Aurel Stein has been able to trace over a long distance the remains of the old *Limes*, or defensive wall, erected at the end of the second century B.C., no doubt the ancient representative of our modern line of block-houses.

The reasons for the abandonment of such a route as this, and the evacuation of the former inhabitants, give rise to interesting speculation. Was it due to an alteration of climate bringing about excessive desiccation, or to some other cause? From the high state of preservation of records, and other materials, discovered by Sir Aurel, it would appear that the climate has not altered and is much the same now as it was 2000 years ago, for such friable objects as were found could only stand the test of time in an extremely arid climate. If, then, the occupation of this area was independent of climate, it must have been dependent on irrigation, which in its turn was dependent on water derived from glacier-fed rivers.

The lecturer discovered several abandoned sites of human habitation which had been maintained by irrigation, but are now far distant from the terminal point of the river which formerly supplied them with water. This seems to show, as originally suggested by Sir Sidney Burrard, that the evacuation of this area was due to the diminution of water in the rivers and not to an alteration of climate, of which there is no evidence. The glaciers, probably the remains of a former ice-age, are known to be shrinking, with the consequent decrease of the water-supply, which probably accounted for the abandonment of this important commercial highway.

('Innermost Asia: its Geography as a Factor in History,' *Nature*, vol. CXIV, 1924, p. 806.)

Baluchistan, Sistan, Afghanistan

Of Alexander we have already spoken, and have shown that he marched through a desert country quite as miserable and sterile as we now know on the coast of Ichthyophagi.

Of the Garmsel, or the country on the lower Hilmend, Bellew writes: "The valley everywhere bears the marks of former prosperity and population. Its soil is extremely fertile and the command of water is unlimited. It only requires a strong and just government to quickly recover its lost prosperity and to render it a fruitful garden, crowded with towns and villages in unbroken succession all the way from Sistan to Kandahar. The present desolation and waste of this naturally fertile tract intensify the aridity and heat of its climate. But with the increase of cultivation and the growth of trees these defects of the climate would be reduced to a minimum, and the Garmsel would then become habitable, which in its present state it can hardly be considered to be."

Politics, wars, religions, all have been affected by the geography of the country, and everywhere we detect the influence of the great desert on the life of the people. As long as records go back, the great desert has lain where it still lies, and, on the whole, with the same characteristics as to-day.

(Sven Hedin, *Overland to India*, Macmillan and Co., Ltd., London, 1910, vol. II, pp. 225, 227, 234.)

Central America

A study of the ruins of Yucatan, in 1912, and of Guatemala in 1913, as is explained in *The Climatic Factor*, has led to the conclusion that the climate of those regions has changed in the opposite way from the changes which appear to have taken place in the desert regions farther south. These Maya ruins in Central America are in many cases located in regions of such heavy rainfall, such dense forests, and such malignant fevers that habitation is now practically impossible. The land cannot be cultivated except in especially favourable places. The people are terribly weakened by disease and are among the lowest in Central America. Only a hundred miles from the unhealthful forests we find healthful areas, such as the coasts of Yucatan and the plateau of Guatemala. Here the vast majority of the population is gathered, the large towns are located, and the only progressive people are found. Nevertheless, in the past the region of the forests was the home of by far the most progressive people who are ever known to have lived in America previous to the days of Columbus.

(Ellsworth Huntington and S. S. Visher, *Climatic Changes, their nature and causes*, p. 95.)

Europe

The chronicles of the 6th, 13th and 14th centuries contain a series of climatic disasters which cannot be explained as accidental variations in the *weather*. They seem rather to denote an epoch of violent disturbance in the climate, the effects of which may be traced through centuries to our own days. Sudden transitory variations in the climatic conditions are not infrequent in this century, but these are of small importance when compared with what the chronicles relate of the warm and cold winters of the 13th and 14th centuries, or of the encroachment of the sea upon the coasts of the North Sea and the Baltic during the storm-floods in winter.

Nowadays it is not uncommon that a cold winter succeeds a warm one and we experienced quite a series of such two-year periods in the decade 1870–1880. Several times in recent years mild and stormy winters have been succeeded by dry and hot summers. But it never happens nowadays that the seasons change their character or become displaced, as was the case in the years 1302 and 1328, when the fruit-trees blossomed in January, the vine in April, and the ripe corn was harvested in May, while the vintage in Germany began on July 25. Similar climatic conditions occurred in 1384, when the corn and vine were in flower by May. No less than three such years were recorded during the 14th century, and they were, as a rule, succeeded by extremely cold ones. After the year 1302, when the vine blossomed in May, there followed an intensely cold winter. The Rhine and the Doubs were frozen up, and vineyards were destroyed by frost at the end of May. In other years, as in 1391, after a warm summer, the cold set in with such severity in September that the grapes were frozen as hard as stones and had to be crushed with clubs before pressing.

Part of the 13th and the whole of the 14th century show a record of extreme climatic variations. In the cold winter the rivers Rhine, Danube, Thames, and Po were frozen for weeks and months. On these cold winters there followed violent floods, so that the rivers mentioned inundated their valleys. Such floods are recorded in 55 summers in the 14th century. There is, of course, nothing astonishing in the fact that the inundations of the great rivers of Europe were more devastating 600 to 700 years ago than in our days, when the flow of the rivers has been regulated by canals, locks, etc.; but still the inundations in the 13th and 14th centuries must have surpassed everything of that kind which has occurred since then. In 1342 the waters of the Rhine rose so high that they inundated the city of Mayence and the Cathedral "usque ad cingulum hominis." The walls of Cologne were flooded so that they could be passed by boats in July. This occurred also in 1374 in the midst of the month of February, which is of course an unusual season for disasters of the kind. Again in other years the drought was so intense that the same rivers, the Danube, Rhine, and others, nearly dried up, and the Rhine could be forded at Cologne. This happened at least twice in the same century.

There is one exceptional summer of such evil record that centuries afterwards it was spoken of as "the old hot summer of 1357."

If a meteorologist of our age should try to account for the anomalous atmospheric conditions of the mediaeval age, I presume that he would have to choose between two alternatives. Either he could suggest the possibility that there had been in certain years, or groups of years, rainfalls of unequalled magnitude in Europe followed by alternating periods of drought, which could be caused by fluctuations in the Atlantic warm water current, or he could ascribe the climatic vicissitudes to variations in the intensity of solar activity, which in that case ought to have manifested itself by a greater frequency of sunspots. If we should try to discuss the question whether either of these conjectures is borne out by historical facts, we should have to come to the rather surprising conclusion that there is sufficient evidence for the validity of both conjectures.

There existed in those centuries a more intense oceanic circulation and a greater frequency of sunspots.

> (Professor Otto Pettersson, 'The Connection between hydrographical and meteorological Phenomena,' *Q. J. Roy. Meteor. Soc.* vol. XXXVIII, 1912, p. 174.)

Northern Africa

In a masterly paper in the *Transactions* of the K. K. Geographischen Gesellschaft in Wien, 1909, Dr Hermann Leiter deals with the question of the change of climate during historical times in North Africa. He cites the indications of climate to be gathered from classical authors with abundant references, and provides a detailed account of temperature and rainfall at the present time, with a curve showing the rainfall in Algiers, year by year, from 1838 to 1908. After a careful review of the evidence he decides against the suggestion of a change of climate in the following words: "Die häufig behauptete Zunahme der Temperatur und Minderung der Niederschläge in Nordafrika während geschichtlicher Zeit lässt sich also nicht beweisen, eher lassen sich Spuren vom Gegenteil beobachten."

Palestine and California

In a paper with the interrogative title 'Is the Earth Drying Up?' before the Royal Geographical Society of London (1914) from which our first quotation was extracted, Professor J. W. Gregory examined the evidence for change of climate and, relying especially on that obtained from the references to vegetation in early historic times in Palestine, came to the conclusion that the isothermal line of 69° had not moved and that the climate of the Mediter-

CLIMATIC CHANGES

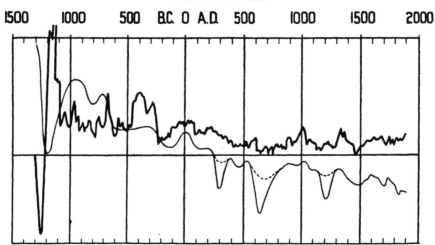

Fig. 10. Changes of climate (rainfall) as deduced from the rings of trees in California—thick full line—and changes of climate in Eastern Mediterranean regions as inferred from the study of ruins and of history—thin full line—with subsequent corrections indicated by the dotted lines.

(From Huntington and Visher, *Climatic Changes*, Yale University Press, 1922. The discussion is continued in 'Quaternary Climates,' see *Nature*, Feb. 13, 1926, p. 238.)

ranean regions is to all intents and purposes the same now as it was in the period before the Christian era. Professor Ellsworth Huntington, who has devoted himself especially to the study of climatic changes and their causes, endorses Professor Gregory's conclusion as regards temperature but cites a considerable amount of circumstantial evidence for fluctuations in rainfall. A very notable addition to the voluminous literature of the subject is A. E. Douglass's examination of the marks of growth in cross-sections of the sequoias of Arizona and California which showed rings ranging in number from 250 to nearly 3250 and therefore provided a record extending back to more than a thousand years before the Christian era. The rings of growth of these trees were connected with rainfall by an examination of the rings of trees within the period of regular observations of rainfall. The climate of the

Eastern Mediterranean is brought into relation with the curve of tree-growth
by examining the rainfall of Jerusalem in comparison with the modern growth
of sequoias in California. Jerusalem is in a similar position to that of California
as regards latitude and some other climatic features and the conclusion of
the inquiry is that "The curve of tree-growth in California seems to be a true
representation of the general features of climatic pulsations in the Mediterranean region[1]." We reproduce the curve for reference. It shows a considerable decline of rainfall after 400 B.C., a secondary increase from 200 B.C.
to A.D. 100, then a decline to A.D. 650 after a slight recrudescence from
A.D. 400 to A.D. 600. Fluctuations with marked maxima at A.D. 1000 and
A.D. 1300 and subsequently such fluctuations as we are accustomed to in our
ordinary climatic tables bring the record down to the present time.

THE GENERAL CIRCULATION OF THE ATMOSPHERE

This sequence of events is in good agreement with the indications derived
from other sources. Besides variations of short period which show very little
regularity there are variations of about eleven years in relation with sunspots
which have been the subject of many investigations. The facts so far as
actual observations of rainfall are concerned are set out very effectively by
Sir Gilbert Walker[2]. Professor E. Brückner[3] made it clear that there have
been fluctuations of the climate of Europe since 1700 with an irregular
period of about 35 years. Dr Otto Pettersson[4] has traced back the incidents
through a series of maxima and minima with special stress upon a remarkable
period of violent extremes of weather, hot and cold, flood and drought,
about the fourteenth century. Sir William Beveridge[5] has in like manner
traced back the fluctuations in the price of corn for more than three centuries.

All these show that there are small but important fluctuations of climate
always in operation which can be included under the general description of
changes in the general circulation of the atmosphere. We shall be in a very
much better position for studying this question when we have made out a
satisfactory account of the general circulation of the atmosphere as it exists
at present and the true meaning of what we have here called a change in the
general circulation. It follows at once from the fundamental idea of circulation
that a change noted in one locality cannot be of the same character all over
the world; it must be accompanied by changes of a complementary character
elsewhere. We cannot really trace the life-history of one aspect of the circulation without reference to other aspects. When we really know the general
circulation and its changes we may be able to select a suitable index of
atmospheric activity that can be used quantitatively with greater confidence

---

[1] Ellsworth Huntington and S. S. Visher, *Climatic Changes, their nature and causes*, p. 76.
[2] G. T. Walker, 'Correlation in seasonal variations of weather. IV. Sunspots and Rainfall,'
Memoirs of the Indian Meteorological Department, vol. XXI, No. 10.
[3] E. Brückner, *Klimaschwankungen, etc.*, Wien, 1890.
[4] Reference on p. 89.
[5] Sir William Beveridge, *Journal of the Royal Statistical Society*, vol. LXXXV, part 3, 1922,
pp. 412-78.

than the local rainfall or any other meteorological element for a single locality Huntington uses as an index storminess in the United States, investigated by Kullmer, and pressure-gradient over the North Atlantic Ocean evaluated in a peculiar manner. Neither is completely satisfactory but serves until the general circulation has been properly scheduled from the physical and dynamical points of view.

VARIATIONS AND THEIR CAUSES

Allowing however that there are fluctuations which can be represented by changes in the general circulation we are assured by geologists that there have also been from time to time glacial epochs, or ice-ages, during which "permanent ice" has covered vast areas which are now entirely free from ice, and on the other hand the existence of coal in Greenland, of fossil plants and warm-water species of shells in other Arctic regions is held to prove that these parts of the world have certainly been warmer than they are now and have been able to support growth of a character which is now impossible. Biologists do not seem to think themselves called upon to allow a change in the biological capacities of the recognised species even after the millions of years which geological changes require, and it is apparently left to geologists and meteorologists to accept an alteration in climate and if necessary to find reasons for it.

Accepting the situation we have accordingly set out at the head of this chapter the principal climatic time-marks in the past up to the last recognised ice-age at 18,000 B.C. as indicated in C. E. P. Brooks's *Evolution of Climate*. They may not be generally accepted in detail but they give a sufficiently definite idea of the changes which have to be faced.

They allow for the sequence of changes indicated by the tree-rings and other evidence up to 1500 B.C. and connect from that point to the ice-age by means of relatively moist peat-bog periods and relatively dry forest periods. Such catastrophic epochs as ice-ages are of great importance because an ice-age is very destructive of animal species and the advent of another might put an end altogether to our version of the human race, or at least a very large part of it, and enable a new start to be made on the recovery of milder conditions by the survivors, whether anthropic, anthropoid or simian.

We do not propose to enter into the detailed discussion of the circumstances which have been assigned as causes for these most important changes. We have already attributed the minor fluctuations, such as are indicated by tree-growths, to changes in the general circulation of the atmosphere that, like the ordinary seasons' changes, may be directly dependent upon variations in solar activity with or without the assistance of such circumstances as the variation in the eccentricity of the earth's orbit or the increase or diminution of carbonic acid gas or of volcanic dust in the earth's atmosphere; and we have drawn therefrom an urgent summons for the effective study of the circulation of the atmosphere in its length and breadth and thickness, and an effective expression of changes as observed in modern times. The greater changes which are connected with the cataclysmic changes in the crust of the earth, or the wandering of the continents with consequential shifts in the position of the polar axis are beyond our present scope. The changes in historic times cannot be classed as cataclysmic, though they may include considerable oscillations in the rainfall of the Mediterranean regions which have resulted in very serious changes in the conditions of civilisation, as a result vast areas that were once thickly peopled have been depopulated and the balance of power has been transferred from one region of the earth to others.

Climate and Civilisation

We remark only upon three points in this connexion. The first is that comparatively slight variations of rainfall may have very disastrous consequences upon the centres of civilisation as pointed out by Sir John Eliot[1] in a note on the famines of India at the Chicago Convention in 1893. Crops can be grown with an assured supply of 500 mm of rainfall, or if the distribution is favourable even with 250 mm, but with a rainfall below 250 mm the climate is arid and between 250 mm and 500 mm (10 and 20 inches) the ordinary fluctuations may easily cause disaster. A curious feature of the growth of crops is that the optimum of rainfall for crops may be very near to the irreducible minimum. It may perhaps be a general biological characteristic that life is most prolific when it is precarious. The particular evidence that we have in mind in this connexion is based upon R. H. Hooker's examination of the relation of the wheat-crop to weather in the Eastern counties of England. There the rainfall is about 500 mm and it is quite clear from the correlation coefficients that it is beyond the optimum. On the other hand we could probably not grow wheat with less than 250 mm. Hence we must look for the optimum somewhere between the irreducible minimum of 250 mm and the undesirable amount of 500. The crops seem to improve with the dryness of the seasons (except the spring) until the minimum of rainfall is nearly approached and then suddenly fail. Hence in a country of optimum rainfall a few years below the average might mean rainfall below the necessary minimum; where irrigation is not practicable, that may mean starvation from which there is no recovery, or migration to some other locality.

Sand

There are two other agencies which may be called indifferently geographical or geological or meteorological. One of them is the gradual disintegration of the rocks in arid regions and the consequent formation of sand which is distributed over vast tracts of country by the wind. The other is the washing away of soil as detritus from the higher levels down to the plains or out to sea with the consequent formation of deltas. Both these processes are

---

[1] 'Droughts and Famines in India,' *Report of the International Meteorological Congress held at Chicago, Ill.*, Aug. 21–24, 1893, published as Bulletin No. 11 of the U.S. Dept. of Agriculture, Weather Bureau, Part II, p. 444.

cumulative and irreversible. The first is a very old story. The formation of
"loess" which forms the soil of steppes is part of its work.

One of the most remarkable formations associated with glacial deposits consists of
vast sheets of the fine-grained, yellowish, wind-blown material called loess. Somewhat
peculiar climatic conditions evidently prevailed when it was formed. At present
similar deposits are being laid down only near the leeward margin of great deserts.
The famous loess deposits of China in the lee of the Desert of Gobi are examples.
During the Pleistocene period, however, loess accumulated in a broad zone along
the margin of the ice-sheet at its maximum extent. In the Old World it extended
from France across Germany and through the Black Earth region of Russia into
Siberia. In the New World a still larger area is loess-covered. In the Mississippi
Valley, tens of thousands of square miles are mantled by a layer exceeding twenty
feet in thickness and in many places approaching a hundred feet. Neither the North
American nor the European deposits are associated with a desert. Indeed, loess is
lacking in the western and drier parts of the great plains and is best developed in
the well-watered states of Iowa, Illinois, and Missouri. Part of the loess overlies the
non-glacial materials of the great central plain, but the northern portions overlie the
drift deposits of the first three glaciations.

(Ellsworth Huntington and S. S. Visher, *Climatic
Changes, their nature and causes*, p. 155.)

Both loess and sand must be subject to similar meteorological laws because
both are wind-blown deposits. Geologists will no doubt wish to draw a
distinction between them, but it is not their difference but their similarity
which is of importance in meteorology.

In *The Times* of November 1, 1923, is a description of the discovery of the
nest and eggs of a dinosaur as well as the skeleton of the dinosaur herself
suddenly overwhelmed by blown sand, at an epoch estimated as being ten
million years ago, as completely and surely as the Psylli and the Persians of
Herodotus' story[1], and sand has been presumably accumulating in the
Eurasian continent during the interval which has elapsed since that event.
In general soil is accumulating in places where it can rest, ground-floors are
gradually becoming basements, Roman causeways in Britain made to be
walked upon as ground-floors, seventeen or eighteen hundred years ago, are
now eight feet below the surface in St Albans and eighteen to twenty feet
below in London. The Royal City of Jerusalem is some forty feet below
the present ground-level.

The importance of the gradual accumulations of blown sand was noticed
briefly by Mascart in a communication printed in the Hann Band of the
Meteorologische Zeitschrift. The formation of sand goes on at a steady rate
and its gradual increase over cultivated areas is inevitable because there is
no means effectively available for its removal by any natural process. Sand
may be converted into cultivable soil by the admixture of other suitable
ingredients as in Egypt by the periodic mixture of the mud of the Nile: but
if the rate of increase of sand is greater than the corresponding supply of
the other ingredients which human activity can provide, progressive deterioration of the soil, encroachment upon the rivers and their ultimate submergence,

[1] See p. 70.

are inevitable. Human agency can prevail against the gradual encroachment of sand for a time; but if the process is continued, any relaxation of effort means losing ground, and ultimately human agency is overcome and cultivation becomes hopeless, irrespective of any change of climate. Under existing geological conditions the process is irreversible.

Soil

In like manner the gradual removal of soil from the higher levels by the natural alternations in the flow of water due to heavy rain-storms, which are characteristic of the coastal climate of the Mediterranean, seems equally inevitable and irreversible. The soil of the uplands may be regarded as a strictly limited deposit of capital left by the preceding ice-age and the moist conditions which would follow it. Soil may accumulate locally by the action of vegetable growth; but on balance there must be less at the end of an ordinary year than at the beginning. As one passes through the high upland valleys of Europe one cannot fail to note the patches of cultivable soil, very limited in extent, that will serve to support a certain number of persons and no more. Assiduous labour might conserve the soil, but once washed away it can never be replaced. The process is irreversible and in course of time the population must migrate.

These aspects of the inevitable course of natural history have to be considered before we appeal to the third process, the alternative which the advocates of a change of climate uphold, a change in the course or character of the general circulation of the atmosphere.

CONCLUSION

Hence we may conclude that in spite of modifications of the atmospheric circulation which have resulted in diminished rainfall and consequently had great effects from time to time upon civilisation in the Mediterranean regions, the general features of the circulation of air over the region from 22° N to 52° N have not changed and the sketch of the climate which we gave in chapter II, with modifications in normal values, may be taken as expressing the relative characteristics of the circulation known to the ancients, the seasons are the same and the crops are still mainly the same and require the same cycle of seasons, though the area over which they can be profitably cultivated may have been considerably reduced and some of the region may have been transformed from habitable land into inhospitable desert.

We have still a similar circulation. The circulation is conditioned by a distribution of temperature over sea and land, and that in its turn by the energy received from the sun on the one hand and radiated into space on the other, and by the distribution of land and water. None of these can be pronounced invariable, but at the moment the quantitative evidence of change is of the nature of minor fluctuations rather than of sweeping modifications.

These things have to be considered and in our ignorance of the realities of the atmospheric circulation we have indeed hardly begun to think with real lucidity about the physical causes of local changes of climate on the earth's surface. The 25 inches of snow in Jerusalem in February 1920 may possibly point the way to views which deserve consideration. If so great a fall of snow could occur at one time it could certainly be repeated if the conditions were favourable, and sufficient repetition would mean a change of climate for the whole of Eastern Syria. The winter gales and rains of the Mediterranean region are doubtless due to invasions of cold air coming from the highlands of Central Asia and finding its way to the shores of the Mediterranean East or West as mistral, tramontana or bora[1]. A little additional coldness of the Asiatic highlands would stimulate the circulation. We have already seen that the distribution of pressure in the winter is a long trough of low pressure over the sea between the vast continental high pressure with its cold winds from the East and the vast oceanic high pressure with its warm winds from the West. The play between the two keeps the balance with a succession of rainy depressions between them. If the continental high pressure should be exaggerated or the oceanic pressure enfeebled the trough would be pushed Southward and the influence of the cold enhanced. A little less cloudiness in winter in the Central Asian region might make all the difference between an adequate and an inadequate rainfall in Eastern Syria. What regulates the cloudiness of the highlands of Asia we have still to discover—it is by no means an insoluble problem. It is simplified by the fact that only the winter conditions need be scrutinised, the summer months are rainless even in well-watered Mediterranean countries.

Having completed our survey of the intricacies of the problem of the change of climate we may return almost to the starting-point by pointing out that the ancients themselves had views similar to those of modern investigators, and in support of that view we cite a quotation from Plato's *Critias* to which attention has been called by Dr E. G. Mariolopoulos, of the National Observatory of Athens[2], in a paper which after full consideration pronounces against any notable change within historic times.

The consequence is, that, in comparison of what then was, there are remaining only the bones of the wasted body, as they may be called, discernible in small islands; all the richer and softer parts of the soil having fallen away, and the mere skeleton of the land being left. But in former days, and in the primitive state of the country, what are now mountains were only regarded as high hills; and the plains, as they are termed by us, of Phelleus were full of rich earth, and there was abundance of wood in the mountains. Of this last the traces still remain, for although some of the mountains now only afford sustenance to bees, not so very long ago there were to be seen roofs of timber cut from trees growing there, which were of a size sufficient to cover the largest houses; and there were many other high trees, bearing fruit and abundance of food for cattle. Moreover, the land enjoyed rain from heaven year by year, not as now losing the water which flows off the bare earth into the sea, but, having an

[1] E. G. Mariolopoulos, 'Sur la formation des dépressions locales méditerranéennes et la théorie norvégienne du "polar front,"' *Comptes rendus*, Tome CLXXVII, 1923, pp. 597–600.
[2] *Étude sur le Climat de la Grèce*, Les Presses Universitaires de France, Paris, 1925.

abundance in all places, and receiving and treasuring up' in the close clay soil the streams which descended from the heights, it let them off into the hollows, providing everywhere abundant fountains and rivers, of which there may still be observed indications in ancient sacred places, where fountains once existed; and this proves the truth of what I am saying.

Such was the natural state of the country, which was cultivated, as we may well believe, by true husbandmen, who did the work of husbandmen, and were lovers of honour, and of a noble nature, and had a soil the best in the world, and abundance of water, and in the heaven above an excellently tempered climate.

> (*The Dialogues of Plato*, translated into English with analyses and introductions by B. Jowett, M.A., vol. III, *Critias*, p. 691, 2nd ed. Oxford, Clarendon Press, 1875.)

While these pages are passing through the press we learn of a new discovery that supports the hypothesis of a definite and permanent change of climate within the last five hundred years in the case of Greenland.

Recent excavations near Cape Farewell, described by William Hovgaard in the *Geographical Review* for October, 1925, throw light on the fate of one at least of the early Norse colonies in Greenland.... The colony referred to is the "Eastern settlement," just west of Cape Farewell, and the interesting finds come from the cemetery, where they have been preserved by being permanently frozen into the ground, a condition which must have persisted for at least five hundred years. When the bodies were buried, however, the soil must have thawed, at least at midsummer. The costumes and many of the coffins, even the deepest-lying, are pierced and matted by the roots of plants, which would not have happened if the ground was permanently frozen.

> (*Meteorological Magazine*, vol. LXI, February, 1926, p. 13.)

Greenland was discovered by Norsemen in 982 and flourishing colonies were established. But in 1400 trade with Europe had virtually ceased. In these circumstances, which included the change of climate, the colonists were overcome by the Eskimos; the new excavations disclose "the last Norseman... lying dead and unburied by his desolate and deserted dwelling, and holding in his hand the emblem of the cultural superiority of the European, the iron knife, which had been ground and ground to the verge of possibility."

The reader will not fail to remark that, in these discussions, the changes of climate which are noted are mostly for the worse. The implications of that view are rather depressing. Aristotle however seems to have taken the more cheerful attitude that deterioration in one locality may be accompanied by improvement in others.

[This chapter ought not to close without a reference to Baron de Geer's discoveries of the deposits of former glaciers of southern Sweden which show the traces of successive years after the close of the last ice-age, and form a record of greater antiquity but with the same eloquence as Douglass's tree-rings. With these advantages "palaeometeorology" has developed an attraction for scientific enterprise that current weather only commands with difficulty. See *Geogr. Ann.* 1930, 'The finiglacial sub-epoch in Sweden, etc.' by Gerard de Geer, Stockholms Högskolas Geokronol. Inst. and other papers of the same Institute.]

CHAPTER VII

FROM ARISTOTLE TO THE INVENTION OF THE BAROMETER: WEATHER-LORE, ASTROLOGY AND ALMANACS

The hollow winds begin to blow,
The clouds look black, the glass is low,
The soot falls down, the spaniels sleep,
And spiders from their cobwebs peep.
Last night the sun went pale to bed,
The moon in halos hid her head.
The boding shepherd heaves a sigh,
For, see! a rainbow spans the sky.
The walls are damp, the ditches smell,
Closed is the pink-eyed pimpernel.
Hark! how the chairs and tables crack,
Old Betty's joints are on the rack;
Her corns with shooting pains torment her
And to her bed untimely send her.
Loud quack the ducks, the peacocks cry,
The distant hills are looking nigh.
How restless are the snorting swine!
The busy flies disturb the kine.
Low o'er the grass the swallow wings,
The cricket, too, how sharp he sings!
Puss on the hearth, with velvet paws

Sits wiping o'er her whiskered jaws.
Through the clear stream the fishes rise
And nimbly catch th' incautious flies.
The glow-worms, numerous and bright,
Illumed the dewy dell last night.
At dusk the squalid toad was seen
Hopping and crawling o'er the green.
The whirling dust the wind obeys,
And in the rapid eddy plays.
The frog has changed his yellow vest
And in a russet coat is dressed.
Though June, the air is cold and still,
The mellow blackbird's voice is shrill.
My dog, so altered is his taste,
Quits mutton bones on grass to feast.
And, see yon rooks, how odd their flight,
They imitate the gliding kite,
And seem precipitate to fall,
As if they felt the piercing ball—
'Twill surely rain—I see with sorrow
Our jaunt must be put off to-morrow.

(Erasmus Darwin, also attributed to Edward Jenner)

GREEK WEATHER-LORE.

The physical and dynamical theories of Aristotle's *Meteorologica* found little application as practical meteorology. The physical processes of the atmosphere were not effectively explained. The conflict between religion and science which had been the subject of Aristophanes' humour was still unsettled after the four centuries of experience of weather that lie between Aristophanes and Horace.

> Parcus deorum cultor et infrequens,
> insanientis dum sapientiae
>   consultus erro, nunc retrorsum
>   vela dare atque iterare cursus
>
> Cogor relictos: namque Diespiter
> igni corusco nubila dividens
>   plerumque, per purum tonantes
>   egit equos volucremque currum.

(Horace, *Odes*, Liber I, Carmen xxxiv.)

The crucial case of lightning out of a clear sky propounded by the jestful personation of Socrates had in the experience of Horace gone against the philosopher. Even in the early days the philosopher was commonly regarded as having got out of his depth in his explanation of nature. The Greeks coined a special word μετεωρολέσχης for a babbler about things sublime. Meanwhile the poets and philosophers concurred in dealing with the weather by observing the winds or the signs in the sky or the behaviour of plants and animals, whatever views they held as to ultimate causes; and such science as is contained in weather-lore was the practical guide for the farmer, the sailor and others who had to take the natural risks of weather in the course of their daily life.

An elaborate exposition of weather-lore was in fact drawn up by Theophrastus in his *Book of Signs*. He gives some eighty different signs of rain, forty-five of wind, fifty of storm, twenty-four of fair weather and seven signs of the weather for periods of a year or less. To guide those interested in future weather there are upwards of two hundred maxims altogether, many of them duplicated it is true and some contradictory. Many are quite familiar to us though derived perhaps from other sources and are common at least to Europe; others are less well-known.

By way of illustration we give a number of quotations from the translation by J. G. Wood:

The plainest sign is that which is to be observed in the morning, when, before the sun rises, the sky appears reddened over; and it indicates rain, either on the same day, or generally within three days; and the other signs show the same; for rain is indicated, if not sooner, within three days at the most, by a reddened sky at sunset also, but less certainly than when it is seen in the morning. And if, either in winter or spring, the sun goes down into a thin cloud, it generally indicates rain within three days; and so also if there are streaks of clouds from the southward; but these same appearances from the north are less certain. And if the sun, as it rises, has a dark mark on it, and if it rises out of clouds, rain is indicated; and if, as it is rising, rays stretch upward before it actually rises, this is a sign alike of rain and of wind. And if, as the sun is going down, a cloud comes under it so that the rays are thereby divided, it is a sign of storm. And whenever the sun is fiery [burning] at its rising, or setting, unless the wind rise, it is a sign of rain. The same is indicated by the moon as it rises at the full; but less by the crescent moon. If it be fiery, it indicates that the month will be windy; if hazy, that it will be wet. And whatever the crescent moon indicates, it indicates when it is three days old.

(*Theophrastus of Eresus on Winds and on Weather Signs*, translated by J. G. Wood, M.A., LL.B., F.G.S. E. Stanford, 1894, §§ 10-12.)

If shooting stars are frequent, they are a sign either of rain or wind; and the wind or rain will come from the quarter whence they proceed. And if, while the sun is either rising or setting, numerous rays arise therefrom it is a sign of rain. And when during sunrise the rays retain a colour as if the sun were being eclipsed, it is a sign of rain. And when the clouds are like fleeces of wool, it indicates rain. An unusual number of bubbles on the surface of the rivers indicates great rain. (§ 13.)

Black spots on the sun and moon indicate rain: red show wind. (§ 27.)

Whenever there is a rainbow it is a sure sign of rain. If many occur, it indicates a great deal of rain. And so also in many cases when a burning sun breaks forth from a cloud. (§ 22.)

Whenever there is fog, there is little or no rain. (§ 52.)

An ass shaking its ears indicates storm; and so do sheep and herds fighting for their food more than usual. (§ 41.)

Cattle eating more than usual and lying down on the right side indicate storm. (§ 41.)

If birds which do not live on the water wash themselves, it indicates either rain or storms. (§ 15.)

The common saying about flies is true; for when they bite vigorously it is a sign of rain. (§ 23.)

Snuffs in the lamp indicate either wind or rain. (§ 34.)

If the crow caws twice quickly and then a third time, it indicates a storm. (§ 39.)

There are many other signs from the behaviour of animals some of which are very widely known and we find the wealth of berries used, as is still customary, as an indication of the future though it is undoubtedly a record of the past.

If the scarlet oak be full of berries there will be very many storms. (§ 45.)

And again:

They generally indicate a severe winter; but they say that sometimes drought follows. (§ 49.)

Almost all the prognostics that are known to us will be found in the collection of Theophrastus. We stretch the space which can be allotted to quotations in order to include some of a much more physical character than the behaviour of animals.

Hurricanes [or cloud-bursts] occur when winds conflict with each other principally in late autumn and next in spring. (§ 37.)

If there is a great deal of bright weather in the late autumn, the spring generally is cold; but if the spring is late and cold the early autumn is late, and the late autumn is generally close and hot. (§ 48.)

Whenever there is much snow, a fruitful season generally follows. (§ 24.)

If there is much rain in the winter the spring is generally dry. (§ 24.)

If the winter is dry the spring is rainy. (§ 24.)

In a repetition, however, of a similar maxim in regard to storms the spring is said to be fair (§ 44). This correlation of wet winter and dry spring is specially interesting to us because we have found a somewhat similar suggestion in the winters and springs of Britain[1]. "On seventeen occasions out of the twenty-one a deficient autumn rainfall has been followed by excess of rainfall in the spring or *vice versa*....On the other hand, on sixteen occasions out of the twenty-one the deviation of the winter rainfall from the normal has been in the same direction as that of the autumn rainfall."

And still more noteworthy is Theophrastus' remark about Kaikias, the wind from the North East or East North East. In discussing the signs of winds he gives the characteristics which are included in Aristotle's description and which we have quoted there (chapter v, p. 78). Then he adds:

The E.N.E. [Kaikias] chiefly, and then the W.S.W. [Lips] makes the sky dense, and covers it with clouds. All other winds drive the clouds before them; the E.N.E. alone draws them towards itself. (§ 36.)

A similar note appears also in the book *On Winds*:

[1] *The Air and its Ways*, Cambridge University Press, 1923, p. 178.

The most striking peculiarities in fact are those of the E.N.E. wind (Kaikias) and the W. wind (Zephyros). For the E.N.E. wind (Kaikias) alone attracts the clouds towards itself as the proverb says:

> To himself he gathers alway,
> As doth Kaikias the clouds.... (§ 37.)

The reason [for the peculiarity just mentioned] in the case of the E.N.E. wind (Kaikias) is, that it is its nature to move in a curved line, of which the concave side is towards the sky, and not extended over the earth, as in the case of other winds; for this wind blows from below; and *blowing in this way towards its commencement,* it attracts the clouds towards itself. For towards whatever point the current is, thence also is the movement of the clouds. (§ 39.)

We may remark that drawing the clouds towards himself must mean that the clouds are travelling in an upper current in the direction opposite to the North Easter which is found at the surface—that is a peculiarity of North Easterly winds over the whole of the Northern Hemisphere most commonly known in connexion with the North East trade-winds but applicable also to the North East winds of other countries.

There is another example of wind, in the work on that subject, which is not so easily interpreted:

There is also a sort of rebound of the winds, so that they blow back against themselves, when they fail to surmount the places against which they blow by reason of the superior height of such places. Thus it happens that clouds are sometimes borne, by an undercurrent, in directions contrary to the winds. As for instance in Ægaea of Macedonia, the clouds are carried towards the north, while the north wind is blowing. The reason of this is that, the hills around Olympus and Ossa being high, the winds fall on and do not surmount them, but are turned back in the opposite direction; so that the clouds, being at a lower level, are also carried in the contrary direction. The same thing happens in other places as well. (§ 27.)

Theophrastus attributes this phenomenon to the rebound of a Northerly current from the high mountains which lie to the Southward of Ægaea in Macedonia. He supposes that the impact of the air on the high land causes an "undercurrent" which carries the clouds Northward below the Northerly wind. It seems much more likely that the clouds were in an upper layer.

It is peculiarly interesting to find such things noticed in the earliest meteorological work, and to note at the same time the importance attributed to "streaks of cloud from the southward" and of "clouds like fleeces of wool."

In this remarkable manner Theophrastus introduced a compilation of weather-lore to the world as a subject of study: little of importance was added to his representation of it during the two thousand years which followed the writing of his book on signs, though the literature of the subject is very voluminous. Aratus, as we have seen, adopted the study of weather-signs as a legitimate object of concern for the Greek who was still loyal to the religious idea. We have given some extracts from his presentation of the subject in a previous chapter.

ROMAN WEATHER-LORE

By far the most notable version of the subject in Latin is that of Virgil in the first book of the *Georgics* in which he gives instructions to husbandmen about agricultural pursuits, the signs of the seasons and the times for ploughing, sowing, etc. We quote his description of the means of anticipating changes in the weather.

And these things that we might avail to learn by sure tokens, the heats and the rains and the winds that bring cold weather, our Lord himself hath ordained what the moon in her month should foreshadow, at what sign the south wind should drop, what husbandmen should often mark and keep their cattle nearer the farmyard. Straightway when gales are gathering, either the seaways begin to shudder and heave, and a dry roaring to be heard on the mountain heights, or the far-echoing beaches to stir, and a rustling swell through the woodland. Even in that hour the rude surge spares not the curving hull, when gulls fly swiftly back from mid ocean and press screaming shoreward, or when sea-coot play on dry land, and the heron leaves his home in the marshes and soars high above the mist. Often likewise when a gale is toward wilt thou see shooting stars glide down the sky, and through the darkness of night long trails of flame glimmer in their track: often light chaff and fallen leaves flutter in air, or floating feathers dance on the water's surface. But when it lightens from the fierce northern regions, and when Eurus and Zephyrus thunder through their hall, the whole countryside is afloat with brimming ditches, and every mariner at sea furls his soaking sails. Never is rain on us unwarned: either as it gathers in the valley bottoms the crane soars high in flight before it; or the heifer gazing up into the sky snuffs the breeze with wide-opened nostril, or the shrill swallow darts circling about the pond, and the frogs in the mire intone their old complaint. Often likewise the ant carries forth her eggs from her secret chambers along her narrow trodden path, and a vast rainbow drinks, and leaving their feeding ground in long column armies of rooks crowd with flapping wings. Then seafowl many in sort, and birds that search the fresh pools round the Asian meadows of Caÿster, eagerly splash showers of spray over their shoulders, and thou mayest see them now ducking in the channels, now running up into the waves, and wantoning in their bath with vain desire. Then the villain raven calls full-voiced for rain, and stalks along the dry sand in solitary state. Nor even to girls who ply their spinning nightlong is the storm unknown, while they see the oil sputter, and spongy mould gather on the blazing lamp.

And even thus sunlight after rain and cloudless clearness mayest thou foresee and know by sure tokens. For then neither is the keen edge of the starlight dulled to view, nor does the moon rise flushed by her brother's rays, nor are thin woolly fleeces borne across the sky; neither do kingfishers beloved of Thetis spread their plumage to the sun's warmth upon the shore, nor unclean swine remember to shake out their litter and toss it with their snout. But the mists gather lower down and settle on the flats, and, constant to sunset, the night-owl from the roof-top keeps vainly calling through the dark. Aloft in the liquid sky Nisus is in sight and Scylla pays the debt of that purple hair: wheresoever her pinions cleave the thin air in flight, lo, hostile, fierce, loud-swooping down the wind, Nisus is upon her; where Nisus mounts into the wind, her hurrying pinions cleave the thin air in flight. Therewithal rooks repeat three or four times a clear thin-throated cry, and often where they sit aloft, happy in some strange unwonted delight, chatter together among the leaves, glad when rains are over to look to their little brood and darling nests once again; not, to my thinking, that their instinct is divine or their dower of fate a larger foresight into nature: but when the weather veers about and the saturated air shifts, and under dripping skies of the south what was rare but now condenses and what was dense expands, their temper changes its fashion, and other motions stir within their breasts than stirred while the clouds drove on before the wind; hence the birds make such chorus in the fields, and the cattle are glad, and the rooks caw in exultation.

If indeed thou wilt regard the hastening sun and the moon's ordered sequences, never will an hour of the morrow deceive thee, nor wilt thou be taken in the wiles of a cloudless night. When the moon first gathers her returning fires, if she clasp a dark mist in her dim crescent, drenching rain will be in store for husbandman and seafarer; but if a maiden flush suffuse her face, wind is coming: wind always flushes the gold of the moon: while if at her fourth rising (for that is surest of warrant) she travel through the sky with clear sharp-cut horns, both that whole day and those that shall dawn after it till the month be done will be rainless and windless, and sailors preserved will pay their vows to Glaucus and Panope and Melicertes son of Ino.

The sun likewise, both in his arising and when he sinks into the waves, will issue signs; most sure are the signs that attend the sun, yielded with morning or at the ascending of the stars. When at dayspring he is dappled with spots and sunk in a mist, and his orbed centre retires, mistrust thou of showers; for a gale is bearing hard from seaward, ill-ominous for trees and crops and herds. Either when towards daybreak spreading shafts struggle out between thick clouds, or when Dawn springs pale from Tithonus' saffron bed, alas! weak defence will the vine-tendril be then to the mellow cluster, so heavily the rough hail dances rattling on the roofs. This likewise, when he has run his race and is now sinking from the sky, will be of yet more service to remember; for often we see shifting colours fluctuate on his face; green presages rain, flame-colour east winds; but if spots begin to mingle with fiery red, then wilt thou see all a single riot of wind and storm-clouds; not on such a night at any persuasion would I voyage through the deep or part moorings from land. But if his circle be bright alike when he brings the day and buries the day he brought, vain will be thy terror of rain-clouds, and thou shalt discern the forests waving in a clear wind from the north.

Lastly, what burden evenfall carries, whence the wind chases clear the clouds, what the dripping South broods over, the sun will signify to thee; who shall dare to call the sun untrue?

(*The Eclogues and Georgics of Virgil*, translated from the Latin by J. W. Mackail, London, Longmans Green and Co., 1915.)

Except that the pig makes its appearance among the animals that are endowed with weather-wisdom the whole might have been taken from Theophrastus. Such a wholesale adoption of previous ideas upon a subject of practical interest like the weather could hardly have been possible unless the maxims were in reasonably good accord with the experience of the poet and his prospective readers, however "far-fetched" the signs may be.

MEDIAEVAL AND MODERN LORE

R. Inwards has compiled the proverbs and maxims of all periods in a volume of *Weather-lore*[1]. The additions of importance during mediaeval times to the classical list of prognostics are those connected with the kalendar such as St Swithin on July 15 and the ice-saints of May, St Luke's summer and St Martin's summer. The legend of St Swithin's day, that rain on July 15 is the commencement of a long rainy spell, obviously belongs to the countries that have a summer rainfall and not to the Mediterranean; and the ice-saints of the second week in May are not mentioned by Theophrastus. With additions of that kind we still find his signs in use. Apart from the solitary exception of "the glass is low" the expressive lines of Erasmus Darwin, quoted at the head of this chapter, might be regarded as a metrical version of Theophrastus' signs of rain.

[1] London, Elliot Stock, 1898.

Nearly all places have a local weather-lore based on the relation of clouds to the hills of the locality and there are occasional suggestions of common proverbs being coloured locally.

> Mony hips and haws
> Mony frosts and snaws

is a Scottish version of the many-berried ilex of the Mediterranean, but

> If the sun goes pale to bed,
> 'Twill rain to-morrow, it is said

is two millenniums old.

A wet summer almost always precedes a cold stormy winter because evaporation absorbs the heat of the earth. As a wet summer is favourable to the growth of the blackthorn, whenever this shrub is laden with fruit a cold winter may be predicted.

(Inwards, Professor Boerne's Latin MS., 1677–1799.)

The attitude towards the many indications of the apparent prescience of animals has changed little in the centuries Virgil writing at the beginning of the first century rightly regarded the behaviour of animals as reminiscent rather than prescient. Joseph Taylor in the *Complete Weather Guide* at the beginning of the nineteenth writes:

In general the senses of men, who in their way of life deviate from the simplicity of nature, are coarse, dull, and void of energy. Those also, who are distracted by a thousand other objects, scarcely feel the impression of the air, and if they speak of it to fill up a vacuum in their miserable and frivolous conversation, they do it without thinking of its causes or effects, and without ever paying attention to them. But animals,—which retain their natural instinct, which have their organs better constituted, and their senses in a more perfect state, and besides are not changed by vicious and depraved habits,—perceive sooner, and are more susceptible of the impressions produced in them by variations in the atmosphere, and sooner exhibit signs of them.

He adds, with a real nineteenth-century touch, the appeal to electricity:

Until the discovery of animal electricity, little attention was paid to those signs, which were consequently ascribed to a certain natural prescience.

ASTROLOGY—THE VOICES OF THE STARS

But if little or nothing was added to our knowledge of the physics of the atmosphere during the long period between Aristotle and Torricelli, there was a remarkable development on the part of the astronomers. Derived originally from the Chaldaeans the study of the heavens had been extraordinarily successful in keeping people informed about the order of the seasons: as a contribution to climatology it was very impressive; the hot season, the rains and the harvest-time followed in the order in which they were predicted. Even eclipses of the sun could be foretold.

Associated with the earlier forms of religion astronomy was a specialised form of science which must have required training in accurate observation and careful drawing. It may therefore be regarded as natural that the experts in the subject, the astronomer-priests as they are sometimes called, should endeavour to extend their method from the regularities of the heavenly bodies and the Mediterranean climates to the irregularities of the weather, and to find in the periodic migrations of the wandering stars, which we call planets, an anticipation of the prominent features of weather and of the other vicissitudes of life. From the attribution of the details of weather to the powers of the air it is a natural psychological transition to the influence of the stars which had become personified with the names of the deities. We thus pass from the action of the gods to the influence of the heavenly bodies associated with the gods. Stellar influences having been effectively traced in the cycle of the seasons they might be found to have some application in the details of practical life by a close study of astronomy with the aid of a lively imagination and an ingenuity that might, under more favourable circumstances, have been scientific genius.

The use of ingenuity to peer into the future is indeed a very ancient cult. The Greek oracles appear to have been as skilful in making a little knowledge go a long way as the most able of the political correspondents of the newspapers of our day, and the Roman augurs and haruspices must have required a good deal of judicious acumen to maintain their reputation and position. Their efforts were different from magic and witchcraft; those claimed to control the vicissitudes of life, whereas the augurs and haruspices only claimed to foresee them, or more simply to guide the conduct of citizens or the state in cases of doubt. So the astronomers in developing their knowledge in the direction of astrology claimed to interpret and not to control the sequence of events by studying the influence of the stars and planets. They acquired official positions of great influence in many countries.

The practice of astrology as a means of anticipating future events or as a guide to conduct was followed throughout the dark and middle ages and is even now pursued to a considerable extent. It has produced a considerable impress upon our language. The word "influence" itself is reminiscent of it and there are many other examples. We take the following from *Urania*, a modern Journal of Astrology, Meteorology and Physical Science.

When the fierce Puritanic spirit swept over England, the sublime science of astrology was for a long period destined to suffer an almost total eclipse. Its followers could only study its truths in secret, for derision and persecution were the rewards of those who professedly disseminated its doctrines, which were branded as savouring of unholy practices. It is almost incredible that a science believed in and pursued by the greatest minds of antiquity should suffer such unmerited wrong, and that men of culture and education should be unable to separate truth from falsehood, and see that, stripped of the absurd superstitions which had become incorporated with it, astrology was indubitably pure and truthful.

Notwithstanding its almost total suppression for so long a period, it is interesting to note how deep a hold its teachings had taken on the national mind, as is evident from a careful study of many words and phrases in the language and popular usages and beliefs.

As it was known that the planet in the ascendant at birth impressed on the native its own peculiar properties, certain descriptive words came into use. An open, generous, laughter-loving nature is spoken of as *jovial*; a bold, energetic, warlike spirit as *martial*; a changeable, active temperament as *mercurial*; and a dull, serious, solemn deportment as *saturnine*.

What more telling words than these of astrologic import could be employed to describe the niceties of individual character?

Then we have the popular exclamations of—"My stars!" "Bless my stars!" and "O Gemini!" (the ruling sign of London), or, as Winifred Jenkins had it, "O Jiminy!" Shakespeare makes Othello address the dead Desdemona as, "O ill-starred wench!"

A lucky personage is said to be born under a fortunate star, and should it be the fate of another to suffer discredit or loss, he is described by a certain class of writers as one whose "star is on the wane."

"When the planets in evil mixture to disorder wander," and sorrow and misfortune are the consequences, what more apt than the word "*disaster*" (which literally means the blow of an evil star) to express the result.

The days of the week and the months of the year have undoubtedly an astrologic origin. The French have retained the literal names of the planets for the former, while ours have come to us through the Saxon speech.

Then again, the most prominent decoration in ancient orders of merit was a star. . . .

As the science of medicine was for so many ages wedded to astrology (for the greatest physicians were also the greatest astrologers of the age), numerous are the relics of astrologic observances within its pale.

The physician to the present day, before writing his prescription, makes a cabalistic mark, supposed to mean Recipe, but which is really the astrologic symbol of Jupiter the Healer— ♃ .

The power of the moon was held to be great on the mind and body. Her phases were carefully watched in disease. Hence such words as lunatic, moonstruck, etc.

. . . .

All these point to the close connection of astrology with religion, for the religions of the world were all based on astrology, and we commend to the student to note how full is the Bible of astrologic allusions, which proves that the science had a Divine origin.

Let us hope that the long night of oblivion is passed, and the day is at hand when the celestial science shall again be honoured, and take its fitting rank with those which engage the attention of the wise and learned of our age.

('Relics of Astrologic Teaching,' *Urania*, January 1880, p. 24.)

Even now it might be possible to find a scheme of weather-sequence for a year in advance made out by the study of the positions of the heavenly bodies and their supposed influence upon the pressure and temperature of the atmosphere.

ALMANACS

Astrological prediction received its greatest development with the invention of printing. So long as communication was limited to the horizon of speech and manuscript an astrologer must have been dependent upon a wealthy patron or a local clientèle, but the possibility of unlimited multiplication by the printing-press was as important for astrologers as the invention of telegraphy at a later date was for meteorologists. It provided the opportunity for Nostradamus, or Michel de Notredame, who achieved a great reputation as an astrologer and published a volume of predictions entitled *Centuries*, originally in 1555, which have been very celebrated. Born in 1503 of Jewish descent, he practised at Agen and afterwards at Salon near Aix.

Printing made it possible to include astrological predictions in the information given in almanacs which were designed to give the ephemerides, or data of position of the heavenly bodies for each day of the year.

The association of meteorological notes with almanacs or kalendars may be traced back through Roman times and the Greek parapegmata even to Babylonian practices. In Roman times it was the duty of the pontifices to indicate exceptional occurrences such as cold, drought, hail-storms, floods, etc. A good example of a collection of weather-notes is given in the *Calendarium rusticum* of Columella. Many of the Roman *fasti* in later times contained a weather-kalendar which announced the weather of the year[1]. With printing the custom received a notable extension and in that form astrological prediction became so popular in France at the end of the sixteenth century that almanacs had to be prevented by law from making political prophecies.

In England, by grant of James I, prophetic almanacs were made a monopoly of the Universities and the Stationers' Company, under the licence of the Archbishop of Canterbury. Such almanacs still survive in the various forms of *Moore's Almanac*, which has been carried on for some centuries, or *Zadkiel's Almanac*.

Astrological prediction is however not now regarded as a practical science by the ordinary scientific societies. The almanacs which find a place among our recognised books of reference, as the *Nautical Almanac* or *Whitaker's Almanac*, confine themselves to specifying the positions of the heavenly bodies according to calculations without expressing an opinion as to their influence upon mundane affairs.

LUNAR WEATHER-LORE

Associated with the practice of astrology, if not entirely dependent upon it, is the supposed influence of the moon on the weather. In the past it has certainly been regarded as of primary importance, and the notion of associating weather-changes with lunar changes is probably still regarded by the majority of people in this country as being at least founded on fact.

Of course the prognostics derived from the colour of the moon at rising or setting, or her position in a halo or corona, or the nautical proverb that "the full moon eats clouds" are ordinary atmospheric weather-lore and with these may be associated Virgil's prognostic about the points of the new moon being dull, but when weather-influences are attributed to the points of the new moon being upward or downward and people speak of the new moon lying on her back as a prognostic of wet weather, we are back in astrology, because what is visible of the moon depends entirely upon her position relative to the sun. It is not surprising that the proverbs are contradictory.

> The bonnie moon is on her back;
> Mend your shoes and sort your thack

is a Scottish view, but on the Welsh border, "It is sure to be a dry moon if it lies on its back, so that you can hang your hat on its horns."

---

[1] H. Leiter, *K. K. Geog. Gesell. in Wien*, Band VIII, No. 1, 1909.

The most categorical statement of relationship between moon and weather is a table of the weather associated with the time of entry upon the phases of the moon[1].

Hour of new or full moon or of entry into first or last quarter	Summer	Winter
12 Noon	Very rainy	Snow and rain
Between 2 and 4	Changeable	Fair and mild
,, 4 and 6	Fair	Fair
,, 6 and 8	{Fair if wind NW {Rainy if S or SW	{Fair and frosty if N or NE {Rainy if S or SW
,, 8 and 10	do.	do.
,, 10 and Midnight	Fair	Fair and frosty
,, Midnight and 2	Fair	Hard frost, unless wind S or SW
,, 2 and 4	Cold with frequent showers	Snow and stormy
,, 4 and 6	Rain	Snow and stormy
,, 6 and 8	Wind and rain	Stormy
,, 8 and 10	Changeable	Cold, rain if W., snow if E
,, 10 and Noon	Frequent showers	Cold, with high wind

We may mention further that at the same time Luke Howard in his *Barometrographia*, an elaborate work with provision for tracing graphically the relation between pressure-changes and weather, is careful to introduce the moon's positions into his diagrams. But since the introduction of the weather-map a direct relation between the weather and the phases of the moon has been definitely ignored and regarded as outside the pale of scientific discussion. Still more recently Professor S. Chapman[2] has examined the effect of the moon's position upon the atmospheric pressure with the result that it must be regarded as quite too small to produce any appreciable influence upon weather.

THE ACCEPTANCE OF PROVERBIAL PHILOSOPHY

We can therefore only regard "vox stellarum," the supposed influences of the stars or the moon upon the sequence of weather, involving as they do a very extensive knowledge of astronomy, as being other examples of the same kind as the ordinary proverbs and weather-lore which are the simple expression of what is regarded as direct experience. In this respect the predictions of astrology may be paralleled in a very remarkable manner by some of the most elaborate and most modern mathematical reasoning as applied to the study of weather. With astrology, as with mathematical physics, when once a formula has been obtained the deductive reasoning is rigorous and conclusive, however complicated it may be. Its practical value depends upon the assumptions upon which the original formula is based; if the assumptions represent reality the conclusions have practical application, but not otherwise. We are now sufficiently certain that the astrologers made an erroneous assumption when they thought they could specify the course of a man's life

---

[1] Joseph Taylor, *The Complete Weather Guide*, London, 1814, p. 58. The table is attributed to Dr [Sir William] Herschel (*Europ. Mag.*, vol. LX, p. 24).

[2] *Q.J. Roy. Meteor. Soc.*, vol. XLIV, 1918, p. 277; vol. XLV, 1919, p. 113; vol. XLVIII, 1922, p. 246; vol. L, 1924, p. 99.

as the influence of the heavenly bodies in the positions which they happened to occupy on the day of his nativity. The knowledge of the positions and relation of the heavenly bodies at any time, which is required to draw a horoscope, is a wonderful performance, but the supposed influence of each planet or star is pure fiction and the result is consequently valueless—not much less than pure fiction are some of the assumptions which form the ultimate basis of a good deal of modern mathematical reasoning about meteorological problems, and to that cause we must also attribute the lack of recognition which a good deal of meteorological reasoning has to deplore.

However that may be, we have to face the position that for fifteen centuries of our era the world allowed its mind to rest in the accounts which Aristotle had given of the physical processes of the air, and was content to make use of the statements to be found in weather-lore or astrological writings, as the best available guides to conduct in face of the uncertainties of weather. And even now people are to be found who consider that the weather-lore which is characteristic of the middle ages and which is .dependent upon personal experience and not upon instruments and maps is still the highest expression of human intelligence on the subject.

We propose to give an example which carries the title of *The Shepherd of Banbury's Rules*, not perhaps dating actually from before the invention of the barometer but making no reference to that instrument nor any other.

Who the shepherd of Banbury was, we know not; nor indeed have we any proof that the rules called his were penned by a real shepherd: both these points are, however, immaterial: *their truth is their best voucher*. Mr Claridge (who published them in the year 1744) states, that they are grounded on forty years' experience, and thus, very rightly, accounts for the presumption in their favour. "The shepherd," he remarks, "whose sole business it is to observe what has a reference to the flock under his care, who spends all his days, and many of his nights in the open air, under the wide-spread canopy of Heaven, is obliged to take particular notice of the alterations of the weather; and when he comes to take a pleasure in making such observations, it is amazing how great a progress he makes in them, and to how great a certainty he arrives at last, by mere dint of comparing signs and events, and correcting one remark by another. Every thing, in time, becomes to him a sort of weather-gage. The sun, the moon, the stars, the clouds, the winds, the mists, the trees, the flowers, the herbs, and almost every animal with which he is acquainted, all these become, to such a person, instruments of real knowledge."

The rules enumerated are typical of all rules based on experience of the weather; what of truth or error there is in them the reader may judge; they are as follows:

    I. SUN.—*If the sun rise red and fiery*—Wind and rain.
    II. CLOUDS.—*If cloudy, and the clouds soon decrease*—Certain fair weather.
    III. *Clouds small and round, like a dapple-grey, with a north-wind*—Fair weather for two or three days.
    IV. *If small Clouds increase*—Much rain.
    V. *If large Clouds decrease*—Fair weather.
    VI. *In Summer or Harvest, when the wind has been South two or three days, and it grows very hot, and you see Clouds rise with great white Tops like Towers, as if one were upon the Top of another, and joined together with black on the nether side*—There will be thunder and rain suddenly.

VII. *If two such Clouds arise, one on either hand*—It is time to make haste to shelter.

VIII. *If you see a Cloud rise against the Wind or side Wind, when that Cloud comes up to you*—The Wind will blow the same way that the Cloud came. *And the same Rule holds of a clear Place, when all the Sky is equally thick, except one Edge.*

IX. MIST.—*If Mists rise in low Grounds, and soon vanish*—Fair Weather.

X. *If Mists rise to the Hill-tops*—Rain in a Day or two.

XI. *A general Mist before the Sun rises, near the full Moon*—Fair Weather.

XII. *If Mists in the New Moon*—Rain in the Old.

XIII. *If Mists in the Old*—Rain in the New Moon.

XIV. RAIN.—*Sudden Rains never last long: but when the Air grows thick by degrees and the Sun, Moon and Stars shine dimmer and dimmer*, then it is like to rain six Hours usually.

XV. *If it begin to rain from the South, with a high Wind for two or three Hours, and the Wind falls, but the Rain continues*, it is like to rain twelve Hours or more, and does usually rain till a strong North Wind clears the Air. *These long Rains seldom hold above twelve Hours, or happen above once a year.*

XVI. *If it begins to rain an Hour or two before Sunrising*, it is likely to be fair before Noon, and to continue so that day; *but if the Rain begins an Hour or two after Sunrising*, it is likely to rain all that day, except the Rainbow be seen before it rains.

XVII. WINDS.—*Observe that in eight Years' Time there is as much South-West Wind as North-East, and consequently as many wet Years as dry.*

XVIII. *When the Wind turns to North-East, and it continues two Days without Rain, and does not turn South the third Day, nor Rain the third Day, it is likely to continue North-East for eight or nine Days*, all fair; *and then to come to the South again.*

XIX. *After a Northerly Wind for the most part of two Months or more, and then coming South*, there are usually three or four fair Days at first, and then on the fourth or fifth Day comes Rain, *or else the Wind turns North again*, and continues dry.

XX. *If it turns again out of the South to the North-East with Rain, and continues in the North-East two Days without Rain, and neither turns South nor rains the third Day*, it is likely to continue North-East two or three months.

XXI. *If it returns to the South within a Day or two without Rain, and turns Northward with Rain, and returns to the South in one or two Days as before, two or three times together after this sort*, then it is like to be in the South or South-West two or three Months together, *as it was in the North before.*

The winds will finish these turns in a fortnight.

XXII. *Fair Weather for a Week with a Southern Wind* is like to produce a great Drought, *if there has been much Rain out of the South before. The Wind usually turns from the North to South with a quiet Wind without Rain; but returns to the North with a strong Wind and Rain. The strongest Winds are when it turns from South to North by West. When the North Wind first clears the Air, which is usually once a Week*, be sure of a fair Day or two.

XXIII. SPRING AND SUMMER.—*If the last eighteen Days of February and ten Days of March be for the most part rainy*, then the Spring and Summer Quarters are like to be so too; and I never knew a great Drought but it entered in that Season.

XXIV. WINTER.—*If the latter End of October and Beginning of November be for the most part warm and rainy*, then January and February are like to be frosty and cold, except after a very dry Summer.

XXV. *If October and November be Snow and Frost*, January and February are likely to be open and mild.

The correlation coefficients for one or two of the above rules have been worked out, but they are disappointing:

XXIII. For 38 years, S.E. England, between rainfall of last 18 days of February and first 10 days of March, and spring rainfall, the correlation coefficient is + 0·14; between the rainfall for the same period and summer rainfall, + 0·07.

XXIV. For 64 years at Greenwich, between October-November temperature and that of the following January-February, + 0·05; between October-November temperature and that of the following December-January-February-March, + 0·25.

It will be noticed that these rules are much more reminiscent in their form and structure of the teaching of Theophrastus than of the rhymes or phrases of the mediaeval weather-lore; they suggest at least careful observation and compilation. We may remember that the practice of keeping diaries of weather had already taken root. The earliest extant is that of the Rev. William Merle of Oxford which is entitled *Consideraciones temperiei pro 7 annis* 1337–1344, and contains daily observations partly at Oxford and partly at Driby in Lincoln-shire of which parish Merle was vicar. The diary was reproduced in facsimile through the activity of G. J. Symons (London, 1891) and is also included in the collection of reprints of meteorological classics edited by G. Hellmann and enumerated on p. 146; the list also includes a number of other contributions of a somewhat similar kind to meteorology in the early and middle ages.

The practice of keeping notes of the weather was of course much encouraged by the spread of printing and by the use of kalendars and almanacs, and as in these the astronomical data or ephemerides were frequently supplemented by astrological predictions of the weather, the seasons and other notable events, the practice of recording weather came into association on the one side with the astrological basis of weather-study and on the other side with the weather-lore of common experience. Astrology was regarded as a scientific method by all students of natural philosophy; even the celebrated astronomer Kepler, who takes the highest rank among inductive philosophers for his formulation of the motion of the planetary system, was himself an astrologer. His prognostics for 1604 and 1624 have recently come to light and have been published by W. van Dyck.

With this great wealth of weather-lore and astrological prediction that was generally accessible and much sought after we have to inquire why the general meteorological problem was not solved and why first the invention of the barometer and secondly the invention of the weather-map were, each in turn, hailed as the beginning of a new era in weather-study, and confidently looked to as a new step towards the certainty that is so striking a feature of astronomical predictions. We must conclude that in spite of the number of signs, and perhaps on that account, and in spite of repeated assertions, people have never yet been satisfied with the weather-prediction of their own times. It is not only that those who are scientifically-minded desire to place the explanation of natural phenomena upon a scientific basis, practical folk want practical assurance about the prospects of weather and have not yet found it. They have tried weather-lore and astrology and both have been found insufficient and yet each has found vigorous adherents and defenders.

The grounds for dissatisfaction can be easily explained by the consideration that there is no reciprocity about the arguments. Theophrastus to begin with gives eighty signs of rain and Erasmus Darwin gives thirty-three indications of a wet day in June; but, stated in the other way, that if rain is coming all these things will happen first is quite a different proposition. Aratus suggests that the signs reinforce one another and the probability becomes greater with the accumulation of signs; but that gives away the whole position. There will always be ample ground for dissatisfaction until we can find reciprocal relations like astronomical formulae—on such a day and such an hour there will be an eclipse visible at such a place and *vice versa* on other days there will be no eclipse visible anywhere.

THE POPULAR APPRECIATION OF WEATHER-SIGNS

We may therefore turn to the more interesting question of the reasons for accepting the various signs as genuine. Here we may remark upon an attitude of the human mind with reference to promises and predictions. One success produces as a rule a deeper and more lasting impression than many failures, though in both cases they may have been equally unforeseen. We say, as a rule, because we remember a striking case the other way when a forecast of a fine day for a presidential inauguration that turned out to be dolefully wet produced a most painful impression in the Eastern United States and some adverse criticism of the official system that has not even yet quite lost its pungency.

Further the specifications of weather-signs are as a rule remarkably vague. If one takes the red sky in the evening as a sign of fine weather and in the morning as a sign of wet—we are not at all clear as to what a red sky means. Does it mean clouds overhead, or nearly so, suffused with red light from the West or from the East as the case may be? or does it mean clouds fringed with red or with fiery yellow in the Western or Eastern horizon?

Or again with the changes of the moon there is quite a pleasurable vagueness, is new moon the middle day of no moon? or is it as some authors seem to suggest the period of seven days between the last sight of the old moon and the first sight of the new? So first quarter is from three days before to three days after the semicircle and so on. With the wider interpretation of the latter suggestion it is clear enough that the moon's period being divided into four weeeks every event and every change of weather must occur at one of the quarters and in that case it is more a matter of temperament than anything else that decides whether the facts when they arrive agree with the prediction. And if on the occurrence of any event any appropriate sign can be remembered no further evidence of success is necessary until something more definite turns upon it.

It seems remarkable that we have so few records of any attempts to verify by actual comparison the various signs that have been quoted. Any such attempts would have led to greater precision in the statements and better guidance to the judgment.

In modern days statistical methods have been introduced which can be used to check the accuracy of the predictions when they are sufficiently defined and in that way the rules XXIII and XXIV of the Shepherd of Banbury have been examined and the result is that so far as Eastern England is concerned they fail and ought to be removed from the canon.

In the meantime explanations of many of the signs have been found by Abercromby and Marriott[1] in the properties of travelling depressions as disclosed in the weather-map, and Professor W. J. Humphreys[2] has given an explanation of others on the basis of recognised physical processes.

THE "CAUSE" OF ABNORMALITIES IN WEATHER

The popular acceptance of the validity of weather-proverbs may be related to the equally common practice of seeking or expecting an immediate cause of any notable abnormality of weather. Every flood, or spell of rainy weather, a heavy thunderstorm, a "heat wave" or a "cold wave," provokes many inquiries as to "the cause." The habit is doubtless in some way connected with the philosophical dictum that "for every effect there must be a cause." But the dictum cannot be applied in its simplicity to the sequence of weather. We cannot assume that any condition of weather, normal or abnormal, can be ascribed to a single cause. Meteorology is not in a position to assign a definite cause for normal weather; even that is the result of the interaction of many conditions all of them liable to variations, the nature and extent of which are at present undefined.

A modern writer has pointed out that the cause of any phenomenon for any inquirer is that account of the phenomenon in which his mind "rests." The popular mind is generally content to rest in a description of the phenomena of abnormal weather which offers no solution to the student who has always in mind the natural complexity of the ordinary events of weather. Hence the endeavour to find a cause of any abnormal weather in some single circumstance such as a change in the Gulf Stream, or ice in the Atlantic, the prevalence of wireless waves, the discharge of explosives or the proximity of Mars is likely to afford satisfaction to those only who do not appreciate the necessity for seeking an explanation of weather that is not abnormal enough to interfere with the pursuit of ordinary avocations.

To many people the reality of one or other as the "cause" is just a matter of the opinion of this or that expert.

So, in like manner, the mind easily rests in a weather-proverb as the assertion of direct relation between notable events such as a heavy crop of hips and haws and the coldness of the succeeding winter, or the significance of the phases of the moon, if the assertion can be supported by some telling reminiscences.

The science of meteorology must however look at such matters differently. Its mind cannot rest in any description of the phenomena which is not com-

---

[1] 'Popular Weather Prognostics,' *Q. J. Roy. Meteor. Soc.*, vol. IX, 1883, pp. 27–43.
[2] *Weather Proverbs and Paradoxes*, Baltimore, 1923.

plete in all essential particulars. It must regard the whole complex of normal and abnormal as being within the scope of its inquiry and must therefore seek its explanation of the abnormal in the examination of the normal in all its details.

It is not the unscientific mind exclusively that is liable to otiose acceptance of assertions that will not bear critical examination. The history of meteorology affords an abundance of examples of the assertion of quite inadequate causes. An inherent difficulty of the science is that an assertion about any phenomenon cannot be immediately tested by repeating the phenomenon in the same way that a physicist can repeat an experiment. The repetition must await the natural occasion, or similar occasions must be sought for in the available records. The practice that makes the selection of appropriate occasions possible is a special form of training.

A NEW ASTROLOGY

We conclude this chapter, which leads from the Chaldaean astrology to the direct measurement of the physical properties of the atmosphere, by noting that after some thirty centuries of wandering in an astrological wilderness a new era of what may legitimately be called astrology, or astrometeorology, appears now to be dawning. The close watching of the sun at the Solar Physics Observatory of the Smithsonian Institution of Washington has disclosed a relation between the intensity of solar radiation and the number of sunspots visible on the sun's disc, and these in their turn are found to be related to exceptional seasons on the earth. A further development is to find a cause for the irregularly periodic variations in the sunspot activity. The more recent suggestion of these perturbations of the sun is that they are due to the surrounding planets, and so we come back again to the planets controlling our weather, not however directly as the old astrologers erroneously assumed, nor yet through the gravitational action of the planets on the sun, but indirectly through the thermionic emission through space of electrons from all bodies in the universe.

The agency through which the planets influence the solar atmosphere is not yet clear. The suggested agencies are the direct pull of gravitation, the tidal effect of the planets, and an electromagnetic effect. In *Earth and Sun* the conclusion is reached that the first two are out of the question, a conclusion in which E. W. Brown acquiesces. Unless some unknown cause is appealed to, this leaves an electromagnetic hypothesis as the only one which has a reasonable foundation. Schuster inclines to this view.

(Huntington and Visher, *Climatic Changes*, Yale University Press, 1922, p. 244.)

Just as the "emanation of air" from certain parts of the earth might have been a prophetic anticipation of the meteorological science of the future, so if the early astrologers had been fortunate in their grouping of the planetary influences they might have anticipated the latest theory of the subject.

CHAPTER VIII

THE REIGN OF THE BAROMETER AS WEATHER-GLASS: PIONEERS IN THE SCIENCE OF WEATHER

Torricelli's barometer of argento vivo, 1643   Variation of temperature with height, cir. 1784
Determination of g [1644] 1657   Newton's law of gravitation, 1665–6
Boyle's Law, 1662   General equations of motion with moving axes, 1758
Mariotte's Law, 1676   Latent heat of water, Black, 1761
Charles's Law, 1802   Laplace's law of variation of pressure with height, 1805
Gay Lussac's Law, 1802   Hydrodynamics. General equations of motion, 1755
Dalton's laws of the pressure of vapour, 1801–3   Variation of gravity with latitude, 1672

Law of conservation of momentum, 1687
        ,,   ,,   angular momentum, cir. 1742
        ,,   ,,   mass, cir. 1782
        ,,   ,,   energy, 1844

Une première période s'étend de la Renaissance à la Révolution française. C'est la période des origines. La physique qui, depuis Aristote, constituait une branche de la philosophie, par opposition à la métaphysique, se détache peu à peu de la souche commune et tout d'abord ne se distingue pas de la mécanique proprement dite[1].

METEOROLOGY AS A SCIENCE

THE invention of the barometer and thermometer marks the dawn of the real study of the physics of the atmosphere, the quantitative study by which alone we are enabled to form any true conception of its structure. We cannot complete the conception without the weather-map, and the balloon. Without the measurements which provide the material of the physics of the atmosphere, we are restricted to such information as our experiences of wind and observations of cloud; we have no more effective means of comparing or co-ordinating conditions in different places than had Herodotus or Callisthenes.

Hence in the period upon which we enter now, the period from the invention of the barometer to that of the weather-map, we are concerned with two important aspects of the history of meteorology, first the work of the laboratory, the discovery of the physical laws which are operative in the atmosphere, and secondly the work of the traveller and the librarian, the earliest steps in the collection of data from different parts of the earth's atmosphere at the surface and in the upper air and their co-ordination to form a mental picture of the state of the atmosphere as a whole. And here we may pause to point out that both these aspects have to be combined before we can begin the real science of meteorology.

The physical and chemical laboratories can supply us with the laws of physics and chemistry which the phenomena of the atmosphere must obey. Weather is indeed a continuous series of physical and chemical processes all of which must be referred to laws and processes established by experiments in a laboratory—the laws of gases and vapours, the laws of dynamics, the laws of heat and of radiation, the laws of optics and of sound, the laws of

---

[1] *La Science Française.* Exposition universelle et internationale de San Francisco. Tome premier, p. 131, Paris 1915.

electricity and magnetism; but the atmospheric processes are different from laboratory experiments in that they are on such a vast scale, both horizontal and vertical; they cannot be imitated though they may be illustrated by laboratory experiments. The science of meteorology *is* essentially the combination of the knowledge which is obtained from the laboratory and the weather-map. A scientific memoir which is a study of the properties of air, a contribution to the science of physics, may be indispensable for the study of meteorology, but it is not a contribution to the special science of meteorology unless it takes into account the variations of pressure and temperature with height and the pressure-differences associated with the rotation of the earth.

Looked at from the historical point of view what is now called physics was included in what Aristotle called meteorology, the natural phenomena of the material universe. In the special period now under consideration it was the natural display of physical processes of the atmosphere which excited attention and claimed the interest of the physicist. The steam-engine, the galvanic battery, the galvanometer, the dynamo, the motor and the electric circuit, had not arrived, the natural phenomena which prompted inquiry were the formation of wind, cloud, rain, snow, hail, radiant heat, light and atmospheric optics; the magnetism was terrestrial magnetism, the electricity if not in reality atmospheric electricity was the so-called static electricity, the electricity of Coulomb and Henry Cavendish to which atmospheric electricity was soon shown to be related. It was the physics of these phenomena which found expression in the laboratories, while the other side of meteorology, the comparative side, was a separate study, although many physicists interested themselves in the keeping of a meteorological register. An official register was indeed kept for the Royal Society itself from 1774 to 1850.

Indeed every physicist was, in a certain sense, a meteorologist and meteorology was generally regarded as a department of physics. But there was no real conception of a general circulation of the atmosphere, with its periodic and temporary changes, to form a basis of reference for the atmospheric processes. No doubt each physicist had some conception in his mind when he wrote of the application of physical laws to the atmosphere, just as Halley, Hadley and others had some working ideas, but they took no account of pressure-distribution over the surface or temperature-distribution in the upper air, and without these no scheme of general circulation can be effective.

Hence while explorers gradually laid the foundation of an atmospheric structure the physicists were dealing with an unexpressed ideal of the atmosphere, and as contributions to meteorology they are of the nature of marginal notes. An accepted scheme of general circulation is required before meteorological science can be said really to exist.

We regard it as important to set out here this view of meteorology as a separate science because the aim of this book is to treat the subject from that point of view, and we shall endeavour to set out the lines of the subject in such a way as to justify it. We are afraid this may cause some disappointment to the reader because many notable papers which deal with the dynamics and physics of air, though monuments of human genius or at least ingenuity, do not comply with the condition that we are only treating of the atmosphere, as distinguished from air, when the treatment has regard to the natural distribution of the elements in space. Many important memoirs on that account do not find a place in the mental equipment of the meteorologist.

In his report on Meteorology to the meeting of the British Association of 1832 J. D. Forbes, subsequently Principal of the University of St Andrews, comments upon the fragmentary treatment of meteorology with the natural severity of a young man of twenty-three years. He remarks:

Very different [from that of astronomy] is the position of an infant science like Meteorology. The unity of the whole, or of the individual greater divisions of which it is composed, is not always kept in view, even as far as our present very limited general conceptions will admit of; and as few persons have devoted their whole attention to this science alone, or the whole exertions which they *did* bestow, to one branch of so wide a field,—no wonder that we find strewed over its irregular and far-spread surface, patches of cultivation upon spots chosen without discrimination and treated on no common principle, which defy the improver to inclose, and the surveyor to estimate and connect them. Meteorological instruments have been for the most part treated like toys, and much time and labour have been lost in making and re-cording observations utterly useless for any scientific purpose....

As this appears to me the place to insist upon a total revision of the principles upon which meteorologists have hitherto very generally proceeded, I shall explain my views a little more particularly.

It is in the first place worthy of remark, that the most interesting views which have been given in this science, and the most important general laws at which it has yet arrived, have for the most part been contributed by philosophers who, in pursuit of other objects, have stepped aside for a moment from their systematic studies, and bestowed upon the science of Meteorology some permanent mark of their casual notice of a subject which they never intended to prosecute, and which they soon deserted for other and more favoured paths of inquiry. Mr Dalton descends for a moment from his chemistry in the abstract, to illustrate the constitution of the atmosphere and the theory of vapour. Laplace, viewing nature with the eye of a master, introduces into his *Mécanique Céleste* an investigation of the mechanical structure and laws of equilibrium of the gaseous envelope of our planet: he applies Meteorology to one of its great objects,—the laws of atmospherical refraction; and gives to the scientific world a new formula for the measurement of heights by the barometer, which greatly exceeds in accuracy those which had previously been proposed. Yet may the speculations of these philosophers, and the discussions to which they give rise, be more important to the science than the labours of a professed meteorologist, who has made, with minute scrupulosity, all the ordinary entries in his Journal, daily for a life-time.

In like manner William Whewell (1795–1866), a celebrated mathematician and mineralogist and Master of Trinity College, Cambridge, who took a good deal of interest in meteorology and invented a self-recording anemometer, omitted meteorology from the list of inductive sciences because in the end it was only the application of the laws of physics to the special case of the atmosphere.

In view of the fact that the contributions of physicists have been isolated contributions to the physics of the air, as J. F. Daniell himself pointed out, we have thought that the reader would get a more real conception of the position of the science between 1643 and 1860 by short biographical notices of those to whom we owe the various steps which have made a rational idea of the structure of the atmosphere possible.

For the greater number of the facts and dates which are included in this chapter we are indebted to the *Encyclopædia Britannica*, *Chambers's Encyclopædia*, ed. 1895, the *Dictionary of National Biography* and *Biographie universelle*, and desire here to express our obligations to those works.

Further details may be found in any good Biographical Dictionary or Encyclopaedia: we have sought only to give sufficient information to enable the reader to appreciate the relation of the several authors to the study of weather.

PIONEERS IN THE STUDY OF WEATHER AND OF THE PHYSICAL PROCESSES OF THE ATMOSPHERE

Francis Bacon, Viscount St Albans, son of Sir Nicholas Bacon, Lord-Keeper of the Great Seal, born in London 1561, died 1626. Educated at Cambridge. The author of the *Novum Organum* and exponent of the inductive method. His work ranges over all branches of science. We quote only one example as an expression of inductive curiosity in respect of the science of weather.

There is a toy, which I have heard, and I would not have it given over, but waited upon a little. They say it is observed in the Low Countries (I know not in what part), that every five and thirty years, the same kind and sute of years and weathers comes about again; as great frosts, great wet, great droughts, warm winters, summers with little heat, and the like; and they call it the prime: it is a thing I do the rather mention because, computing backwards, I have found some concurrence.

(Bacon's *Essays*, Essay LVIII 'Of Vicissitudes of Things.')

Galileo Galilei, eldest son of Vincenzio de' Bonajuti de' Galilei, a Florentine noble in straitened circumstances, and his wife Giulia Ammanati, was born at Pisa on February 18, 1564. He died in Florence on January 8, 1642. At the age of seventeen and a·half he was sent to study medicine and philosophy at the University of Pisa—in the latter we are told not to the satisfaction of his teachers owing to that habit already learnt from his father of examining an assertion to see what it was worth instead of relying on the weight of authority for authority's sake. In consequence of this habit he gained the unfortunate reputation of being imbued with the spirit of contradiction. His eager and constant study of Aristotle, Plato and other ancient authors found no favour in their eyes. To their narrow ideas a philosopher only needed to know Aristotle by heart; to understand him was a secondary consideration; to contradict him a blasphemy: Galileo both understood and contradicted[1].

That seems to be the key to his career, which is too well-known to need repetition. We owe to him the laws of pendulums, the acceleration of gravity, the invention of the thermometer and the telescope, the discovery of Jupiter's satellites as well as his advocacy of the Copernican system of astronomy as opposed to the Ptolemaic system.

[1] *The Private Life of Galileo*, Macmillan and Co., 1870, p. 4.

René Descartes, Chevalier du Perron, the inventor of Cartesian co-ordinates, may be regarded as a French successor to Francis Bacon. His chief works are: (1) *Discours de la Méthode pour bien conduire sa raison*, etc., published at Leyden, 1637, with *La Dioptrique, les Météores, et la Géometrie*; (2) *Des Méditations métaphysiques touchant la première philosophie*, 1644; (3) *Principia philosophiae*, Amsterdam, 1644; (4) *Le Monde*, published posthumously in Paris 1664.

He was born at La Haye in 1596, third son in an ancient Touraine family; from 1604 to 1612 he attended the Jesuit school of La Flèche, and being delicate continued his studies in Paris till the year 1617. He devoted four years to military service, first with Prince Maurice of Orange and then with the Duke of Bavaria. After a good deal of hesitation about his career he migrated to Holland to pursue the study of philosophy and science in a suitable climate, free from disturbance and the kind of interference which had overtaken Galileo. He did not succeed entirely in that object as his pupil Leroy, teaching the new philosophy at the University of Utrecht, got him into trouble. He published his chief works, however, in Holland, but was over-persuaded by Queen Christina of Sweden to migrate to Stockholm in 1649: the climate was too severe for him taken in conjunction with his arduous duties; he died on February 11, 1650.

"Of Bacon's demand for observation and collection of facts he is an imitator"; he wishes (in a letter of 1632) that "someone would undertake to give a history of celestial phenomena after the method of Bacon, and describe the sky exactly as it appears without introducing a single hypothesis."

"The whole field of natural laws excited his desire to explain them."

Evangelista Torricelli was born at Faenza on October 15, 1608. He was educated by the care of his uncle and was sent in 1627 to study science at Rome. His treatise *De motu* inspired by his perusal of Galileo's *Dialoghi delle nuove scienze* led to his introduction to Galileo and to his appointment as his amanuensis, a position which he held until Galileo's death three months later. He was then nominated "grand-ducal mathematician and professor of mathematics" in the Florentine academy.

In 1643 he discovered the principle of the barometer which is always associated with his name. He is also responsible for certain mathematical theorems and for the discovery of important principles of mechanics, and we owe to him improvements in both the telescope and the microscope. He died of pleurisy on October 25, 1647.

Blaise Pascal, one of the best writers and profoundest thinkers France has produced, was born of a good legal family at Clermont Ferrand in Auvergne on June 19, 1623. A religious enthusiast and ascetic, devoted with other members of his family to the cause of the Jansenists of Port Royal, a religious establishment of somewhat advanced opinions. He is best-known for his *Letters written to a Provincial by one of his Friends* 1656 and *Pensées* 1669, but in 1647 he had already distinguished himself by the publication of his

Nouvelles Expériences sur le Vide which gave the variation of the barometer consequent upon the ascent of the Tour St Jacques. "Next year occurred his famous Puy-de-Dôme experiments on atmospheric pressure which may be said to have completed the work of Galileo and Torricelli." He died at his sister's house in Paris on August 19, 1662, "his own house having been given up to a poor family one of whose children had been seized with small-pox[1]."

On sait que les anciens ne s'étaient guère élevés au delà de la statique des corps solides. S'ils étaient en possession du principe d'Archimède, ce n'est pourtant qu'au milieu du XVIIe siècle que notre Pascal énonce le principe, beaucoup plus général, d'où découle toute l'Hydrostatique. Pascal, on le sait, ne sacrifie aux études scientifiques que de rares loisirs. Ce grand initiateur, auquel nous devons encore les célèbres expériences sur la pesanteur de l'air, et, ce qui parut merveilleux à ses contemporains, la première machine à calculer, ne regardait guère la science que comme une distraction. Il ne la fit progresser, pour ainsi dire, que malgré lui.

(*La Science Française*. Paris, Ministère de l'Instruction publique et des Beaux-Arts, 1915, p. 131.)

Robert Boyle was born at Lismore Castle, County Cork, in 1627, the seventh son and fourteenth child of Richard Boyle, first Earl of Cork; at eight years of age he went to Eton for nearly four years, and, after reading with private tutors, was sent to Geneva for accomplishments. He visited Florence and studied the work of Galileo. He was unable to return on account of financial embarrassments until 1644. He took no part in the stormy politics of the time, but joined the "Invisible" club and devoted himself to experimental philosophy. In 1654 he settled in Oxford, at that time the centre of an association of philosophers. There he had a laboratory and made his famous experiments "touching the spring of air," produced the compressing air-pump, and gave the law of relation of volume to pressure which is known by his name, the first step in the dynamics of the atmosphere. The first English thermometer based upon the expansion of a liquid in hermetically sealed glass was also made under his direction. He left Oxford for London in 1668 and thereafter lived with his sister Lady Ranelagh in Pall Mall. He devoted his time to the Royal Society and to religious enterprises until his death in 1691.

Boyle had with him as his laboratory assistant at Oxford *Robert Hooke*, who was born at Freshwater, Isle of Wight, in 1635, his father being minister of the parish. He was educated at Westminster School, and as a chorister at Christchurch, Oxford. He subsequently became curator of experiments at the Royal Society, and Professor at Gresham College. He was an extremely ingenious experimenter. "He invented thirty several ways of flying." We owe to him not only Hooke's law "ut tensio sic vis," but also the invention of the wheel barometer. With that were soon associated the legends "Rain," "Change," "Fair," etc., which formed the basis of popular ideas of scientific meteorology for two hundred years and are still engraved on wheel barometers even to the present day. Hooke also invented the double barometer and a marine barometer described in *Phil. Trans.*, vol. XXII, p. 791. He died in 1703.

[1] *Chambers's Encyclopædia*, s.v. Pascal.

Edmé Mariotte, a French physicist, whose name is associated in French scientific literature with the discovery of Boyle's law of the spring of air, was born in Burgundy during the first half of the seventeenth century and died in Paris May 12, 1684. He was Prior of St Martin sous Beaune, and one of the earliest members of the Academy of Sciences at Paris, wrote on percussion, the nature of air and its pressure, the movements of fluid bodies and of pendulums, the nature of colour, etc.

Sir Isaac Newton, the celebrated natural philosopher, born in 1642 at Woolsthorpe, Lincolnshire, died at Kensington in 1727. It is hardly necessary to say that Newton's work supplied all the sciences with mathematical principles and the methods of the differential and integral calculus.

His interest in physics as related to meteorology is expressed in a scale of temperature, a law of cooling, the law of the square of velocity in fluid resistance, pendulum observations, his work on optics, the theory of colours, and the theory of transmission of sound.

Denis Papin, still known for his "digester" which provides for increasing the temperature of water beyond the normal boiling-point, was born in Blois on August 22, 1647; he studied medicine from 1661 at the University of Angers and graduated in 1669. Later he migrated to Paris and assisted Huygens in his experiments with the air-pump; from there he proceeded to London and worked for some years with Robert Boyle. He made many improvements in the air-pump, and "constructed a model of an engine for raising water from a river by means of pumps worked by a water-wheel." He was admitted to the Royal Society in 1680, and after a visit to Venice he was appointed "temporary curator of experiments" at the Royal Society. From 1687 to 1696 he occupied the chair of mathematics in the University of Marburg. In 1707 he resolved to return to London, but on his arrival he found himself without resources. He died in total obscurity in 1714.

William Dampier, sailor and buccaneer, was a pioneer in the other side of meteorology, namely, the collection of information about the distribution of the wind over the oceans, for the perfecting of which the Meteorological Department of the Board of Trade was subsequently founded. He was born in 1652 near Yeovil, the son of a tenant farmer, and after many adventures died in London in 1715. His first buccaneering expedition was in 1679 to the Isthmus of Darien, when he got as far south as Juan Fernandez, getting or building a ship on the other side, as so many travellers in the Spanish Main seem to have done. Later on Alexander Selkirk, the prototype of Robinson Crusoe, was one of his men. In 1683 he was out again to Chile, Peru, and Mexico. Not being popular with his men he was put ashore on the Nicobar Islands in 1688, but managed to find his way home by 1691, and in 1697 published his remarkable book *A Voyage round the World*, about which all that need be said here is that the discourse on the winds is so excellent that Captain Basil Hall, known as a climber of the Peak of Tenerife, writing to J. F. Daniell, speaks of Dampier's book as the authority on details of winds though it had then been published more than a hundred years. It includes a description of a typhoon which is given in chapter XIV. Here is his own account of his work:

And thus have I finished what my own experience or relations from my friends have furnished me with on this useful subject of Winds, Tides, Currents, &c., which I humbly offer, not as a complete and perfect account but as a rude and imperfect beginning or specimen of what may be better done by other hands hereafter and I hope this may be useful so far as to give a few hints to direct the more accurate observations of others.

The facts about the winds of the oceans must have been largely a matter of common knowledge in those days because our next meteorologist, *Edmund Halley*, the father of dynamical meteorology, published his account of the trade winds and monsoons, with an excellent map, in the year before Dampier's book was published. Halley was indeed a pioneer in meteorology in that he first endeavoured to connect the general circulation of the atmosphere with the distribution of the sun's heat over the earth's surface. Born at Haggerston in 1656, died at Greenwich in 1742, he was educated at St Paul's School, and Queen's College, Oxford; he was a great natural philosopher, a great traveller in search of the facts of terrestrial magnetism, a great secretary of the Royal Society, for he it was who persuaded Newton to publish the *Principia*, and a great Astronomer-Royal.

René Antoine Ferchault De Réaumur, physicist, was born at La Rochelle on February 28, 1683, and studied at Poitiers and at Bourges. At the age of twenty he published three geometrical memoirs, five years later he was elected a member of the Academy of Sciences and was appointed to superintend for the government the work *Description des divers Arts et Métiers*. During the period which followed he carried out various researches in natural history and also discovered a method of producing steel from iron and of tinning iron. For these and other researches he received the sum of 12,000 livres from the government, which he desired should be used by the Academy of Sciences for experiments on improved industrial processes. He died on October 17, 1757, by falling from a horse.

He is perhaps best-known for his thermometric researches on air and on mixtures of fluids with fluids or solids which led to the invention of the thermometer which bears his name. His original thermometer was of alcohol, the freezing-point was taken as 0° and the degree was one-thousandth of the volume contained by the bulb and tube up to the zero mark.

John Theophilus Desaguliers, natural philosopher, was born on March 13, 1683, at La Rochelle. On the revocation of the Edict of Nantes in 1685 his father, a pastor of a Protestant congregation, fled to England with his son concealed in a barrel. He assisted his father in the management of a school in Islington. On his father's death he took his degree at Oxford, entered into deacon's orders in 1710, and lectured on hydrostatics, optics and mechanics at Hart Hall; he continued his lectures in Channel Row, Westminster, on his removal there in 1713. He was elected fellow of the Royal

Society in 1714 and became their demonstrator and curator and later was awarded the Copley medal. George I rewarded him for a lecture he delivered before him by appointing him to a benefice in Norfolk. Subsequently he held a living in Essex and was chaplain to the Prince of Wales. He died on February 29, 1744.

He is said to have been the first to deliver lectures on learned subjects to the general public. He invented a machine for determining the distances of the heavenly bodies according to the systems of Newton and Copernicus. His works include papers on physics, natural philosophy, the history of Freemasons, astronomy, electricity, mechanical and experimental philosophy, etc. He also contributed papers on light, colours, variations of the barometer, to the *Philosophical Transactions*.

George Hadley, a younger brother of the John Hadley of the sextant, was by profession a lawyer. Born in London in 1685, he entered Pembroke College, Oxford, in 1700, joined Lincoln's Inn in 1701, and was "called" in 1709. He was one of the coterie of philosophers that gathered at the Royal Society. He first showed how to make allowance for the rotation of the earth in the explanation of the trade-winds in papers before the Royal Society in 1735. His theory ultimately found its way into every text-book of physical geography, though it occurred to John Dalton independently sixty years after it was published. The same theory has been claimed in certain quarters as originating with Immanuel Kant; it was urged that Kant was hardly likely to have seen the *Philosophical Transactions* which had only been published thirty years before.

Besides his papers on theory Hadley contributed also abstracts of meteorological diaries for the years 1729 and 1730, and subsequently a paper on the meteorology of 1731–35.

I have said elsewhere and still hold the opinion that the theories of the trade-winds which Halley started and Hadley improved belong to the fairy tales of science because they explain the complexity of nature by a simplicity which is suggestive of a fairy's wand. They are none the less attractive on that account. Every theory of the course of events in nature is necessarily based on some process of simplification of the phenomena and is to some extent therefore a fairy tale.

Gabriel Daniel Fahrenheit, physicist, was born at Danzig on May 14, 1686. He lived chiefly in England and Holland where he studied physics, apparently supporting himself by the manufacture of meteorological instruments. He died in Holland on September 16, 1736. He made valuable improvements in the construction of hygrometers and thermometers and introduced the scale of temperature, known by his name, which is still in use among most English-speaking peoples.

Pierre Bouguer, the originator of the exponential law of absorption, a French mathematician. He was born on February 16, 1698 and died on August 15,

1758. In 1713 he was appointed to succeed his father as professor of hydro-graphy at Croisic in Lower Brittany, and in 1730 he held a similar position at Havre. His writings are chiefly concerned with the theory of navigation. To meteorologists, however, he is known as the author of *Essai d'optique sur la gradation de la lumière* published in 1729, which is concerned with the diminution in the intensity of light as it passes through a given thickness of atmosphere. He also invented a heliometer and made some of the earliest measurements of photometry. In later life he joined an expedition to Peru to measure a degree of the meridian near the equator.

Anders Celsius, Swedish astronomer, was born at Upsala on November 27, 1701; he died there on April 25, 1744. He occupied the chair of astronomy at Upsala from 1730 to 1744, and during that period travelled in Germany, Italy and France. He published a collection of observations of the aurora borealis and also took part in an expedition organised for the measurement of an arc of the meridian in Lapland. His name will long be remembered for his paper on the centigrade thermometer which was read before the Swedish Academy of Sciences in 1742. As originally designed the scale read down-wards from a zero at the boiling-point to 100 at the freezing-point.

Benjamin Franklin was born on January 17, 1706, in a house in Milk Street, Boston, Mass.; he was the tenth son of Josiah Franklin and was sent when he was "eight years old to Boston grammar school, being destined by his father for the church as a tithe of his sons." At the age of ten he was taken from school to assist his father in the business of a tallow-chandler and soap-boiler. Two years later he was apprenticed to his half-brother as printer and it was his work as a printer that led to his first visit to England. After many vicissitudes he published in 1732 the first of his almanacks known as *Poor Richard's Almanack*; these were issued for 25 years and the annual sale averaged 10,000 copies. He took a leading part in the public life of Philadelphia and held many civil appointments; in 1753 he was put in charge of the post-service of the colonies, and later was sent by the assembly of Pennsylvania on a diplomatic mission to England. Subsequently he played a very prominent part as a diplomatist during and after the American War of Independence. "His services to America in England and France rank him as one of the heroes of the American War of Independence and as the greatest of American diplomats." Even during his lifetime his services to science were recognised by his contemporaries and he was a member of every important learned society in Europe. At the present day he is perhaps best-known for his invention of the lightning-rod, for which he was awarded the Copley medal of the Royal Society, and for his classical experiments with kites in 1752 by which he demonstrated conclusively that lightning is an electrical phenomenon; his scientific writings include papers on the causes of earth-quakes and on waterspouts and whirlwinds. The versatility of his genius may be illustrated by the fact that he invented the Franklin stove, was much engaged in remedying smoking chimneys, studied the temperature of the

Gulf Stream, made many experiments with oil on stormy waters and was also much interested in agriculture, aeronautics, medicine and mathematics.

Leonhard Euler, mathematician and author of the equations of motion with reference to moving axes that bear his name, born at Basel, April 15, 1707, pupil and friend of the Bernoullis whom he followed to St Petersburg whither they had been called by Catherine I in connexion with the founding of the Academy, migrated to Berlin in 1741 on the invitation of Frederick the Great but returned to St Petersburg in 1766 and died there on September 18, 1783. His *Lettres à une Princesse d'Allemagne* are an exposition of contemporary physics. "Euler was of an upright, amiable and religious character and a man of wider general culture than might have been looked for in one who pursued his special studies so keenly."

Jean le Rond d'Alembert, French mathematician and philosopher, was born in Paris in November 1717. He was left a foundling near the church of St Jean le Rond where he was found on November 17, and he was entrusted to the wife of a glazier who lived close by. The identity of his parents was afterwards made known. His father, without disclosing himself, settled an annuity on him and he was sent first to a boarding-school and later to the Mazarin College under the Jansenists. On leaving college he lived for thirty years at the house of his foster-mother, and began at first to study for the law and later for medicine, but his natural inclination proved too strong for him and he decided to give his whole time to mathematics. "His knowledge of higher mathematics was acquired by his own unaided efforts after he had left the college."

His writings include many valuable contributions to mechanics and dynamics and hydrodynamics; the principle with which his name is associated was first enunciated in 1742. In 1747 he published his *Réflexion sur la cause générale des vents*, which he dedicated to Frederick the Great.

In association with Diderot he was connected with the preparation of the *Dictionnaire Encyclopédique* to which he contributed several literary and mathematical articles.

In spite of many inducements he continued to lead a quiet and frugal life until his death in Paris on October 29, 1783, and out of a small income "he contrived to find means to support his foster-mother in her old age, to educate the children of his first teacher, and to help various deserving students during their college career."

James Hutton, "the first great British geologist," who was born in Edinburgh in 1726 and educated at Edinburgh, Paris, and Leyden as a physician, devoted his life to natural philosophy and natural history; he died in his native city in 1797. He is said to have given us the wet-and-dry-bulb thermometer. To him we owe an attempt to explain the physics of the formation of rain. It is contained in a work published in 1784 on the Theory of Rain and its applications, according to which rain is produced by the mixing of masses of air with different temperatures and different amounts of water-vapour. The precipitation of water by mixing different specimens of air is dependent upon the fact that the capacity of a mass of air to contain moisture in the gaseous form increases more rapidly than in direct proportion with its temperature; so that when two masses, both saturated, are mixed the mixture is too cold for it to hold all the moisture in the gaseous form, consequently some of it must be deposited as cloud or rain. Since Hutton's time we have come to regard the dynamical cooling of air, the idea of which seems to have originated with Dalton, as having more significance than mixing, in the formation of a shower of rainfall which may represent a travelling waterfall of something like 1000 million horse-power[1]; but the insight into the effects of mixture is true, and is the more remarkable because the laws of saturation of air as subsequently expounded by Dalton were not then understood.

Jean André Deluc, geologist and meteorologist, was born at Geneva on February 8, 1727. Well educated in mathematics and the natural sciences, commerce occupied the first forty-six years of his life, except for scientific excursions among the Alps in conjunction with his brother. He also took a prominent part in politics. In 1773 owing to reverses in business he left his native town; he thus was free to carry on scientific pursuits; he visited England, was made a fellow of the Royal Society and appointed reader to Queen Charlotte, a position which he held for forty-four years. He died at Windsor on November 7, 1817.

His researches in meteorology deal chiefly with the moisture of the atmosphere; he showed that water was more dense at 40° F than at the freezing-point, and he was the originator of the theory that the quantity of aqueous-vapour contained in any space is independent of the presence of any other élastic fluid. He also devoted much attention to experiments on moisture and evaporation, and invented a new hygrometer; he gave the first correct rules for measuring heights by the barometer. The discovery of the dry pile or electric column has been regarded as his best achievement; the paper describing it was presented to the Royal Society but it was so little in accord with the opinions of the day that it was considered inadvisable to print the paper in the *Transactions*.

Joseph Black, the originator of the doctrine of "latent heat," born in 1728 in Bordeaux where his father was engaged in the wine-trade, was of a Belfast family of Scottish descent on both sides. His thesis for medical graduation at Edinburgh in 1756 disclosed the relation of caustic lime and alkalis to their carbonates through the action of "fixed air," subsequently called carbonic acid by Lavoisier in 1784. Black was a busy practising physician and professor, first of anatomy and chemistry, and then of the institutes of medicine in Glasgow, and from 1766 of medicine and chemistry at Edinburgh. His name is still celebrated there for "the elaboration of his lectures in which he aimed at the utmost degree of perspicuity and with perfect success....It occasioned however some disappointment that one so capable of

---

[1] Bjerknes, *Q. J. Roy. Meteor. Soc.*, vol. XLVI, 1920, p. 129.

enlarging its territory made no further contribution to chemistry." He died on December 6, 1799. The doctrine of latent heat dates from 1756–61.

Johann Heinrich Lambert, German physicist, mathematician and astronomer, was born in Mulhausen on August 26, 1728. The son of a tailor, the slight education he received at the free school of his native town was supplemented by his private reading. He served in turn as book-keeper at Montbéliard iron-works, as secretary to a newspaper editor and as private tutor. The latter post gave him access to a good library and opportunities for pursuing his scientific studies. He toured with his pupils in Utrecht, Paris, Marseilles and Turin, and in 1759 resigned his post and after brief periods of residence in Augsburg, Munich, Erlangen, Coire and Leipzig, he settled in Berlin in 1764, where ten years later he edited the Berlin *Ephemeris*. He died on September 25, 1777.

His work was concerned chiefly with mathematics and their application to practical questions, and he made valuable contributions to astronomy, geometry, conics and trigonometry. His most important work is *Pyrometrie*, a treatise on heat which includes records and a discussion of his own experiments. His contributions to the Berlin Academy include papers on the resistance of fluids, comets, probabilities and meteorology, etc.

Richard Kirwan may be called a "practical" meteorologist. He was born in 1733, son of Martin Kirwan, Esq., of Cregg, County Galway. He entered the Jesuit novitiate of St Omer in 1754, but quitted it in 1755 and married, to find himself arrested on his wedding-day for his wife's debts. He lived in London from 1777 to 1787, chiefly interesting himself in the Royal Society, of which he was made a fellow in 1780 and Copley medallist in 1782. His work is remarkable for an estimate of the temperatures of different latitudes (London, 1787) designed to pave the way for a theory of the winds. We thus get a glimpse of the meteorology of the globe based upon ascertained facts of weather and climate. Kirwan left London in 1787, and spent the rest of his life in Dublin, where he died in 1812. "He was consulted as a weather prophet by half the farmers of Ireland."

Joseph Louis Lagrange, of French extraction, was born at Turin on August 25, 1736. His earliest tastes were literary rather than scientific; inspired however by a tract of Halley's he turned his attention to analytical methods and entered upon a two years' course of study. At the age of nineteen he communicated with Leonhard Euler his ideas on problems which later became known as the Calculus of Variations, and Euler withheld from publication his own further researches on the subject until Lagrange should have time to complete his invention.

In 1754 he was appointed professor of geometry in the royal school of artillery and in 1758 he founded a society which became the Turin Academy of Sciences. In 1766 he was appointed director of the mathematical department of the Berlin Academy in succession to Euler. In 1787 he removed to

Paris and was warmly patronised by Marie Antoinette; he remained in Paris during the revolution and in 1799 sat on the commission for the construction of the metric system and largely contributed to its adoption. On the foundation of the École Polytechnique Lagrange returned to the teaching of mathematics, and he was appointed head of the section of geometry on the establishment of the Institute. He was one of the first members of the Bureau des Longitudes, and in 1791 was made a foreign member of the Royal Society. He died on April 10, 1813.

His name is specially associated with the general equations of motion, the calculus of variations, planetary perturbations and the stability of the solar system; he took a leading part in the advancement of almost every branch of pure mathematics. "It was his just boast to have transformed mechanics into a branch of analysis and to have exhibited the so-called mechanical 'principles' as simple results of the calculus."

Sir (Frederick) William Herschel, English astronomer, was born at Hanover on November 15, 1738. His father was a musician in the Hanoverian guard, and in 1752 he himself joined the band and three years later visited England with his detachment. His health suffered so severely during the hardships of the Seven Years' War that he was removed from the regiment in 1757 and sent to England where he endeavoured, at first with but little success, to obtain a living by teaching music. In 1766 he was appointed organist at the Octagon Chapel in Bath, and while there he was led to the study of astronomy. At the outset of his work, owing to the difficulty of obtaining a satisfactory telescope and the exorbitant prices charged for the instruments, he set about the construction of a telescope of his own and on its completion in 1774, with the assistance of his brother and sister, he began his survey of the heavens.

His early work was concerned with the phenomena of variable stars and sunspots. In 1781 he discovered the planet Uranus (at first called Georgium Sidus) and several of its satellites. In 1782 he was appointed private astronomer to the king at a salary of £200 per annum, to which £50 per annum was added for the astronomical assistance of his sister; in consequence of the appointment he removed to Datchet and later to Slough. In 1781 he was awarded the Copley medal of the Royal Society and was knighted in 1816. He was the first president of the Royal Astronomical Society. He died at Slough on August 25, 1822, at the age of eighty-three.

His work was chiefly astronomical; he was the first to detect the existence of binary stars and in 1802 he announced his discovery that the laws of gravitation which bind our solar system are operative also among the distant stars. His famous reflecting telescope of 40 feet focal length and 4 feet aperture was completed in 1789. Apparently it was he who made the table of p. 108 giving the association of the weather with the times of day of the moon's changes. It is also of interest to meteorologists that in 1801 he examined the question as to whether there was any relation between sunspots (regarded as an index of solar activity) and the varying seasons of the earth as exhibited by the price of corn; his reply was inconclusive.

Horace Bénédict de Saussure, géologue et physicien suisse, fils de Nicolas de Saussure, agronome suisse, né à Couches près Genève le 17 février 1740, mort à Genève le 22 janvier 1799....L'Université de Genève lui confia en 1762 une chaire de philosophie; il n'avait que vingt deux ans et dès ses premières leçons il montra cet esprit de méthode qui contribua si puissamment plus tard à assurer les résultats de ses découvertes scientifiques.

* * *

Il visita la Suisse, la France, l'Angleterre, l'Italie et traversa quatorze fois les Alpes par huit passages différents. Le 3 août 1787 il s'éleva jusqu'au sommet du Mont Blanc où n'étaient encore parvenu que deux habitants de Chamonix, Balmet et Paccard, dont l'ascension s'était effectuée le 8 août de l'année précédente. Sa dernière course fut celle du Mont Rose en 1789. Les observations de Saussure portèrent principalement sur les minéraux. A ces recherches sur la géologie, but définitif de ses travaux, il unit les sciences qui s'y lient nécessairement, la physique, la météorologie et la botanique.

* * *

Pour la plupart de ses recherches il manquait d'instruments ou n'avait d'abord que des instruments imparfaits: "Il perfectionna le thermomètre, pour mesurer la température de l'eau à toutes les profondeurs; l'hygromètre, pour indiquer l'abondance plus ou moins grande des vapeurs aqueuses; l'eudiomètre pour déterminer la pureté de l'air et savoir s'il n'y a point autre chose que les vapeurs dans les causes de la pluie; l'électromètre pour connaître l'état de l'électricité, qui influe si puissamment sur les météores aqueux; l'anémomètre pour donner à la fois la direction, la vitesse et la force des courants d'air et inventa enfin la cyanomètre et la diaphanomètre pour comparer les degrés de la transparence de l'air aux différentes hauteurs."

Saussure garda sa chaire de philosophie à Genève jusqu'en 1786; il fut nommé en 1798 professeur d'histoire naturelle à l'école centrale du département de Leman, formé lors de la réunion de Genève à la France. Il avait été frappé de trois attaques successives de paralysie; il mourut à l'âge de cinquante neuf ans et après quatre années de souffrance.

Il fut l'auteur d'un *Essai sur l'hygrométrie*—un des plus beaux ouvrages, dit Cuvier, dont la science se soit enrichie à la fin du dix-huitième siècle. C'est là que Saussure fit connaître son importante découverte que l'air se dilate et devient spécifiquement plus léger à mesure qu'il se charge d'humidité[1].

Antoine Laurent Lavoisier, a French chemist, was born in Paris on August 26, 1743; he was educated at the Collège Mazarin and was encouraged by his father in his taste for natural science. Among his early work were papers on thunder and on the aurora. He held many posts in which his administrative abilities found full scope; he was nominated to a committee on agriculture, was chosen a member of the provincial assembly at Orleans, and formulated a scheme of taxation for the national assembly. In 1790 he was secretary

[1] *Nouvelle Biographie Générale*, Firman Didot Frères, Paris, 1863.

and treasurer of the commission to secure uniformity of weights and measures. His membership of the "farmers-general" which was suppressed in 1791 caused him to become an object of suspicion and on May 8, 1794, he was guillotined with twenty-seven other savants. "Il ne leur a fallu qu'un moment pour faire tomber cette tête, et cent années peut-être ne suffiront pas pour en reproduire une semblable."

His name is associated with the overthrow of the phlogistic doctrine, and with his theory of combustion which he showed was the union of the combustible substance with atmospheric oxygen.

James Capper, meteorologist, born on December 15, 1743, was educated at Harrow and entered the Hon. E. India Company's service in which he attained the rank of colonel. On his retirement he lived for some years in South Wales and devoted much of his time to meteorology and agriculture. He died at Ditchingham Lodge in Norfolk on September 6, 1825.

In addition to papers on travel and on agriculture he wrote the following meteorological memoirs: *Observations on the winds and monsoons, Meteorological and miscellaneous tracts applicable to navigation, gardening and farming*, with kalendars of flora for Greece, France, England and Sweden.

Jean Baptiste Antoine de Monnet, Chevalier de Lamarck was born at Bazentin in Picardy August 1, 1744, and died in comparative obscurity December 18, 1829. He was a great biologist especially in relation to the *invertebrata* and one of the first evolutionists. Since Darwin's time his views have been reasserted by Samuel Butler and meet with a good deal of support. In meteorology he was a pioneer in weather-mapping. In association with Lavoisier, Laplace and others he established a *réseau* of stations. In 1800 he commenced the publication of a series of *Annuaires Météorologiques* and continued them until 1810. They were given up because Napoleon at a public reception rudely told him to keep to Natural History. His pioneer work in the study of weather earned him the reputation of "spending much time in fruitless meteorological prediction." His biological work was apparently less appreciated at the time than it is now.

Pierre Simon Laplace was born in Normandy on March 28, 1749; the son of a small farmer, he owed his education to "the interest excited by his lively parts in some persons of position." When eighteen years of age he endeavoured to approach d'Alembert; the letters however remained unnoticed until he addressed one on the principles of mechanics which received immediate recognition and he obtained an appointment as professor of mathematics in the École Militaire of Paris. His subsequent work on planetary perturbations and the stability of the solar system earned him the title of "The Newton of France."

Subsequently he devoted his activity "to offer a complete solution of the great mechanical problem presented by the solar system, and to bring theory to coincide so closely with observation that empirical equations should no longer find a place in astronomical tables." The result is set out in *Mécanique céleste* which ranks second only to the *Principia*. He became a member of the Academy of Sciences in 1785 and was associated with most of the leading scientific societies of Europe. He became president of the Bureau of Longitudes and aided in the organisation of the decimal system. His ambition seems to have been towards politics; in 1803 he became Chancellor of the Senate and in 1817 he attained the dignity of a marquisate.

During the later years of his life he retired to a country seat at Arcueil. He died on March 5, 1827. Laplace will be remembered for the barometrical determination of height which is included in the *Mécanique céleste*, his removal of the discrepancy between the actual and the Newtonian velocity of sound, his experiments with Lavoisier on specific heats and his invention of the ice-calorimeter, and on the development of electricity by evaporation. He is also well-known for his introduction into the theory of analysis of "Laplace's coefficients" and the potential function; he laid the foundations of the mathematical sciences of heat, electricity and magnetism, and brought nearly to perfection the science of the theory of probabilities which had been initiated by Blaise Pascal and P. de Fermat.

Pierre Prévost, Swiss philosopher and physicist, son of a Protestant clergyman, was born in Geneva on March 3, 1751, and was brought up for a career in the church. He left this for the law, and later devoted himself to education and travel. In 1780 he was appointed by Frederick II of Prussia professor of philosophy in Berlin, and by acquaintance with Lagrange was led to the study of physical science. He worked for some years on the principles of economy and the fine arts, and later returned to Geneva and turned his attention to magnetism and heat. He died at Geneva on April 8, 1839. His name is chiefly known for his work on radiation and for the theory of exchanges which he enunciated.

The theory of dew was set out in detail by *Charles Wells*, Physician to St Thomas's Hospital. He was born at Charleston, South Carolina, in 1757. His father was the printer of a local newspaper. He went to Edinburgh University when he was thirteen, and was back in Carolina apprenticed to a doctor in 1771. He left the revolted colony in 1784 and settled in London. He died in 1817. His work on dew introduces to us the study of radiation as a meteorological agency. By numerous and careful experiments described in the essay which was published in 1814 he made it clear that the deposition of dew was to be accounted for by the cooling of the objects exposed to the sky on clear nights. His theory of dew has been critically examined by the perspicacious John Aitken, who raised the question whether the dewdrops were derived from the air, as Wells would have allowed, or from the earth. We may, in consequence, have to picture to ourselves dewdrops as exuded by plants when evaporation is inhibited instead of being condensed from the passing air, and the deposited moisture as coming from the local earth; but the process of cooling by which these operations are brought about is correctly ascribed to radiation, and we are introduced to one of the real causes of the meteorological phenomena of the land as distinguished from those of the sea.

Sir John Leslie, well-known as the originator of Leslie's cube, for experiments on radiation, a dry-and-wet-bulb-hygrometer, and a differential air-thermometer, was born at Largo, Fife, April 16, 1766, and died November 3, 1832, at his estate of Coates in Fife near his birthplace. Traveller as tutor to young Americans, experimental researcher and translator of Buffon's Natural History of Birds, he published in 1804 his *Experimental inquiry into the nature and propagation of heat*, for which the Rumford Medal was awarded by the Royal Society. In 1805 he became professor of mathematics in Edinburgh and in 1819 professor of natural philosophy. In 1810 he invented the process of artificial refrigeration. He is remembered for the experimental illustrations of his university lectures.

John Dalton, the distinguished chemist and physicist, the true founder of the physics of a mixture of air and water-vapour, was a very enthusiastic meteorologist. Dalton was all his life a student and a natural philosopher, dependent upon teaching for his maintenance. He was born at Eaglesfield, near Cockermouth, in 1766, the son of "a poor weaver undistinguished either for parts or energy who married in 1755 Deborah Greenup, a woman of strong character and, like himself, a member of the Society of Friends." He set up school at Eaglesfield in 1778 at the age of twelve, and, after two years, at Kendal in 1781. He began a meteorological journal with the observation of a remarkable aurora on March 24, 1787, and maintained it for the rest of his life. In 1793 he became professor of mathematics in New College, Manchester, and in Manchester he continued his researches until his death in 1844. For his atomic theory his name is known wherever science is pursued, and his investigations of the laws of pressure of aqueous vapour in air are the foundation of one of the most difficult conceptions which every student of meteorology must make his own. But he was also interested particularly in the study of the aurora. He estimated the height of one of these remarkable appearances in 1793 at 150 miles, and gave the height later on in 1826 from further investigations at 100 miles, which agrees fairly well with the results of the most recent measurements of the present day. His contributions to the physics of the atmosphere include a paper read on June 27, 1800, "on experiments and observations on the Heat and Cold produced by the mechanical condensation and rarefaction of air," and in that he crossed the threshold of the physical explanation of countless phenomena of the atmosphere.

In the last eleven years of his life Dalton received a pension from the Civil List. He never had time to marry but he got through an astonishing amount of scientific work.

Jean Baptiste Joseph Fourier, the creator of what Lord Kelvin called "harmonic analysis," born at Auxerre on March 21, 1768, was the son of a tailor. Left an orphan at an early age, he was educated by the kindness of a friend at the military school in his native town under the direction of the Benedictines of St Maur. In 1795 he was appointed to teach in the

École Normale at Paris and afterwards was attached to the École Poly-technique. In 1798 he accompanied Bonaparte to Egypt and undertook both political and scientific duties, serving for a time virtually as governor of half Egypt. He returned to France in 1801 and the following year was nominated Prefect of Isère, a position which he held for fourteen years, during which period he conducted his famous investigations on the conduction of heat. In 1826 he became a member of the French Academy and in the following year was appointed President of the council of the École Polytechnique. He died at Paris on May 16, 1830. Although remarkably successful as a politician Fourier is chiefly remembered for his scientific work on the theory of heat and for his mathematical researches chiefly connected with the theory of equations.

Friedrich Heinrich Alexander Baron von Humboldt, one of the greatest of naturalists, born at Berlin September 14, 1769, died in Berlin May 6, 1859. Traveller, especially in the Spanish settlements in America and the Indian Ocean, with a French friend Aimé Bonpland 1799 to 1804, lived in Paris 1807–27, then at Berlin, and again a traveller in 1829 with two friends Ehrenberg and Rose, exploring the north of Asia, Ural and Altai mountains, Chinese Dzungaria and the Caspian Sea. Author of *Cosmos* and celebrated as a meteorologist for his survey of the distribution of temperature and pressure over the globe.

Luke Howard, like Dalton, was a member of the Society of Friends and was a successful man of business. He was born in London in 1772. His father, Robert Howard, was a manufacturer, the chief introducer of the argand burner. Luke Howard was educated at a private school in Burford, Oxon, and subsequently apprenticed to a chemist at Stockport. In 1793 he took up the business of a chemist in London, near Temple Bar, with chemical works in Essex. He spent the remainder of his life at Tottenham, whither he removed in 1812, or on his estate at Ackworth in Yorkshire. He died in 1864. He became a fellow of the Royal Society in 1821. His scientific interests were much developed by the Askesian Society (from ἄσκησις, exercise), which was instituted in March 1796 to hold meetings every fortnight at the house of William Allen, F.R.S., Plough Court, Lombard Street, and sub-sequently at the other members' houses, the last being Dr Babington's in Aldermanbury. It lasted about ten years. Howard became sufficiently famous, principally on account of clouds, to correspond with Goethe and receive a short poem of acknowledgment entitled "Howards Ehrengedächtniss." His best-known works are on the forms of clouds and on the climate of London. He was as assiduous as Dalton in keeping a meteorological register which was continued till 1830.

His *Barometrographia*, a book of some magnificence, gives a series of curves of the variation of the barometer during a number of years obtained from the "Barometer Clock," a self-recording barograph. In the book are large diagrams in circular form representing the barometer's changes with other particulars of weather. Howard devoted a good deal of attention to the study of the relation of the weather to the phases and the apsides of the moon and wrote about a *cycle* of 18 years in the seasons of Britain.

Thomas Young, born at Milverton, Somersetshire, June 13, 1773, also of Quaker parents, physician and physicist, professor of natural philosophy at the Royal Institution 1801. His course of lectures (1807) expounded the doctrine of interference which established the undulatory theory and explained a method of estimating the size of particles in a cloud by measuring the diameter of the coronal rays, with a special instrument, the eriometer, for measuring the diameters of fibres. His work was so severely criticised by Lord Brougham in the *Edinburgh Review* that though he published a defence one copy only was sold and his discovery was completely disregarded until the subject was taken up independently by Fresnel. Young gave up scientific pursuits and devoted himself to Egyptology in which also he was disappointed. He died May 10, 1829.

Jean Baptiste Biot was born in Paris on April 21, 1774; he served for a time in the artillery. In 1797 he was appointed professor of mathematics at Beauvais, in 1800 he became professor of physics at the College of France, and in 1803 he was elected a member of the Academy of Sciences. In the following year he accompanied Gay-Lussac on the first balloon ascent undertaken for scientific investigations. He was made commander of the Legion of Honour in 1849 and in 1856 was elected member of the French Academy. He died in Paris on February 3, 1862. His researches extended to almost every branch of physical science and included many geodetic determinations, but his name is most closely associated with his optical work, especially that on the polarisation of light.

Sir Francis Beaufort, Rear-Admiral and Hydrographer of the Navy, known for the Beaufort Scale of Wind Force as indicated by the behaviour of a ship at sea, and the Beaufort notation for weather, whereby its different phases are indicated by letters of the alphabet. He was born in 1774, son of the Rev. Daniel Augustus Beaufort, Rector of Navan, County Meath. He entered the Navy in 1787. In an interval of unemployment at sea in 1803–4 he assisted in establishing a line of telegraphs from Dublin to Galway. In June 1805 he was in command of the *Woolwich* and surveyed Rio de la Plata, and from the next year his scale of wind force is dated. He surveyed the Greek Archipelago and combined the duty with suppressing the pirates of the Levant. He was Hydrographer of the Navy from 1829 for 26 years, and died in 1857. He was a fellow of the Royal Society. He made no contribution to its publications, but his contributions to the organisation of meteorological observations at sea have been of the greatest value to meteorologists for more than a hundred years on account of the practical advantages of definition and of brevity in meteorological records.

Johann Karl Friedrich Gauss, after whom the practical c.g.s. unit of magnetic force is named, was born at Brunswick on April 30, 1777, and died at Göttingen on February 23, 1855. At an early age he published an important work on the theory of numbers, and in 1809, while director of the Observatory at Göttingen and professor of mathematics, he published his *Theoria Motus Corporum Caelestium* which contains the new methods he had invented for the calculation of the orbits of planets. He was appointed by the government of Hanover to conduct the trigonometric survey and the measurement of an arc of the meridian; during the course of the work he invented the "heliotrope." His researches in geodesy were extraordinarily successful. Later, with no less success, he turned his attention to the study of terrestrial magnetism, especially the principles of absolute magnetic measures. He was instrumental in founding a magnetic association. He also generalised the passage of light through a series of lenses, wrote papers on probability, the method of least squares, and many problems of mathematical analysis.

Heinrich Wilhelm Brandes, mathematician and meteorologist, who was the first to construct a series of daily charts, was born on July 27, 1777, at Groden near Ritzebüttel; from 1801 to 1811 he served as Dyke inspector of the Weser, then professor of mathematics in the University of Breslau, and subsequently of Leipzig in 1826. He died at Leipzig on May 17, 1834.

Louis Joseph Gay-Lussac, chemist and physicist, born December 6, 1778, at St Léonard (Haute Vienne), assistant to Berthollet at the chemical works at Arcueil, made a series of original researches on the dilatation of gases, the "tension" of vapours, the improvement of thermometers and barometers, the density of vapours, hygrometry, evaporation and capillary action. "Next, first with Biot, and a month later alone, he made two balloon ascents for the purpose of investigating the temperature and moisture of the air and the laws of terrestrial magnetism. Along with Alexander von Humboldt he analysed the properties of air brought down from a height of nearly 23,000 feet, and their joint memoir to the Academy of Sciences (read October 1, 1804) contained the first announcement of the fact that oxygen and hydrogen unite to form water in the proportion of one volume of the former to two volumes of the latter." In 1809 he was appointed professor of chemistry at the Polytechnic School, and from 1832 also filled the corresponding chair in the Jardin des Plantes. In 1815 he succeeded in isolating Cyanogen and made other chemical discoveries. From 1816 he was editor of the *Annales de Chimie et de Physique* in association with Arago. He died in Paris on May 9, 1850.

Clément Désormes[1], Chevalier of the Legion of Honour, the originator of a well-known laboratory experiment, professor of chemistry at the Conservatoire des Arts et Métiers, was born about 1778, and died in Paris on November 21, 1840. His experiments on the gaseous oxide of carbon, published in the year 1801, refuted the last argument of the partisans of phlogiston. To him science is indebted for the theory of the production of sulphuric acid, the determination of the absolute zero of temperature, and for an able and elaborate investigation of the mechanical force of heat. In

---

[1] *Pharmaceutical Transactions*, quoted in the *Quarterly Journal of Meteorology and Physical Science*, edited by J. W. C. Gutch and W. H. White, London, 1843, vol. 1, part 1, p. 114.

that last memoir was promulgated the important fact that the quantity of heat contained in a given weight of vapour is constant for all temperatures and all pressures, if the space occupied by the vapour be saturated therewith.

Friedrich Wilhelm Bessel, born at Minden on July 22, 1784, was placed at the age of fifteen in a counting-house at Bremen, where he was led to study navigation, mathematics and astronomy by his desire to obtain a situation as supercargo on a foreign voyage. His calculation of the orbit of Halley's comet brought him into prominence and his masterly investigation of the comet of 1807 led to his being summoned to superintend the erection of a new observatory at Königsberg of which he acted as director until his death on March 17, 1846.

"Modern astronomy of precision is essentially Bessel's creation. Apart from the large scope of his activity, he introduced such important novelties as the effective use of the heliometer, the correction for personal equation and the systematic investigation of instrumental errors." "In pure mathematics he enlarged the resources of analysis by the invention of Bessel's Functions."

James Pollard Espy[1], meteorologist, comes next. He was born in Pennsylvania on May 9, 1785, and died on January 24, 1860, in Cincinnati. At the age of eighteen he became a student in Transylvania University at Lexington, and after taking his degree taught in a school in Xenia and studied law. Although he completed his law-studies his love of teaching caused him to adopt that rather than law as his profession. In 1817 he became teacher in the classical department of the Franklin Institute and, impressed by the writings of Dalton and Daniell on meteorology, he began to observe the phenomena and to experiment. His enthusiasm became so strong that he gave up teaching in order to carry on his researches. We shall deal with his theory of storms in a subsequent chapter (chap. XIV).

François Jean Dominique Arago, born at Estagel near Perpignan February 26, 1786, died in Paris October 3, 1853, astronomer, geodesist and physicist; secretary and from 1830 director of the Observatory of Paris and meanwhile professor of analytical geometry at the École Polytechnique. With Biot he completed the measurement of an arc of meridian, he confirmed the truth of the undulatory theory of light, and "may be said to have proved the relation between the aurora borealis and magnetic variations."

Augustin Jean Fresnel, born at Broglie, Eure, May 10, 1788, died July 14, 1827, at Villa d'Avray near Paris, engineer, ultimately head of the department of Ponts et Chaussées.

La théorie de l'émission se heurtait aux phénomènes de diffraction dont elle n'avait pu fournir une interprétation satisfaisante. L'Académie française des Sciences propose cette question à la sagacité des jeunes savants prescrivant d'ailleurs de faire usage de la théorie de l'émission. Un inconnu nommé Fresnel, bien qu'à peu pres dénué des

[1] *Pioneers of Science in America*, 1896, edited by W. J. Youmans.

ressources indispensables à une expérimentation délicate, a l'audace et le bonheur de résoudre la question proposée de la manière la plus complète: mais son explication repose sur la théorie des ondulations. Poisson, l'un des commissaires de l'Académie, observe que les théories de l'auteur entraînent cette conséquence paradoxale, non explicitée par Fresnel, que le centre de l'ombre géométrique d'un petit disque opaque doit se trouver éclairé. Contre toute vraisemblance, l'expérience tentée de suite, donne le résultat prédit. Le prix fut décerné à Fresnel.

(*La Science Française*, Paris, 1915.)

Sir Edward Sabine was in his ninety-fifth year when he died. His long life was largely devoted to the meteorology and magnetism of the globe as dependent upon co-ordinated observations in all countries, organised by Government in association with recognised scientific institutions. He was born in Dublin in 1788, educated at the Royal Military Academy, Woolwich, became a fellow of the Royal Society in 1818, joined in arctic explorations in the same year with [Sir] J. Ross and in the next year with [Sir] Edward Parry, and was engaged upon pendulum observations in 1821–29; was Secretary of the Royal Society in 1828–29; undertook a magnetic survey of the British Isles in 1834; General Secretary of the British Association for the Advancement of Science from 1839 for 20 years, with the exception of 1852, when he was President. In 1840, with the aid of these scientific bodies, he obtained the establishment of magnetic observatories at Toronto (Canada), St Helena, Cape of Good Hope, Hobart (Tasmania), and stations in India determined by the East India Company, many of which are still in operation; meteorological observations were initiated there also. A naval expedition to the Antarctic was also arranged for the purpose of a magnetic survey. The Kew Observatory, Richmond, was obtained from the Crown by the British Association for use as a physical observatory in 1842, and maintained by that body until 1871, when it was provided for partly by the Meteorological Office, then administered by a Committee of the Royal Society of which Sabine was Chairman, and partly by a private benefaction entrusted to the Royal Society by John Peter Gassiot, the Chairman of the Kew Committee of the British Association. To the movement of which Sabine was at least the most prominent figure we owe the observatories of Mauritius and Hongkong and the beginnings of the meteorological services of Canada and India. In 1852 he secured the establishment of meteorological observations at all the foreign and colonial stations of the Royal Engineers, which were under the general superintendence of Sir Henry James, Director of the Ordnance Survey at Southampton, where a base station was maintained. Under Sir Henry's direction the book of *Instructions* for Meteorological Observations was drawn up which appears in modern form as the *Observer's Handbook* of the Meteorological Office. The foreign and colonial stations of the Army Medical Department were brought to take part in the meteorological work, and in 1862 the whole of that work was taken over by the Army Medical Department.

Sabine undertook the reduction and publication of the magnetic observations himself with a small clerical staff at Woolwich and continued it for about twenty years, and indeed after the provision for the clerical staff was withdrawn. He had also in hand the repetition of the magnetic survey of the British Isles, 1858–61. He was President of the Royal Society from 1861 until 1871, and within that time pendulum observations were initiated in connexion with the Trigonometrical Survey of India, an enterprise started in 1864.

It was in 1867 that the Meteorological Office was detached from the Board of Trade, and the grant was entrusted to a Committee of the Royal Society, of which Sabine was Chairman. The arrangement was altered in 1876 and Sabine withdrew. He died at Richmond on June 26, 1883.

John Frederic Daniell, professor of chemistry in King's College, London, from the foundation of that institution in 1831 until his sudden death in 1845, was born in Essex Street, Strand, in 1790. His father was a bencher of the Inner Temple. He went into business as a sugar refiner, but left it to become a fellow of the Royal Society at the age of 23. He began his meteorological contributions in the same year. He is known as the inventor of the Daniell cell, and almost equally well as the inventor of the Daniell hygrometer for the determination of the dew-point. It was used by James Glaisher for many thousands of observations of dew-point in comparison with the dry and wet bulbs, and it is on these observations that Glaisher's table of factors for obtaining the dew-point from the readings of the dry and wet bulbs is based and thence the tables for the vapour pressure and relative humidity which were used for many years in British meteorology.

Daniell also interested himself in particulars of the climate of London of which Howard had made a special study. He was awarded the Medal of the Horticultural Society for an essay on artificial climate in which stress was laid upon the appropriate degree of humidity for greenhouses. His *Meteorological Essays* first appeared in 1823, a second edition in 1827. Among other meteorological experiments he set up in 1832 a water barometer in the rooms of the Royal Society, at that time at Somerset House.

Claude-Servais Mathias Pouillet[1], French physicist, and member of the Institute, was born at Cuzance on February 16, 1791, and in 1811 entered the École Normale; he held successively the posts of professor of physics at the Collège Bourbon, instructor in physics to the Duke of Chartres and later to the other sons of Louis-Philippe, professor of physics and sous-directeur of the Conservatoire des Arts et Métiers, professor at the École Polytechnique in succession to Dulong. He was obliged through ill-health to retire from this last post and was appointed director of the Conservatoire and professor at the Faculty of Sciences in Paris. He sat in the Chamber of Deputies but retired from political life after the Revolution of February, 1848, and from his duties in the university after the coup d'état of December 2, 1851, devoting himself to his researches and to the publication of his papers. He was an active member of the Academy of Sciences and in 1845 the Legion

[1] *Dictionnaire des Contemporains*, Vapereau, 1870.

of Honour was conferred upon him. He died on June 15, 1868. His most important works are *Éléments de physique expérimentale et de météorologie* and *Notions générales de physique et de météorologie à l'usage de la jeunesse.* He wrote papers on many subjects connected with the science of heat and is especially commemorated for his work on solar heat, the radiative and absorptive powers of the atmosphere, and the temperature of space. He also wrote on the height, velocity and direction of motion of clouds, and devised a method of determining photographically the height of clouds.

The names of *William Reid*, Major-General R.E., born at Kinglassie in 1791, eldest son of the minister there, and of *Henry Piddington*, who was born at Uckfield in 1797, bred in the Mercantile Marine, for some time commander of a ship in the East India and China trade and subsequently curator of a museum in Calcutta, will be referred to in greater detail in chapter XIV on 'The Circulation and the Cyclone.'

Sir John Frederick William Herschel, English astronomer and only son of the celebrated Sir William Herschel, was born at Slough on March 7, 1792. He was educated partly at Eton and partly by a private tutor, and entered St John's College, Cambridge, at the age of seventeen; he graduated as Senior Wrangler in 1813. He entered his name at Lincoln's Inn the following year but his interests soon turned to the study of optics and astronomy. For his work on double stars in continuation of that of his father he received many scientific awards; he was closely connected with the Royal Astronomical Society and for several years served as secretary and subsequently as president of that body. In 1833 he set out for the Cape in order to complete his father's survey of the heavens by exploring the southern skies. He returned to England in 1838 and was created a baronet by the Queen. His later life was spent in the collation of his father's catalogues of nebulae and double stars with his own observations and those of other astronomers; he was also engaged on the translation of the *Iliad* into verse. He held the office of Master of the Mint for five years. He died at Collingwood, near Hawkhurst, Kent, on May 11, 1871, and was buried in Westminster Abbey.

Herschel is said to have become an astronomer from a sense of duty, his natural inclination being towards chemistry; he was accomplished in that science and made discoveries of great importance to photography and also contributed valuable researches to the undulatory theory of light. He wrote articles on meteorology, physical geography and the telescope for the eighth edition of the *Encyclopædia Britannica*. He wrote also *A preliminary discourse on the study of Natural Philosophy.* While at the Cape he "gave an impulse to the science of meteorology, having the merit of having suggested the scheme for taking meteorological observations simultaneously at different places."

John Thomas Romney Robinson, Irish astronomer and physicist, born in Dublin on April 23, 1792; he studied at Trinity College, Dublin, and obtained a fellowship in 1814; he acted as deputy-professor of natural philosophy until he obtained the college living at Enniskillen in 1821. Two years later he became astronomer at Armagh Observatory and while residing at the observatory he also held the living of Carrickmacross. He died on February 28, 1882. His work is chiefly concerned with researches in astronomy and physics. He is known to meteorologists as the inventor of the cup-anemometer for recording the velocity of the wind, which dates from 1846.

Lambert Adolphe Jacques Quetelet, Belgian astronomer, meteorologist and statistician, was born in Ghent on February 22, 1796. He was educated in the town and in 1819 became professor of mathematics at the athenaeum of Brussels. After holding other posts in that town he was, in 1828, appointed director of the new royal observatory which had been founded chiefly at his instigation. The building was completed in 1832 and since 1835 observations have been made and published in the *Annales* of the Observatory. In 1834 he was appointed perpetual secretary of the Brussels Academy. On his death on February 17, 1874, he was succeeded in the directorship of the observatory by his son Ernest Quetelet.

Quetelet interested · himself chiefly in meteorology and statistics and "organised extensive magnetical and meteorological observations and in 1839 he started regular observations of the periodical phenomena of vegetation especially the flowering of plants." The results are given in various memoirs and in his works, *Sur le climat de la Belgique* and *Sur la physique du globe.* His name is perhaps even better known for his statistical investigations.

Sadi Nicolas Léonhard Carnot was born in Paris on June 1, 1796. He was admitted to the École Polytechnique in 1812 and left late in 1814 with a commission in the Engineers. In 1819 he obtained a lieutenancy in the staff corps. He died of cholera, following an attack of brain-fever, in Paris on August 24, 1832. "He was one of the most original and profound thinkers." "He devoted himself with astonishing ardour to mathematics, chemistry, natural history, technology and even political economy. He was an enthusiast in music and other fine arts; and he habitually practised as an amusement, while deeply studying in theory, all sorts of athletic sports, including swimming and fencing." "The only work he published was his *Réflexions sur la puissance motrice du feu et sur les machines propres à développer cette puissance.*" Extracts from his manuscript, subsequently appended to a reprint of the *Réflexions*, show that he had realised the true nature of heat and had noted down for trial many of the best methods of finding its mechanical equivalent; Kelvin's experiment on the passage of air under pressure through a porous plug is also given. Carnot's principle is fundamental in thermodynamics and is the basis of the second law of that subject.

Macedonio Melloni, Italian physicist, was born at Parma on April 11, 1798. He was professor in his native town for seven years but in 1831 he escaped to France, having taken part in the revolution. In 1839 he went to Naples and was appointed director of the Vesuvius Observatory, a post which he held until 1848. He died of cholera at Portici near Naples on August 11, 1854.

Melloni's scientific work was chiefly connected with the thermopile and its application to the study of radiant heat, including the power of transmitting dark heat possessed by various substances, and the changes produced in the heat rays by passage through different materials. He also "studied the reflexion and polarisation of radiant heat, the magnetism of rocks, electrostatic induction, daguerrotypy, etc." He was awarded the Rumford medal of the Royal Society and was a correspondent of the Paris Academy and a foreign member of the Royal Society.

Ludwig Friedrich Kämtz was born at Treptow (Pomerania) on January 11, 1801, the son of a simple farmer: he died on December 8/20, 1867, at St Petersburg, a member of the Academy and director of the Central Physical Observatory of that capital. His biography is typical of the meteorology of the first half of the nineteenth century, the early days of professional meteorology. After a restless school-experience he entered the University of Halle in 1819 to study law but was attracted to the study of the classical languages and antiquities instead; however, he graduated in philosophy with a mathematical dissertation in 1822. Impressed by Biot's text-book he became, as Privatdocent, the first exponent in Germany of Fresnel's work; then, as ordinary professor, himself the author of a *Lehrbuch der Experimental-physik* in 1839. Meanwhile the physical properties of the atmosphere had claimed his attention which was further impressed by the experience of a sudden storm on the Baltic and in consequence he produced a *Lehrbuch der Meteorologie* in three volumes (1831–36) which has been recognised as a classical attempt to present meteorology as a co-ordinate system with due regard to recent advances in physical science. "Es gibt kaum ein einziges, irgendwie bemerkenswerthes Werk über physikalische Geographie oder Klimatologie, das im Laufe der letzten vierzig Jahre erschienen wäre und sich nicht mehr oder weniger auf die Arbeit von Kämtz stützen[1]." Dissatisfied with the experimental facilities of Halle he migrated to Russia in 1841 to occupy a chair of physics at Dorpat. Thence he was called to St Petersburg two years before his death. Kämtz's *Repertorium für Meteorologie* is one of the best known periodicals devoted to the science.

Heinrich Gustav Magnus, German chemist and physicist, the son of a wealthy merchant, was born in Berlin on May 2, 1802. He studied in Berlin, Stockholm and Paris, and in 1831 he obtained a post in Berlin as lecturer in technology and physics in the University, in which capacity he showed remarkable ability. During the course of his career he held many important posts in the University of Berlin and represented the government at several conferences and missions, among others that for introducing the metric system in Germany. He died on April 4, 1870.

[1] *Zeitschr. der Öst. Gesell. für Met.*, vol. III, 1868, p. 182.

Between 1827 and 1868 he was occupied with researches at first chiefly chemical but later he dealt with the vapour-pressures of water, among other physical subjects; in the last ten years of his life he interested himself in the question of diathermancy in gases and vapours, especially dry and moist air, and in the thermal effects of condensation of moisture on solid surfaces.

Heinrich Wilhelm Dove, born in 1803 at Liegnitz in Silesia, physicist and meteorologist, professor of natural philosophy at Berlin from 1845 until his death on April 4, 1879, director of the Royal Prussian Meteorological Institute from 1848, an indefatigable organiser of meteorological work, the author of *The Law of Storms* to whom reference will be made in chapter XIV.

Sir David Brewster, the distinguished physicist and biographer of Newton, one of the originators of the British Association for the Advancement of Science, president of St Leonard's College at St Andrews, was born in 1781 and died in 1868; he is known to meteorologists as drawing attention in 1820 to the difference between the geographical pole and the poles of greatest cold in the Northern hemisphere.

Johann von Lamont, Scottish-German astronomer and magnetician, was born at Braemar, Aberdeenshire, on December 13, 1805. He was educated from the age of twelve at the Scottish monastery in Regensburg, and apparently never afterwards returned to his native country. In 1827 he was admitted to the new observatory at Bogenhausen near Munich and after the death of his chief he was appointed director of the observatory in 1835. In 1852 he became professor of astronomy at the University of Munich and he held both posts until his death on August 6, 1879.

Lamont was a member of many foreign scientific societies and corporations. He made valuable contributions to astronomy and was instrumental in initiating the equipment of a magnetic observatory at Bogenhausen; his magnetic work includes comprehensive magnetic surveys, the discovery of the magnetic period of eleven years and the discovery of earth-currents.

Matthew Fontaine Maury, American naval officer and hydrographer, was born near Fredericksburg in Virginia on January 24, 1806. He entered the navy as midshipman in 1825 and sailed round the globe in the *Vincennes* during a cruise of four years 1826–30. He served as master of the *Falmouth* and in other vessels from 1831–34, and in the following year was engaged on a treatise on navigation. In 1839 he met with an accident which rendered him unfit for active service. Maury worked hard to collect observations of winds and currents and distributed to captains of vessels specially prepared log-books. In the course of nine years he collected enough to give the equivalent of about 500,000 days' observations. The result of his work was to show the necessity for the co-operation of the maritime nations with regard to ocean meteorology, and his efforts led to an international conference in Brussels in 1853, the result of which was to benefit both navigation and meteorology. In his endeavours to organise meteorological work on land he met with little encouragement from the government. His most popular work, *The Physical Geography of the Sea*, was published in London in 1855, and in New York in 1856. On the outbreak of the American Civil War in 1861 he became head of coast, harbour and river-defences, and in connexion with his work in that capacity he invented an electric torpedo. He lost most of his money during the war and after further vicissitudes he settled for a time in England where he was presented with a testimonial raised by public subscription; he returned to America in 1868 and became professor of meteorology in the Virginia Military Institute. He died at Lexington on February 1, 1873.

Henri Victor Regnault, French chemist and physicist, was born at Aix-la-Chapelle on July 21, 1810. As a youth he worked for many years in a drapery establishment, his spare time being occupied in study. He entered the École Polytechnique in 1830 and from there passed to the École des Mines, later he was appointed professor of chemistry at Lyons and while holding that office he spent some time on comparisons of the composition of atmospheric air in different parts of the world. He held successively the posts of professor of chemistry in the École Polytechnique in Paris, professor of physics in the Collège de France, and was a member of the Academy of Sciences. In 1854 he became director of the porcelain manufactory at Sèvres. Most of his later work was destroyed during the Franco-German War in which his son was killed; he never recovered from the shock. He died on January 19, 1878, but his scientific work ended in 1872.

Among the subjects to which he devoted his attention in addition to organic chemistry may be mentioned: the re-determination of the specific heats of solids, liquids and gases and of the coefficient of expansion of gases, the examination of the divergences from Boyle's law, thermometry and hygrometry, including the comparison of an air and mercury thermometer, and the invention of a hygrometer. He is also well-known for his work on the practical treatment of steam-engines.

Elias Loomis, American mathematician, was born at Willington, Connecticut, on August 7, 1811, and educated at Yale College. He studied for a year in Paris, after which he was appointed professor of mathematics and physics first at the Western Reserve College in Ohio and later in New York. He became professor of natural philosophy and astronomy at Yale in 1860. He died on August 15, 1889. He is the author of a large number of scientific manuals which have become classics. His work included papers on algebra, geometry, trigonometry, astronomy and meteorology, particularly the travel of cyclones.

Thomas Andrews, Irish chemist and physicist, was born at Belfast on December 19, 1813. He attended the Belfast Academy and the Academical Institution and in 1828 went to study chemistry at Glasgow whence he migrated to Trinity College, Dublin, and gained distinction in classics as well as in science. He graduated as M.D. at Edinburgh in 1835 and returned to carry on medical practice in his native town, at the same time giving instruction in chemistry. Ten years later he was appointed vice-president of Queen's College, which had recently been established, and professor of chemistry, both of which offices he held until he was compelled through ill-health to retire in 1879. He died on November 26, 1885.

He showed remarkable skill and resourcefulness in experimental work; in 1844 he was awarded the Royal medal for his work on the heat developed in chemical actions; he also carried on important work on ozone, but he is best known for his research on the liquefaction of gases and for his study of the continuity of the liquid and gaseous states and introduction of the conceptions of critical temperature and critical pressure.

Julius Robert Mayer, German physicist, was born at Heilbronn on November 25, 1814; he studied medicine at Tübingen, Munich and Paris, and in 1840 made a journey to Java as surgeon on a Dutch vessel, after which he settled in his native town with a medical post. He is said to have propounded independently the first law of thermodynamics and made many applications of the law to the explanation of both terrestrial and cosmical phenomena. He made out the general conservation of energy and applied the principle with great ability to many physical phenomena. He died in 1878.

William Ferrel, scientist, was born in Bedford, now Fulton, Pa., on January 29, 1817. He studied at Marshall and Bethany Colleges, and in 1857 he became assistant in the office of the *American Ephemeris* and *Nautical Almanac*, a post which he held for ten years. Later he held an appointment in the U.S. Coast Survey for the discussion of tidal observations and in 1882 was made assistant, with the rank of professor, in the signal service bureau. He was made a member of the National Academy of Sciences in 1868 and was also an honorary member of the Austrian, English and German meteorological societies.

His published works include papers on the motions of fluids and solids relative to the earth's surface, researches on tides, meteorological researches, the temperature of the atmosphere and the earth's surface, and recent advances in meteorology[1].

Quentin-Paul Désains, French physicist, born at St Quentin (Aisne) on July 12, 1817, was educated at the Collège Louis le Grand and was admitted in 1835 to the École Normale. After holding many educational posts he was appointed to the chair of physics in the Faculty of Sciences in Paris in 1853. His collaborator in scientific work was *Ferdinand Hervé de la Provostaye*, a French physicist who was born at Redon on February 15, 1812, and died at Algiers on December 28, 1863. Together they published a large number of works on radiant heat, the polarisation of heat-rays and the latent heat of water. Désains is also the author of a *Traité de physique*[2].

[1] *Dictionary of American Biography.*
[2] *Dictionnaire des Contemporains*, Vapereau, 1870.

Angelo Secchi, Italian astronomer, was born on June 29, 1818, at Reggio in Lombardy, and at an early age he entered the Society of Jesus. He became director of the observatory of the Collegio Romano in 1849, four years before the rebuilding of the observatory. He died in Rome on February 26, 1878. He worked with great assiduity on his researches in physical astronomy and meteorology; most of his papers are concerned with the spectrum analysis of the stars and the sun.

The seventy-four names of scientific worthies between the invention of the barometer and the invention of the weather-map have been chosen as representing work which helps to furnish the instrumental, observational and intellectual equipment of the exponents of the modern science of meteorology.

Few of the number could be fairly described as belonging exclusively to the meteorological section of the world of science, most of them made their contributions more or less incidentally after the manner suggested by Principal Forbes and Professor Daniell. Thirty-three were experimenters and designers of physical instruments. De Saussure, Dalton and Howard are also conspicuous as regular observers. Twenty-four are noted for their writings upon meteorological theory. Dampier and Kirwan, the Herschels, Lamarck, Humboldt, Beaufort, Brandes, Sabine, Quetelet, Kämtz, Dove, Maury, as well as the investigators of the cyclone, are included because they had special regard to the co-ordination of observations with the object of forming a view of one or other of the aspects of the general circulation of the atmosphere. The founders and formulators of the principle of the conservation of energy cannot be excluded from any list of inductive philosophers, but they come later than the limit of our list.

Thus we have already evidence of the division of labour in the general meteorological programme which became so conspicuous after the introduction of the weather-map, namely, first the improvement of equipment and extension of the range of observations, secondly the use of equipment to obtain an organised series of observations, thirdly the co-ordination of observations to represent the structure and circulation of the atmosphere leading on to inductive laws for the application of weather knowledge, and fourthly the development of a physical and dynamical theory of the circulation either as a whole or in detail.

This fourfold division of the whole duty of the science can be traced in other ways than the selection of conspicuous names   As another indication we give the titles of a list of representative classics which have been published within the last thirty years under the editorship of Dr G. Hellmann, formerly Director of the Prussian Meteorological Institute and Secretary of the International Meteorological Committee. It is the more necessary to refer to this list because it leads us more especially to consider the gradual extension and co-ordination of observations over the land-areas of the globe, following the example originally set in respect of winds at sea.

LIST OF CLASSICAL METEOROLOGICAL PAPERS SELECTED
FOR REPRINTING BY G. HELLMANN

1. L. Reynman, Wetterbüchlein. Von wahrer Erkenntniss des Wetters. 1510.
2. Blaise Pascal. Récit de la Grande Expérience de l'Équilibre des Liqueurs. Paris, 1648.
3. Luke Howard. On the Modifications of Clouds. London, 1803.
4. E. Halley, W. Whiston, J. C. Wilcke, A. von Humboldt, C. Hansteen. Die ältesten Karten der Isogonen, Isoklinen, Isodynamen, 1701, 1721, 1768, 1804, 1825, 1826.
5. Die Bauern-Praktik. 1508.
6. George Hadley. Concerning the Cause of the General Trade-Winds. London, 1735.
7. Evangelista Torricelli. Esperienza dell' Argento Vivo. Accademia del Cimento. Istrumenti per conoscer l' Alterazioni dell' Aria.
8. E. Halley, A. von Humboldt, E. Loomis, U. J. Le Verrier, E. Renou. Meteorologische Karten, 1688, 1817, 1846, 1863, 1864.
9. Henry Gellibrand. A Discourse Mathematical on the Variation of the Magneticall Needle. London, 1635.
10. Rara Magnetica, 1269–1599. P. De Maricourt, F. Falero, P. Nunes, J. De Castro, G. Hartmann, M. Cortés, G. Mercator, R. Norman, W. Borough, S. Stevin.
11. J. H. Winkler, B. Franklin, T. F. Dalibard, L. G. Le Monnier. Ueber Luftelektricität, 1746–1753.
12. Wetterprognosen und Wetterberichte des XV und XVI Jahrhunderts.
13. Meteorologische Beobachtungen vom XIV bis XVII Jahrhundert.
14. Meteorologische Optik, 1000–1836. Theodoricus Teutonicus, R. Descartes, I. Newton, G. B. Airy, A. De Ulloa, P. Bouguer, J. Hevel, T. Lowitz, J. Fraunhofer, G. Monge, W. Scoresby, Alhazen, J. De Mairan.
15. Denkmäler Mittelalterlicher Meteorologie.

The following list of observations is included under No. 13 of the reprints, the fifteen first in the list are without instruments and the last ten give observations made at sea.

William Merle's Wetterbeobachtungen. Driby (England), 1337, 1343.
Wetterbeobachtungen aus England, 1439.
Martin Biem's Wetterbeobachtungen. Krakau, 1502.
Aventin's Wetterbeobachtungen. München, 1511.
Johann Werner's Wetterbeobachtungen. Nürnberg, 1513.
Wetterbeobachtungen aus Mainz, 1517, 1518.
Andrea Pietramellara's Wetterbeobachtungen. Bologna, 1524.
F. von Žerotin's Wetterbeobachtungen. Mähren, 1533, 1534.
Wolfgang Haller's Wetterbeobachtungen. Zürich, 1552.
Diego Palomino's Wetterbeobachtungen. Jodar (Spanien), 1556–1595.
Tycho Brahe's Wetterbeobachtungen. Uraniborg, 1582.
Kepler's Wetterbeobachtungen. Linz, 1623.
Gleichzeitige Wetterbeobachtungen in Hessen und in Pommern, 1635.
Georg Marggraf's Wetterbeobachtungen. Brasilien, 1640.
Johann Campanius' Wetterbeobachtungen. Nordamerika, 1644, 1645.

Récit des observations faites à Clermont, à Paris et à Stockholm, 1649–1651.
Florentiae, 1655 Januario, Gradus Aquae.
Diario delle mutazioni del tempo. Pisa, 1657.
Diario delle mutazioni del tempo. Firenze, 1658.
Ismael Boulliau, Ad Thermometrum observationes anno 1658, Parisiis.
A register kept by Mr Locke in Oxford, 1666.
A letter from Dr Robert Plot of Oxford, 1684.
R. J. Camerarius, Ephemerides Meteorologicae Tubingenses, 1691.

B. Ramazzini, Ephemerides Barometricae Mutinenses, 1694.
W. Derham, A Register of the Weather, 1697.
F. Hoffmann, Observationes Barometrico-Meteorologicae Hallenses, 1700.

Primer Viage de Cristóbal Colon, 1492–1493.
Pero Lopes de Souza, Diario de Navegação, 1530.
João de Castro, Roteiro de Lisboa a Goa, 1535.
Third Voyage of John Davis, 1586.
Francis Drake's Logbook, 1596.
Second Voyage of Henry Hudson, 1608.
Journaal van Abel Janszoon Tasman's Reise, 1642.
Friedrich Marten's Spitzbergische Reise, 1671.
Second Voyage of Edmund Halley, 1699.
Observations of the Weather, made in a Voyage to China, A.D. 1700, by
    Mr James Cunningham.

The four divisions which have been employed here furnish a guide to the
arrangement of the succeeding chapters of the history which deal with the
development of meteorology after the introduction of the weather-map.
Chapter IX deals with meteorology as an international science, the meteorological library and the co-ordination of observations; chapters X and XI with
the equipment of the meteorological observatory for the surface air and the
upper air, chapter XII with the meteorological laboratory, the study of the
atmospheric heat-engine and the cycle of physical changes in the general
circulation, chapter XIII with the development of arithmetical manipulation,
chapter XIV with the general circulation and the cyclone, and chapter XV with
meteorological theory.

It is curious to note on reading through J. D. Forbes's report on Meteorology in 1832 that the most modern aspects of the science were not disregarded.
"A few observations with balloons have been published by Lord Minto in
the Edinburgh *Journal of Science*," and further on, "Captain Parry found by
means of a kite that the thermometer indicated no diminution of temperature
at a height of 400 feet. It stood at − 24°." As in many other British contributions to the subject the optical phenomena of the atmosphere have not
received the attention which they deserve. Perhaps we owe some of our
backwardness in that subject to the savage treatment which was meted out
to Thomas Young to whom now we turn for the fundamental principles of
atmospheric optics. Forbes's second report of 1840 is, however, a mine of
information on that subject for the curious inquirer.

THE BAROMETER AS WEATHER-GLASS

While all this preparation for the advent of a special science of meteorology
was in progress we have to inquire what was being done in respect of the
practical application of the subject. Before the invention of the barometer,
as we have seen, practical meteorology was limited to weather-lore which
could not be said to have made any real progress since the time of Theophrastus. The answer to this inquiry is a story of what one can only call
disillusionment if we compare the results with the triumphs of astronomical
science.

When the barometer came into use the changes in the height of the mercury naturally aroused interest and were soon found to be intimately associated with changes in the weather. When Hooke produced the wheel-barometer about 1670 the primary relation between the height of the barometer and the weather was soon set out by the inscription of the words Change at 29·5 inches, with Rain, Much Rain and Stormy at each half-inch on the low side, and Fair, Set Fair and Very Dry on the high side. Where these inscriptions came from we do not know, but their universality upon all barometers supplied for common use irrespective of latitude or height above sea-level is evidence of popular confidence in their meaning. The legends seem to be most suitable for a station close to sea-level and correspond pretty closely with London as regards latitude or barometric range.

W. Ellis in a *Brief historical account of the barometer* writes as follows:

The practice of attaching to barometer-scales words indicating the kind of weather to be expected with different heights of the mercury appears to be one of comparative antiquity. At page 135 of the Amsterdam work, bearing date 1688, special directions are given in regard to the manner of placing them, and nine different distinctions of weather, corresponding to nine different heights of the mercury, are enumerated. They are in descending order, "grande sécheresse, très sec, beau confirmé, beau temps, changeant ou variable, pluie ou vent, grosse pluie & grand vent, orage, grande tempeste." Derham in the *Phil. Trans.* for 1698 mentions weather-plates, but says nothing about any words. Amontons in 1705 speaks of a barometer "monté à la manière d'Angleterre," which carried two little plates of copper marked with the different states of weather that might be expected to occur, as "beau temps, changeant, pluie &c." The illustrations of Fitzgerald's wheel-barometers 1761 and 1770 show six different distinctions of weather, whilst on the portable Ramsden barometer, previously described, and on other barometers of the same period, there are seven. In an old barometer by Reballio of Rotterdam, date unknown, ten such indications appear....Changeux proposed a new wording for barometer-scales having regard to different winds (Cotte, *Mémoires sur la Météorologie*, 1788) much in the same way that FitzRoy in more recent times endeavoured to introduce one that should accord better with actual fact.

(Q. J. Roy. Meteor. Soc., vol. XII, 1886, p. 142.)

Whatever the origin of the legends may be, the barometer came to be regarded and was indeed called a weather-glass instead of being looked upon merely as an instrument for indicating atmospheric pressure which it really is. The inscriptions are not without a substratum of truth, for a pressure of 28·0 inches is quite fairly associated with storms and 31·0 inches means very dry weather. A man of great practical experience once explained that the only infallible weather-rule was that it did not rain if the barometer stood above 30·3 inches, and indeed examples to the contrary are extremely rare in Britain; but in the finer distinctions the inscriptions are inevitably of little real help. Strictly interpreted they would imply that the weather is the same at all points along an isobar and the examination of any synoptic chart shows that to be a very imperfect statement.

When the notion of direct relation between barometric height and weather was found to be inapplicable other rules for the interpretations of the readings of the barometer were sought, the variations from time to time came to be regarded as of greater significance than the actual height at any one time. A vast amount of attention was paid to this aspect of the subject before the introduction of the weather-map, and when the study of weather was considered to be of sufficient practical importance for seamen in 1853 to justify a Maritime Conference in Brussels on the initiative of Lieutenant M. F. Maury, and the establishment of a Meteorological Department of the Board of Trade in the following year, Admiral FitzRoy was chosen to be the head of the new department as a man of science and an experienced sailor. He at once turned his attention to the construction of a mercury barometer suitable for use at sea and a code of instructions in the interpretation of the readings.

EXPLANATORY OF WEATHER-GLASSES IN NORTH LATITUDE

In other Latitudes substitute the word South, or Southerly or Southward, for North, etc.

THE BAROMETER RISES	THE BAROMETER FALLS
for Northerly wind, (including from North-west, by the *North*, to the Eastward,) for dry, or less wet weather,—for less wind, —or for more than one of these changes:—	for Southerly wind, (including from South-east, by the *South*, to the Westward,) for wet weather,—for stronger winds,—or for more than one of these changes:—
EXCEPT on a few occasions when *rain* (or snow) comes from the Northward with *strong* wind.	EXCEPT on a few occasions when *moderate* wind with *rain* (or snow) comes from the Northward.
For change of wind towards *any* of the above directions:—	For change of wind towards the *upper* of the above directions:—
A THERMOMETER FALLS	A THERMOMETER RISES

Moisture, or dampness, in the air (shown by a hygrometer) increases BEFORE or with rain, fog or dew.

On barometer scales the following contractions may be useful in *North* latitude:—

And the following Summary may be useful *generally* in *any* latitude:—

RISE for N. Ely. NW—N—E DRY or LESS WIND	FALL for S. Wly. SE—S—W WET or MORE WIND	RISE for COLD DRY or LESS WIND	FALL for WARM WET or MORE WIND
Except wet from N. Ed.	Except wet from N. Ed.	Except wet from cooler side	Except wet from cooler side

FITZROY'S INSTRUCTIONS FOR THE USE OF THE BAROMETER TO FORETELL WEATHER

3. The barometer shows whether the air is getting lighter or heavier, or is remaining in the same state. The quicksilver falls as the air becomes lighter, rises as it becomes heavier, and remains at rest in the glass-tube while the air is unchanged in weight. Air presses on everything within about ten miles of the air's surface, like a *much* lighter ocean, at the bottom of which we live—not feeling its weight because our bodies are full of air, but feeling its currents, the winds. Towards any place from which the air has been drawn by suction, air presses with a force or weight of nearly fifteen pounds on a square inch of surface. Such a pressure holds the limpet to the rock, when, by contracting itself, the fish has made a place without air under

its shell. Another familiar instance is that of the fly which walks on the ceiling with feet that stick.

5. The *words* on the earlier scales of barometers should not be so much regarded for weather indications as the rising or falling of the mercury, for if it stand at *Changeable* 29·5 in. [1000 mb] and then rise towards *Fair* 30 in. [1016 mb], it presages a change of wind or weather, though not so great as if the mercury had risen higher; and, on the contrary, if the mercury stand above *Fair*, and then fall, it *presages a change*, though not to so great a degree as if it had stood lower, besides which, neither the direction nor force of wind, nor *elevation above the sea-level*, are in any way noticed on such scales. It is not from the point at which the mercury may stand that we are alone to form a judgment of the state of the weather, but from its *rising* or *falling*, and from the movements of immediately preceding *days as well as hours*, keeping in mind effects of change of *direction* and dryness or moisture, as well as alteration of force or strength of wind.

6. The barometer is said to be *falling* when the mercury in the tube is sinking, at which time its upper surface, if large, is *sometimes* concave or hollow; or when the hand moves to the left. The barometer is rising when the mercurial column is lengthening, its upper surface being convex or rounded; or when the hand moves to the right.

7. In temperate climates, towards the higher latitudes, the quicksilver ranges, or rises or falls, nearly three inches—namely, between about 30 inches and nine-tenths [1046 mb] and less than *twenty-eight* inches [948 mb] on *extraordinary* occasions; but the usual range is from about thirty inches and a half [1033 mb] to about twenty-nine inches [982 mb]. Near the line, or in equatorial places, the range is but a few tenths, except in storms, when it *sometimes* falls to twenty-seven inches [914 mb].

A fall of half a tenth [1½ mb] or, still more, of a whole tenth [3½ mb] in an hour is a sure warning of storm.

8. The sliding-scale (vernier) divides the tenths into ten parts each, or hundredths of an inch. The number of divisions on the vernier exceeds *or* is less than that in an *equal space* of the *fixed* scale, by one.

15. It should always be remembered that the state of the air *foretells coming* weather rather than shows the weather that is *present* (an invaluable fact too often overlooked), that the longer the time between the signs and the change foretold by them, the longer such altered weather will last; and on the contrary, the less the time between a warning and a change, the shorter will be the continuance of such foretold weather.

16. To know the state of the air, not only barometer and thermometers should be watched, but the appearance of the sky should be vigilantly noticed.

17. If the barometer has been about its ordinary height, say near thirty inches, at the sea-level [1016 mb], and is steady or rising, while the thermometer falls, and dampness becomes less, North-westerly, Northerly or North-easterly wind, or less wind, less rain or snow, may be expected.

18. On the contrary—if a fall takes place, with a rising thermometer and increased dampness, wind and rain may be expected from the South-eastward, Southward, or South-westward.

19. In winter, a fall, with low thermometer, foretells snow.

20. Exceptions to these rules occur when a Northerly wind, with wet (rain, hail or snow), is impending, before which the barometer often *rises* (on account of the *direction* of the coming wind alone) and deceives persons who from that sign only (the rising) expect fair weather.

21. When the barometer is rather below its ordinary height down to near twenty-nine inches and a half, say (at the sea-level) [1000 mb], a rise foretells less wind, or a change in its *direction* towards the Northward, or less wet; but when it has been very low, about twenty-nine inches [982·0 mb], the first *rising* usually precedes, or indicates, *strong* wind—at times heavy squalls—from the North-westward, Northward or North-eastward, *after* which violence a gradually rising glass foretells improving weather—if the thermometer falls. But if the warmth continue, probably the wind

will back (shift against the sun's course), and more Southerly or South-westerly wind will follow, especially if the barometer rise has been sudden.

22. The most dangerous shifts of wind, or the *heaviest* Northerly gales, happen *soon* after the barometer *first* rises from a very low point, or if the wind veers *gradually*, at some time afterwards, though with a *rising* glass.

23. Indications of approaching change of weather and the direction and force of winds are shown less by the *height* of the barometer than by its falling or rising. Nevertheless a height of more than 30·0 inches [1016 mb] (at the level of the sea) is indicative of fine weather and *moderate* winds, *except* from East to North *occasionally*, whence it *may* blow strongly.

24. A rapid rise of the barometer indicates unsettled weather, a slow movement to the contrary; as does likewise a *steady* barometer, which, when continued, and with dryness, foretells very fine weather, lasting for some time.

25. A rapid and considerable fall is a sign of stormy weather, with rain (or snow). Alternate rising and sinking, or oscillation, indicates unsettled and threatening weather.

26. The greatest depressions of the barometer are with gales from SE, S or SW; the greatest elevations with wind from NW, N or NE, or with calm.

27. Though the barometer generally falls with a Southerly, and rises with a Northerly wind, the contrary *sometimes* occurs; in which cases the Southerly wind is usually dry with fine weather, or the Northerly wind is violent and accompanied by rain, snow or hail, perhaps with lightning.

28. When the barometer sinks considerably, much wind, rain (perhaps with hail), or snow, will follow, with or without lightning. The wind will be from the Northward, if the thermometer is low (for the season), from the Southward if the thermometer is high. Occasionally a low glass is followed or attended by lightning only; while a storm is beyond the horizon.

29. A sudden fall of the barometer, with a Westerly wind, is sometimes followed by a violent storm from NW or North or NE.

30. If a gale sets in from the E or SE and the wind veers by the South, the barometer will continue falling until the wind is near a marked change, when a lull *may* occur; after which the gale will soon be renewed, perhaps suddenly and violently, and the veering of the wind towards the NW, North or NE, will be indicated by a rising of the barometer with a fall of the thermometer.

39. Another general observation requires attention, which is that the wind usually *appears* to veer, shift, or go round *with the sun* (right-handed or from left to right) and that when it does not do so, or backs, *more* wind or bad weather may be expected instead of improvement, *after a short interval*.

40. It is not by any means intended to discourage attention to what is usually called "weather wisdom." On the contrary, every prudent person will combine observation of the elements with such indications as he may obtain from instruments; and will find that the more accurately the two sources of foreknowledge are compared and combined, the more satisfactory their results will prove.

41. A barometer begins to rise considerably before the conclusion of a gale, sometimes even at its commencement. Although it falls lowest before high winds, it frequently sinks very much before heavy rain. The barometer falls, but not always, on the approach of thunder and lightning. Before and during the *earlier* part of settled weather it usually stands high and is stationary, the air being dry.

42. Instances of fine weather with a *low glass* occur, however, rarely, but they are always preludes to a *duration* of wind or rain, *if not both*.

43. After very warm and calm weather a squall, or storm with rain, may follow; likewise at any time when the atmosphere is *heated* much above the usual temperature of the season, and when there is, or recently has been, much electric (or magnetic) disturbance in the atmosphere.

44. Allowance should *invariably* be made for the previous state of the glass during *some days, as well as some hours*, because their indications *may* be affected by distant

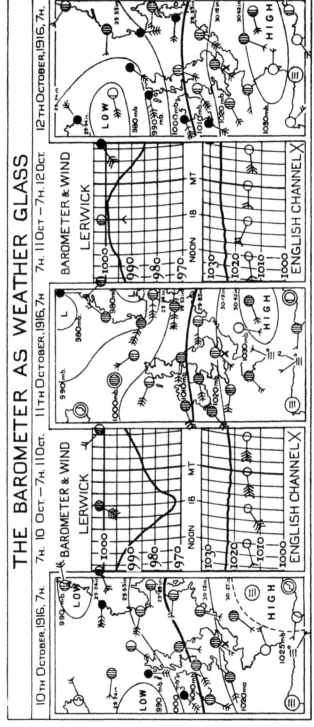

Fig. 11. Charts showing changes from light wind to gale during a period of continuous high pressure at a station marked × in the English Channel in consequence of the passage of the centres of depressions beyond the North of Scotland.

causes, or by changes close at hand. Some of these changes may occur at a greater or less distance, influencing neighbouring regions, but not visible to each observer whose barometer feels their effect.

45. There may be heavy rains, or violent winds, beyond the horizon and the view of an observer, by which his instruments may be affected considerably, though no particular change of weather occurs in his immediate locality.

46. It may be repeated that the longer a change of wind or weather be foretold before it takes place, the longer the presaged weather will last; and, conversely, the shorter the warning, the less time whatever causes the warning, whether wind or a fall of rain or snow, will continue.

47. Sometimes severe weather from the Southward, *not lasting long*, may cause no great fall, because followed by a *duration* of wind from the Northward; and at times the barometer may fall with Northerly winds and fine weather, apparently against these rules, because a *continuance* of Southerly wind is about to follow. By such changes as these one may be misled, and calamity may be the consequence if not duly forewarned.

Some of Admiral FitzRoy's suggestions recall the rhymes which are familiar to sailors.

Thus for numbers 15 and 46:

> Long foretold, long last;
> Short notice, soon past.

And for numbers 21 and 39:

> When the wind backs against the sun
> Trust it not, for back it will run.

And for number 22:

> First rise after low
> Foretells a stronger blow.

To these we may add a quatrain which is not apparently referred to, viz.:

> When the rain comes before the wind
> Your topsail halyards you must mind;
> But when the wind's before the rain
> Hoist your topsails up again.

This maxim finds expression also in many forms and may be associated with the common saying that "it will not rain till the wind drops."

The situation was so altered by the weather-map that those familiar with the newer method will hardly trouble to verify the wording of the rules supposing that in so far as they represent the teaching of the weather-map they are cumbrous and in so far as they do not they are inaccurate.

An obvious drawback to the effective use of the barometer as a weather-glass becomes apparent so soon as we have synoptic charts for reference. A barometer in one place shows only the resultant effect of changes taking place on all sides, and the same barometric height may be the index of an endless variety of barometric situations. To define those situations more accurately is not of course the same thing as explaining them but it is an indispensable step in that direction. We can illustrate the inadequacy of a single barometer as a guide to weather by fig. 11, an example of a series of maps in which the pressure remained practically constant for a week while the changes in the wind and weather were very numerous and notable.

THE THRESHOLD OF MODERN METEOROLOGY

As in this chapter we have brought the reader to the weather-map as the threshold of modern meteorology and in doing so have called attention to certain aspects of what may be called the philosophy of the subject, it will be convenient for us to complete our excursus by pointing out how the introduction of the weather-map led to a curious alienation of the experimental and theoretical physicists from the study of weather.

In the earlier part of this chapter we have given some idea on the one hand of the introduction of new instruments and experiments in the laboratory and thereby the development of physical laws and principles, and on the other hand of the progress of observations, particularly with the barometer, thermometer and rain-gauge, and the importance of a meteorological library with its custodian for the compilation and co-ordination of observations which lead to the identification of the general circulation of the atmosphere, the primary foundation of the real science of meteorology.

We have noted that the exponents of physics used the knowledge which they had acquired in the laboratory to illustrate such natural atmospheric processes as came within their cognisance. With that practice grew the impression that the avenue of progress for meteorology was along the line of new instruments and new experiments, that the compilation of observations except in response to questions suggested by experiment or theory was marking time, if not wasting it. Hence have come endless suggestions for improving the accuracy of observations on the ground that explanation would become easier as greater precision was secured. Pushed to an extreme the attitude of the instrument-designer means that if we are unable to comprehend the variations of the barometer ranging over an inch measured to the hundredth of an inch we shall see our way clearly if we measure the pressure to a ten-thousandth of an inch. So long as meteorological observations were those of a single observer there might be some encouragement to think that greater precision in the specification of physical processes was a possible avenue to solution; but when the weather-map came in and the daily features were presented by a large area of land and sea and not by a single point within the observation of the experimenter, the problem seemed to be withdrawn from the daily avocation of the laboratory altogether.

Various efforts have been made from time to time to explain the variations of the barometric pressure on a physical basis by an appeal to the laws of gases which explain that, other things being equal, warm air is specifically lighter than cold air, or to Dalton's law that moist air is lighter than dry air. These attempted explanations are good examples of the habit of referring the atmosphere to the conditions which obtain in a laboratory where the specimens under investigation are enclosed and at rest, whereas the pressure at any point of the free air is related to the motion of the air as well as to its extension through the whole range of height. Specimens of air of all states with regard to moisture and temperature could be found to range over the whole gamut of the barometer. The determination is a dynamical as well as a thermal question. The warmest or the most moist specimen of air can have any pressure whatever if its environment is so arranged as to produce the necessary compression. Before therefore we embark upon the consideration of the cause of high pressure or low pressure we must be able to form an opinion as to the sequence of changes in the general circulation.

On the other hand the compilers of observations and maps were profoundly conscious that the experiments and theories of the physical laboratories offered no real explanation even of the broadest features of the distribution of pressure and temperature, and that to plunge into the study of minute details, mathematical and physical, when the outlines were an unsolved riddle was equally a waste of time and energy. So there came about a sharp division, physicists on the one side, regarding the efforts of the observers and map-makers as quite unscientific and sometimes suggesting that competent mathematicians should be invited to take the matter up; and meteorologists on the other side, equally firmly convinced that to invite the mathematicians to solve a problem which they could not specify was the same sort of mistake as inviting Newton to solve the problem of the solar system without the previous assistance of Kepler's laws. So sharp has that division become that *La Science Française*, a work already quoted, prepared in 1915 by the French Ministry of Public Instruction and Fine Arts for the Exhibition at San Francisco, with an introduction by Lucien Poincaré, and with separate sections for no less than thirty-three sciences, has nothing to say about meteorology, and in a comprehensive series of bibliographies includes only one meteorological entry "Études sur le climat de la France" under the science of geography.

The present chapter will, it is hoped, be some aid to making the real situation more apparent. For the effective study of meteorology there must be co-ordination between the physical side and the geographical side. Everybody is agreed that the laws applying to atmospheric processes are to be learned by a study of physics in the laboratory; at the same time the conditions under which those processes take place and produce our weather are not the conditions under which experiments are made in the laboratory. The problem cannot be presented for solution with any real hope of success until observations have disclosed to us sufficient evidence of the true structure of the general circulation of the atmosphere and its variations to enable us to make an effective mental picture of the atmospheric problem, just as Kepler's co-ordination made it possible to form a working mental picture of the solar system. The indispensable foundations of knowledge of the atmospheric system are the distribution of land and sea, the distribution of pressure and temperature and water-vapour; any hypothetical arrangement which ignores one or other of these fundamental conditions may be a most valuable exercise but it is not, strictly speaking, meteorology.

CHAPTER IX

METEOROLOGY AS AN INTERNATIONAL SCIENCE: THE METEOROLOGICAL LIBRARY

Some Chiefs of Meteorological Services

Algeria. Service Central Météorologique. [1873] F. Gonnessiat, A. Lasserre.

Argentina. Oficina Meteorológica Central. B. A. Gould 1872–84, W. G. Davis 1885–1915, J. O. Wiggin 1915–.

Australia. Separate observatories were founded at Parramatta 1821, Adelaide (Sir C. Todd 1872), Melbourne (Prof. Neumayer 1858, R. L. J. Ellery, P. Baracchi), Sydney 1859 (H. C. Russell), Hobart 1883 (H. C. Kingsmill), Perth 1896 (W. E. Cooke). In 1887 C. L. Wragge was appointed Government Meteorologist to organise a meteorological bureau for Queensland. The Commonwealth Meteorological Bureau was founded in 1906 and H. A. Hunt was appointed Commonwealth Meteorologist in the following year.

Austria. Zentralanstalt für Meteorologie und Geodynamik, Wien. Karl Kreil 1851–63, Karl Jelinek 1863–76, Julius Hann 1877–97, J. M. Pernter 1897–1908, W. Trabert 1908–15, J. P. Pircher 1916, F. M. Exner 1917–.

Azores. Meteorological Service. F. A. Chaves 1893–.

Belgium. Observatoire royal. A. Quetelet 1826–74, E. Quetelet (acting) 1874–76, J. C. Houzeau 1876–85, F. Folie 1885–98, Ch. Hooreman 1898–99, A. Lancaster 1899–1908, J. Vincent 1908–20, [Institut Royal Météorologique 1913] J. Jaumotte 1920–.

Brazil. Central Meteorological Office. A. P. Pinheiro 1885–96, A. Silvado 1896–1909, H. Morize 1909–21, Sampaio Ferraz 1921–.

Canada. Meteorological Office, Toronto. G. T. Kingston 1871–80, V. Carpmael 1880–94, F. Stupart 1894–.

Chile. [1911] W. Knoche, C. Henriquez.

China. Zi-ka-wei. M. Dechevrens, S.J., L. Froc, S.J., J. de Moidrey, S.J.

Colombia. S. Sarasola, S.J.

Czecho-Slovakia. R. Schneider.

Denmark. Det Danske Meteorologiske Institut, Copenhagen. N. Hoffmeyer 1872–84, A. Paulsen 1884–1907, C. Ryder 1907–23, D. la Cour 1923–.

Ecuador. L. G. Tufino.

Egypt. H. G. Lyons, B. H. H. Wade, B. F. E. Keeling, H. E. Hurst, H. Knox Shaw.

Finland. Institut Météorologique Central, Helsingfors. E. Biese –1906, G. Melander 1907–.

France. Observatoire: Le Verrier 1855–77. Bureau Central Météorologique: E. Mascart 1878–1907, C. A. Angot 1907–20. Office National Météorologique: E. Delcambre 1920–. Strasbourg: E. Rothé 1919–.

Germany. Deutsche Seewarte, Hamburg: G. von Neumayer 1876–1903, D. Herz 1903–11, D. K. Behm 1911–19, H. Capelle 1919–. Meteorologisches Institut, Berlin: W. Mahlmann 1847–48, H. W. Dove 1848–78, T. A. Arndt and G. Hellmann (ad interim) 1878–85, W. von Bezold 1885–1907, G. Hellmann 1907–22, Hugo Hergesell, H. von Ficker 1923–. München: W. von Bezold 1878–85, C. Lang 1885–93, F. Erk 1893–1909, A. Schmauss 1909–. (Other establishments at Aachen, Bremen, Darmstadt, Dresden, Frankfurt, Karlsruhe, Stuttgart, Strassburg: H. Hergesell 1890–1914, K. Wegener and O. Stoll acting directors 1914–18.)

Great Britain. Meteorological Dept. of the Board of Trade: R. FitzRoy (Superintendent) 1854–65, Babington 1865–67. Meteorological Office: R. H. Scott (Director) 1867–76. Meteorological Council 1876–1905. Meteorological Committee: W. N. Shaw 1905–19. Meteorological Office of the Air Ministry: Sir N. Shaw 1919–20 G. C. Simpson 1920–.

Greece. Observatoire National Météorologique: 2nd order station 1847. Observatory: D. Eginitis 1890–.

Holland. Koninklijk Nederlandsch Meteorologisch Instituut, de Bilt. C. H. D. Buys Ballot 1854–89, M. Snellen 1889–1902, C. H. Wind 1902–05, E. van Everdingen 1905–.

Hong Kong. The Observatory. W. Doberck 1883–1907, F. G. Figg 1907–12, T. F. Claxton 1912–.

Hungary. [1870] G. Schenzl, N. von Konkoly, S. Róna.

Iceland. 1920 Th. Thorkelsson.

India. Imperial Meteorological Reporter: H. F. Blanford 1875–89, J. Eliot 1889–1903. Director General of Indian Observatories: G. T. Walker 1904–24, J. H. Field 1924–.

Italy. R. Ufficio Centrale di Meteorologia e Geodinamica, Roma. P. Tacchini 1863–1900, L. Palazzo 1900–.

Jamaica. Weather Service, Kingston. Maxwell Hall 1880–1920, J. F. Brennan 1920–.

Japan. Central Meteorological Observatory, Tokio. K. Kobayashi 1891–99, K. Nakamura c. 1899–1922, T. Okada 1922–.

Java. Nederlandsch Meteorologisch en Magnetisch Observatorium. P. A. Bergsma 1875–82, J. P. van der Stok 1882–1902, S. Figee 1902–05, W. van Bemmelen (acting 1905–11) 1911–21, C. Braak 1921–.

Manila c. 1870. J. Algué, S.J. –1925 (Philippine Weather Bureau 1901).

Mauritius. Royal Alfred Observatory. C. Meldrum 1862–96, T. F. Claxton 1896–1911, A. Walter 1911–26.

Mexico. Central Meteorological Observatory. Mariano Bárcena 1877–99, Pedro C. Sanchez.

New Zealand. Dominion Meteorological Office. D. C. Bates.

Norway. Det Norske Meteorologiske Institut, Oslo. H. Mohn 1866–1913, A. S. Steen 1913–15, Th. Hesselberg 1916–.

Poland. L. Gorczynski.

H. W. Dove   C. H. D. Buys Ballot   M. F. Maury

U. J. J Le Verrier   Edward Sabine   R. FitzRoy

Francis Galton   J. M. Pernter   Richard Strachey

Alexander Buchan   G. J. Symons   H. F. Blanford

INTERNATIONAL METEOROLOGICAL CONGRESS, ROME, 1879

Zenger   Respighi Mendeleef Snellen   Weihrauch
                de Brito Capello Kokides Pittei   Hellmann   Müller
Cecchi   Grassi   Schenzl Weyprecht
Pisati   Tacchini   Auwers   v. Bezold   Mohn   Rubenson
de Rossi   Pujazon   Paugger   Hann   R. H. Scott Bruhns
Lorenz von Liburnau   Henry Smith Denza   Hoffmeyer Mascart
Blaserna   Neumayer   Aguilar   Cantoni Plantamour   Wild
Houzeau

INTERNATIONAL METEOROLOGICAL CONFERENCE, PARIS, 1896

Snellen   Riggenbach   Kesslitz   Biese   Moureaux   Rotch   Fines   Jaubert   Thévenet   Lancaster   Chauveau   Page
                 Mathias   de Fonvielle   Anguiano   Mohn   Paulsen   Angot   Rykatcheff   Hepites   Billwiller   Erk
van Rijckevorsel   Ellis   Hergesell   A. Schmidt   Watzoff   Mascart   von Bezold   Rücker
          Teisserenc de Bort   Hildebrandsson   Scott   Tacchini

James Glaisher

Jean Baptiste Biot

Luke Howard

John Welsh

Samuel Pierpont Langley

Ralph Abercromby

William Henry Dines

Richard Assmann

Portugal. Observatorio do Infante D. Luiz. Dr Pegado 1854–, Fradesso da Silveira 1856–75, J. C. de Brito Capello 1875–1901, A. A. de Pina Vidal 1901–09, C. A. M. de Almeida 1909 (ad interim), J. M. d' Almeida Lima 1910–.

Roumania. Institutul Meteorologic al Romaniei. S. C. Hepites 1884–1908, N. Coculescu 1908–18, E. Otetelisanu 1918–.

Russia. Central Physical Observatory, Leningrad. A. T. Kupffer 1849–65, L. F. Kämtz 1866–68, H. Wild 1868–95, M. Rykatcheff 1896–1913, B. Galizin 1913–15, B. Weinberg, A. Friedmann 1925.

South Africa. Secretaries of the Meteorological Commission. W. L. Blore 1862, C. B. Blore 1870, A. V. Solomon 1874, W. Greathead 1876, J. W. Bailey 1878, W. E. Fry 1879, D. J. May 1890, R. Pillans 1893, C. Stewart 1897.

Spain. Madrid Observatory, 1850. A. Aguilar 1865–82, M. Merino 1882–, A. Arcimis 1890–1910, J. Galbis 1910–21, J. Cruz Condé 1921–25, E. Meseguer 1925–.

Sweden. Statens Meteorologisk-Hydrografiska Anstalt, Stockholm. E. Edlund 1859–75, R. Rubenson 1875–1902, H. E. Hamberg 1902–13, N. Ekholm 1913–18, A. Wallén 1919–.

Switzerland. Schweizerische Meteorologische Centralanstalt, Zürich. R. Billwiller 1880–1904, J. Maurer 1905–.

United States. Weather Bureau, Washington. Chiefs of the Signal Service: A. J. Myer 1860–80, R. C. Drum (acting) 1880, W. B. Hazen 1880–87, A. W. Greely (acting 1886) 1887–91. Chiefs of the Weather Bureau: M. W. Harrington 1891–95, W. L. Moore 1895–1913, C. F. Marvin 1913–.

Presidents of International Congresses and Conferences

Leipzig 1872, C. Bruhns, C. H. D. Buys Ballot, R. H. Scott, H. Wild; Vienna Congress 1873, the same with K. Jelinek; Rome Congress 1879, G. Cantoni; Munich 1891, C. Lang; Paris 1896, E. Mascart; Innsbruck 1905, J. Hann (Président d'honneur), J. M. Pernter; Paris 1919, Utrecht 1923, Sir Napier Shaw.

Presidents of the International Committee

C. H. D. Buys Ballot	Utrecht 1874, London 1876, Utrecht 1878
H. Wild	Berne 1880, Copenhagen 1882, Paris 1885, Zurich 1888, Upsala 1894
E. Mascart	Upsala 1894, St Petersburg 1899, Southport 1903, Paris 1907
W. N. Shaw	Paris 1907, Berlin 1910, Rome 1913, London 1919, Paris 1919, London 1921
E. van Everdingen	Utrecht 1923

THE DISTINCTION BETWEEN THE SCIENCES OF METEOROLOGY AND PHYSICS

IF the subject of this work had been atmospheric physics the equipment of observatories and stations would have claimed the most important section of the historical introduction. The history of physics is controlled by its instruments; new means of measurement are the essential conditions of progress. They in turn are dependent upon the facilities of the workshop which makes it possible to construct improved instruments. That has always been the case. Within the last forty years conditions have developed along those lines to such an extent that modern institutions for physical research are built round power-units, boilers or internal combustion engines. These are necessary not only to supply workshops in which modern instruments of precision are constructed, but also to drive the machinery of electric furnaces or cryogenic apparatus or to maintain vacua or pressures of an order unknown to such physicists as those whose names are recorded in the previous chapter, many of whom worked with apparatus constructed with hand-tools in their own homes.

The first stages of modern physical research deal, as a rule, with the design and construction of new auxiliary instruments; but with meteorology it is essentially different; instruments and measurements are necessary to supply the observations from which the laws and axioms of the science are to be evolved; but they are not in themselves sufficient. An equally essential part of the science and the most difficult from the scientific point of view is the co-ordination of measurements, at different times, in different places and of different kinds. We may have innumerable measurements of pressure, of temperature, of humidity, of wind, of clouds, of solar radiation and of every other element of atmospheric physics, but the co-ordination of the observations in such a manner as to give an insight into the structure of the atmosphere, its changes and their causes, is a separate step requiring other faculties than the power to design or use instruments or to compile observations.

This aspect of the science of meteorology has often been overlooked and a good deal of physical ingenuity has been spent upon the defects and inaccuracies of instruments when the accuracy already achieved was far beyond the capacity of the students of the atmosphere to co-ordinate existing observations in such a way as to account for the features which were not at all in doubt.

Limits of accuracy of meteorological measurements

The writer's earliest effort in meteorology was an inquiry on behalf of the Meteorological Council into the measurement of humidity[1], which in 1878 was regarded as being in a very unsatisfactory state. The starting-point was that expressed in the work of Regnault in the *Annales de Chimie et de Physique*, 1845. On account of the excellence of his apparatus Regnault may be regarded as a pioneer among the workshop-school of physicists; he condemned the hair-hygrometer as a mere hygroscope because it had not a fixed zero. He developed the form of dew-point hygrometer, which bears his name, afterwards modified by Alluard, as an improvement on Daniell's, and devoted a great deal of labour to developing the wet-bulb hygrometer as an instrument of precision. Students of physics taking up meteorology naturally accepted the position, and continued in the attitude towards meteorology which is expressed by Regnault's criticism. But as a matter of fact there is little uncertainty about the measurements of humidity at ordinary temperatures well above the freezing-point of water. For temperatures near or below that point the tables for the reduction of observations of the wet-bulb have not even yet attained the finality of general acceptance. In fact, paradoxical as it may sound, water on a wet-bulb is much less well-defined as a physical entity than a well-prepared hair, or a bundle of them, which makes no trouble about the freezing-point[2]. Meteorological measurements are to a considerable extent interdependent. The primary dynamical measurement, that of the motion of air, is uncertain for reasons which are connected with the structure of the atmosphere and not with the instruments used to measure it. In that as in all other branches of meteorology accuracy is a relative term and the accuracy necessary on any particular occasion depends upon the other elements with which the measurements have to be compared.

[1] *Report on Hygrometric Methods*: First Part, including the Saturation Method and the Chemical Method, and Dew-Point Instruments, by W. N. Shaw. *Quarterly Weather Report of the Meteorological Office for* 1879, Appendix III, reprinted from *Phil. Trans.*, vol. CLXXIX.

[2] *Computer's Handbook*, Section I. Computations based on the physical properties of atmospheric air: Humidity and density, M.O. publication, No. 223, 1916. *A Discussion on Hygrometry*. Report of the Physical Society, London [1922].

The purposes of meteorological observation

Let us therefore in considering the development of modern methods of observation keep in mind the purposes which the observations are intended to serve. Briefly stated those are, first the compilation of observations to be co-ordinated into a faithful representation of the general circulation of the atmosphere and its changes, secondly the identification of the dynamical and physical processes by which the changes are brought about, and thirdly the exposition of the sequence of changes as the result of the physical processes involved therein.

For the first we require primarily a knowledge of the winds defining the motion of air; the pressure at various levels we have seen to be related very closely to the motion of air by Buys Ballot's law and its extension. In mentioning that relation we may pause to point out that the change of pressure along the horizontal which is necessary to balance a wind of the velocity which comes within ordinary observation is very small. Hence we have the curious situation that a measurement of such vagueness as that of wind is associated with the highest precision of meteorological measurement, that of pressure.

After wind and pressure come the elements temperature and humidity—upon which particularly the condensation of water depends—clouds and rainfall as evidences of the condensation which has been accomplished, with the correlative measure of evaporation as nearly as possible in the circumstances operative in nature. To these are joined many non-instrumental observations which have been already mentioned in chapter II.

For the second meteorological purpose, the instruments which are used to identify the nature and extent of the physical processes, in addition to the instruments required for the ordinary routine of a station, include a number for the study of atmospheric electricity, terrestrial magnetism and atmospheric optics which are only indirectly of importance in the study of the general circulation.

And for the third, the physical explanation of the sequence of events, we approach the subject from a different point of view; there we have to regard especially the transformations of energy which take place in the atmosphere and naturally we have to deal primarily with radiation and its influences, both the receipts from the sun and the loss from the earth, and the complicated interchange which takes place between air, earth and sea.

THE EARLY APPRECIATION OF THE METEOROLOGICAL PROBLEM

The barometer was invented by Torricelli in 1643: it is astonishing how soon after that event the study of the atmosphere was formulated as a physical problem to be approached by the co-ordination of observations beginning with those of wind. As a matter of interesting curiosity we reproduce a picture of meteorological instruments (fig. 12) and a specimen table of monthly records (fig. 13). The picture in question accompanies the instructions

A SCHEME

At one View repreſenting to the Eye the Obſervations of the Weather for a Month.

Dayes of the Month and place of the Sun. Remarkable houſe.		Age and ſign of the Moon at Noon.	The Quarters of the Wind and its ſtrength.	The Degrees of Heat and Cold.	The Degrees of Drineſs and Moyſture.	The Degrees of Preſſure.	The Faces or viſible apperances of the Sky.	The Notableſt Effects.	General Deductions to be made after the ſide is fitted with Obſervations As,
14 ♊ 12.46	4 8 12 4 8 12	27 ♉ 9. 46. Perigeu.	W. 2. 3. $3\frac{1}{3}$ W.S W. 1	9 $\frac{3}{4}$ 12 $\frac{1}{2}$ 16 10 $\frac{1}{8}$ 7 $\frac{1}{2}$	2 5 2 8 2 9	29 $\frac{1}{10}$ 29 $\frac{1}{8}$ 29 $\frac{3}{8}$	Clear blew but yellowiſh in the N. E. Clowded toward the S. Checker'd blew.	A great dew. Thunder, far to the South. A very great Tide.	From the laſt Q. of the *Moon* to the Change the Weather was very temperate, but cold for the ſeaſon; the Wind pretty conſtant between N. & W. A little before the laſt great Wind, and till the Wind roſe at its higheſt, the Quick-ſilver continu'd deſcending til it came very low; after wch it began to re aſcend, &c.
15 ♊ 13.40	8 4 6 10	28 ♉ 24.51.	N. W. 3 4 N. 2 1	9 8 $\frac{1}{2}$ 7	28 $\frac{1}{2}$ 2 9 2 10	29 $\frac{1}{16}$ 29	A clear Sky all day, but a little Checker'd at 4. P. M. at Sunſet red and hazy.	Not by much ſo big a Tide as yeſterday. Thunder in the North.	
16 11 14 37	10	N. Moon. at 7. 25 A. M. ♊ 10. 8.	S. 1	10	1 10	28 $\frac{1}{2}$	Overcaſt and very lowring.	No dew upon the ground, but very much upon Marble-ſtones, &c.	
		&c.	&c.	&c.	&c.	&c.	&c.		

Z 2 DI-

Fig. 13. Hooke's scheme of observations. (From Sprat's *History of the Royal Society*, 2nd edition.)

selected over the whole globe, including equatorial, tropical, temperate and polar regions. A selection of stations contributing observations by telegraph for this purpose was advocated by Léon Teisserenc de Bort, who proposed the appointment of a special commission of the International Meteorological Committee for that purpose in 1907. The immediate execution of the proposal was faced with prohibitive expense and the proposal was extended by Professor H. H. Hildebrandsson[1], Secretary of the Commission, to include statistical data compiled from observations of climatological stations. Owing to the death of Teisserenc de Bort and the retirement of Professor Hildebrandsson little progress was made with the enterprise until it was taken up by the President of the International Committee. Data for representative land-stations, at the rate of two stations for each " 10 degree square," were compiled by the British Meteorological Office for the year 1911, and, after some remarkable misadventures, published in 1915. Since that time the publication has been made to include the year 1910, and the years 1912–18 have also been published. This is an important step in advance, but to complete the *réseau mondial* corresponding data are required for the areas covered by the sea and also for considerable districts in the polar regions. The stations already available are established almost entirely for the purpose of compiling a record of weather for economic purposes; the extension to stations in the polar regions and for areas of the sea which are of no economic interest is a matter of great expense and not without difficulty. The Norwegian Government has been very active in maintaining a station in Spitsbergen and Bear Island and initiating a station at Jan Mayen, and for a time at Mygbugten in north-east Greenland—not without some sacrifice. Four stations are now to report from Greenland but the full scheme of a *réseau mondial* is really a world-enterprise and requires an international bureau.

INTERNATIONAL CO-OPERATION AS A MEANS OF PROMOTING METEOROLOGY AS A WORLD-STUDY

A *réseau mondial* has really been the guiding principle of international co-operation between the meteorological establishments of the world for more than fifty years. The idea upon which the co-operation was based was that the several countries of the world should maintain the stations necessary for their own meteorological or economic purposes and exchange the information so acquired, by telegraph in the case of observations necessary for the construction of the daily charts of weather of the respective countries, and by publication in an agreed form for the data for climatological stations. Tentative co-operation of this kind had been developed first in the Palatinate by H. W. Brandes in the early part of the nineteenth century, then for telegraphic information by Le Verrier of the Paris Observatory who instituted the daily Bulletin International in 1863, and by FitzRoy who compiled daily

[1] *Quelques Recherches sur les Centres d'Action de l'Atmosphère*, III, IV and V, by H. Hildebrand Hildebrandsson, Upsala and Stockholm, 1909, 1910 and 1914. *Kungl. Svenska Vetenskapsakademiens Handlingar*, vols. XLV and LI.

weather-maps in London from 1861. Sir Edward Sabine was also very active in promoting international co-operation. For observations at sea international co-operation was sought by a Conference on Maritime Meteorology at Brussels in 1853; but for land-stations the first general effort was a meeting of meteorologists at Leipzig in 1872. This was followed by a congress at Vienna in 1873, the members of which were the official delegates of a large number of countries. A further congress of the same official character was held at Rome in 1879. These congresses drew up resolutions expressing their recommendations as to the instruments to be used, the messages to be transmitted by telegraph and the form of record for observatories and climatological stations. They defined the orders of the stations[1]:

(a) A *Central Office*, or *Central Institute*, is the chief office entrusted by the Government with the management, collection and publication of the meteorological observations of the country.

(b) A *Central Station* is a subordinate centre for the management and collection of observations from a certain province.

(c) A *Station* of the *First Order* is an observatory in which, without the collection of observations from other stations, meteorological observations are conducted on a great scale, i.e. either by hourly readings or by the use of self-recording instruments.

(d) *Stations* of the *Second Order* are the stations where complete and regular observations on the usual meteorological elements, viz., pressure, temperature and humidity of the air, wind, cloud, rain and hydrometeors, etc., are conducted.

(e) *Stations* of the *Third Order*, finally, are the observing stations, where only a greater or less portion of these elements are observed.

During the interval between the two congresses a "permanent committee" under the presidency of Buys Ballot was placed in charge of the details; and after the congress of Rome an International Meteorological Committee was appointed.

The Committee was entrusted with the duty of promoting the adoption of the recommendations of the congresses and of arranging for them to be supplemented from time to time. With this object in view a conference of the Directors of Institutes and Observatories throughout the world was called at Munich in 1891, to which the International Committee reported. The characteristic of the Conference was that it was unofficial, that is to say those who attended it came without special diplomatic powers; for the occasion they carried with them only such authority as they possessed in virtue of their positions as directors of institutes or observatories, and the function of the conference was to ascertain how far co-operation could be made effective within the limits of the powers which their offices gave them. The conference and the committee had no funds at their disposal; for the publication of their reports they depended upon the offices which were interested; the Bureau Central Météorologique published reports in French, an English version was issued by the British Meteorological Office, and a German version by the Prussian Meteorological Institute. A scheme was on foot to arrange for a Spanish version to be issued when the war

---

[1] *Report of the Permanent Committee of the First International Meteorological Congress at Vienna*, 1874, M.O. publication, Non-official, No. 9, p. 67, London, 1875.

interrupted international co-operation in 1914. The special publications *Tables météorologiques internationales* 1890 and *Atlas international des nuages* 1896 were issued from the publishing house of MM. Gauthier-Villars et Fils, and the cost met by subscriptions and sales.

In the course of time a system of co-operation on these informal but effective lines grew up. The ultimate controlling body was the conference of Directors of Institutes and Observatories which met at intervals of about ten years; the executive body was the International Meteorological Committee, consisting of [seventeen] directors of institutes elected by the conference. It met every three years to prepare for a new conference in due time. A third form of organisation was that of Commissions for the study of special questions—terrestrial magnetism, atmospheric electricity, weather telegraphy, solar radiation, scientific aeronautics (the study of the upper air), *réseau mondial*, polar exploration and others. These commissions were initiated by the Conference or by the Committee to continue until a new conference assembled, and then to be dropped or continued as the conference might decide. No formal regulations were drawn up, but in 1907 when Professor Mascart was about to retire from the office of President of the Committee, he assisted in preparing a statement of the regulations that had been acted upon in practice since 1879. Before the Règlement could be submitted to a conference, the war had broken out, and approval was first obtained at Paris in 1919. The approved text was as follows:

RÈGLEMENT DE L'ORGANISATION MÉTÉOROLOGIQUE INTERNATIONALE, 1919

L'organisation météorologique internationale comprend:

1. Les Conférences des directeurs;
2. Le Comité Météorologique International;
3. Les Commissions.

I. *Conférences des Directeurs*

Les Conférences des directeurs ont pour fonction principale de discuter les questions administratives et les moyens d'exécution, de se mettre d'accord sur les méthodes l'observation et de calcul, de décider les travaux communs qui doivent être entrepris t de créer éventuellement les Commissions nécessaires. Les questions purement héoriques ne peuvent pas être comprises dans le programme des Conférences.

Les Conférences sont convoquées par le Comité international; elles doivent avoir ieu au moins tous les six ans; elles peuvent être convoquées extraordinairement par e Comité quand il se présente des questions urgentes à leur soumettre. La convocation st de droit quand elle est réclamée par le quart des membres.

Le bureau du Comité international invite directement aux Conférences, dans chaque ays, tous les directeurs des réseaux ou des observatoires météorologiques officiels et ndépendants les uns des autres. Le bureau s'entend en outre avec les directeurs des ervices officiels des différents pays pour savoir s'il y a lieu d'inviter les directeurs de ertaines institutions privées ou des représentants des sociétés météorologiques. Le ureau du Comité établit la liste des membres des Conférences et publie cette liste evisée chaque année.

Les décisions sont prises à la majorité des voix des membres présents, sauf dans le as où le vote par pays[1] serait réclamé.

---

[1] On entend par ce mot tout pays qui se gouverne lui-même (exemple: Dominion du anada).

II. *Comité Météorologique International*

La Conférence des directeurs nomme le Comité météorologique international dont les pouvoirs prennent fin à la Conférence ordinaire suivante. Tous les membres du Comité doivent appartenir à des pays différents et être directeurs d'une institution météorologique indépendante.

Le nombre des membres du Comité est fixé par la Conférence des directeurs.

Le Comité a le droit de se compléter en cas de décès ou de démission de quelqu'un de ses membres. Il peut aussi éventuellement s'adjoindre, à titre consultatif, des savants dont les conseils paraîtraient utiles. Il nomme son bureau composé d'un président, d'un vice-président et d'un secrétaire.

Le Comité surveille l'exécution des décisions des Conférences, propose toute mesure utile au développement de la science, à l'uniformité de vues, à l'entretien des bonnes relations entre les services des différents pays et prépare les questions à soumettre aux Conférences. Suivant les besoins, il organise des Commissions chargées d'étudier des questions spéciales.

Le Comité se tient en rapport avec le Conseil international de recherches et avec l'Union géophysique créés à Bruxelles en 1919, de façon à assurer la coordination des travaux de ces institutions et des Commissions émanant du Comité.

Le Comité se réunit tous les trois ans au moins en séance plénière. Le bureau du Comité informe par circulaire une année d'avance les membres du Comité et les présidents des Commissions de la réunion du Comité et leur fait désigner par un vote l'époque exacte et le lieu de la réunion.

Pour l'étude des questions qui ne présentent pas un intérêt mondial, le Comité tient, quand il en est besoin, des réunions partielles où sont convoqués seulement les membres appartenant aux régions directement intéressées. A ces réunions peuvent être invités, à titre consultatif, les directeurs des services météorologiques de ces régions qui ne feraient pas partie du Comité.

III. *Commissions*

Un des objets de l'organisation météorologique internationale est d'entreprendre des travaux communs, sans que l'initiative personnelle en éprouve aucune entrave.

Les Commissions instituées dans ce but par la Conférence ou le Comité doivent comprendre au moins un membre du Comité. Pour les Commissions nouvelles dont la création aura été décidée, le président est nommé par la Conférence ou le Comité.

Une fois constituées, les Commissions ont la faculté de se compléter elles-mêmes; elles organisent leurs travaux à leur gré.

Les présidents de Commissions qui ne seraient pas membres du Comité sont invités à assister aux séances de celui-ci et à prendre part aux discussions avec voix consultative. Au commencement de chaque session du Comité, ils présentent un rapport sur les travaux de leur Commission.

Les Commissions se réunissent au moins tous les trois ans. La désignation du lieu et de la date des réunions est faite après entente préalable entre le président du Comité et celui de la Commission.

Les personnes qui veulent proposer une question à la délibération d'une Conférence, du Comité ou d'une Commission, devront préalablement demander au président que cette question soit mise à l'ordre du jour et distribuer aux membres respectifs, deux mois avant la réunion, un court rapport sur cette question.

The Règlement was subsequently revised at a Conference at Utrecht in 1923 and approved with the following changes:

RECENT CHANGES

[The Règlement was revised at a Conference at Utrecht in 1923, and the membership of the Conference restricted to directors of institutions officially responsible for the care of a réseau or network of stations. In 1929 at a Conference at Copenhagen the organisation was extended to provide for an official Secretariat with contributions for its support from the associated Governments, and with the responsibility for the organisation of the meetings of the Conference and its Commissions and the issue of their reports. The Commissions now report their resolutions to the International Meteorological Committee for approval in order that any action recommended may be authorised.]

An International Meteorological Codex, containing the resolutions adopted by the Conferences and Committees which had met since 1872, together with commentaries and explanatory notes, was prepared by H. H. Hildebrandsson of Upsala and G. Hellmann of Berlin and submitted to the Innsbruck Conference in 1905. The German edition was published by the Prussian Meteorological Institute in 1907, an English translation by the British Meteorological Office in 1909, and an Italian translation by Ciro Chistoni for the Società Meteorologica Italiana, was printed at Turin in 1913. A Spanish version translated from the second edition in German was published by the Observatorio Central de Manila in 1913 and includes the resolutions adopted between 1872 and 1910.

It will be noticed that this organisation provides a stimulus and method for the establishment of stations in all parts of the world, but leaves the compilation and publication of the data to the discretion of the Institutes with which they are connected, and makes no provision for the co-ordination of the observations into an organic representation of the meteorology of the globe; that duty has been left to private enterprise, and the science has to recognise especially the services of four distinguished meteorologists: Captain Hoffmeyer of the Danish Meteorological Institute, who commenced a series of daily charts of the Atlantic dating from September 1873 which was subsequently continued by the Danish Office and the Deutsche Seewarte jointly until the publication was stopped by the war; Albert J. Myer, Chief Signal Officer of the Signal Service of the United States, who issued daily charts of the weather of the Northern hemisphere beginning in the year 1877; Alexander Buchan, for many years Secretary of the Scottish Meteorological Society; and Julius von Hann, Director of the Austrian Zentralanstalt für Meteorologie und Geodynamik from 1877 to 1897, Professor of Meteorology in the University of Vienna and editor of the *Meteorologische Zeitschrift*.

The contributions of Buchan to which we refer here are his representation of the pressure, winds and temperature of the globe in the volumes on meteorology in the report of the *Challenger* expedition[1] and the Atlas which he compiled in collaboration with A. J. Herbertson for Bartholomew's Geographical Institute[2].

Those of Hann are of a more varied character; besides his contributions to

[1] *Report on the Scientific Results of the Voyage of H.M.S. "Challenger" during the years 1873–76*, Physics and Chemistry, vol. II, part v, 'Report on Atmospheric Circulation based on the observations made on board H.M.S. "Challenger" during the years 1873–76 and other Meteorological Observations,' by Alexander Buchan, M.A., LL.D., Secretary of the Scottish Meteorological Society: London, Edinburgh and Dublin, 1889.

[2] *Bartholomew's Physical Atlas*, prepared under the direction of J. G. Bartholomew, F.R.S.E., F.R.G.S., vol. III, 'Meteorology,' edited by Alexander Buchan, LL.D., F.R.S.: Archibald Constable and Co., Westminster, 1899.

Berghaus's Hand Atlas, there are two of the best as well as the best-known meteorological text-books, namely *Klimatologie*, which deals with all that was known of the climatology of the world, and the *Lehrbuch der Meteorologie*, which deals with the physical and dynamical aspects of meteorology. But these were only a part of the expression of Hann's interest in the meteorology of the globe. He was editor of the *Meteorologische Zeitschrift* from 1866 to 1920, and in discharge of his duty as editor made a point of bringing within the covers of that Journal all the meteorological data of stations, outside the limits of the *réseaux* of the recognised institutes, that could be collected from any source whatever. It was Hann who prompted the Meteorological Council to publish the results of observations at the stations of the Royal Engineers and Army Medical Department, which had been initiated by Sir E. Sabine, and by his persistence brought to the knowledge of meteorologists much information that would otherwise have been entirely ignored. It is perhaps true that his primary purpose was climatology which separates the world into zones, regions and districts, rather than the meteorology of the globe, which treats of the whole circulation as a unit; but the two are only different aspects of the same general theme and it is no exaggeration to say that Hann's activity in compiling information about all parts of the world was at least an intro-duction into practical meteorology of the idea of a true *réseau mondial*.

THE FUTURE OF INTERNATIONAL METEOROLOGY

The International Meteorological Organisation had no corporate funds, with the solitary exception of sums amounting to 15,626 francs a year, offered as voluntary contributions by the meteorological services specially interested in the study of the upper air to the President of the Commission for Scientific Aeronautics, Dr H. Hergesell, Director of the Meteorological Institute of Strassburg, for the publication in collected form, month by month, of the observations of the upper air made by the participating countries. The organisation had therefore no power to initiate researches apart from the influence which its discussions and resolutions might exert upon the directors of institutes or observatories. Also we may repeat once more that in his relations with the organisation each director was acting only upon the authority of his own institute or observatory. But in the meantime in allied geophysical or cosmical sciences international associations had grown up to which the various Governments made official contributions. Closely related to meteorology is the exploration of the sea which, originating with the investigation of the physical and biological conditions of sea-fisheries, became the care of an International Council with its seat at Copenhagen, supported by contributions from the participating countries. And in the beginning of the year 1914, in consequence of the loss of the s.s. *Titanic*, provision was made for the study of the ice-conditions in the North Atlantic Ocean in the interest of shipping, at the expense of the countries which had a mercantile marine, the work being carried out by the Government of the United States.

And on another side an Agricultural Institute was established at Rome, mainly through the good offices of the King of Italy, which was placed on an international basis, partly with international finance.

These three represent the main practical relationships of meteorology, but there were others which were more definitely associated with the scientific aspects of the science in which the Academies of Science rather than the Institutes were the responsible authorities. These included the International Geodetic Association and the Seismological Association, each of which was in receipt of separate contributions from Government, on the advice in Great Britain of the Royal Society. The funds thus provided were disposable by the Association for the purpose of the science. In like manner an International Astronomical Association was also recognised.

Towards the close of the war it was urged that the old associations could not resume their activity on the old terms and that steps should be taken to form new ones. This suggestion appealed to different countries for different reasons and resulted in the formation of an International Research Council, with its seat in Brussels, and a number of International Unions under the general control of the International Research Council, the participation in which should be subject to national committees organised by the respective National Academies of Science. Each country adhering to one of the Unions contributes to the corporate funds an annual sum fixed according to population and the Union disposes of its funds at its discretion.

Among these Unions is the Union Géodésique et Géophysique Internationale, which has seven sections, Geodesy, Seismology, Meteorology, Terrestrial Magnetism and Atmospheric Electricity, Physical Oceanography, Land Hydrology, and Vulcanology.

Two results follow—first there is a sum of money available for the development of the science of meteorology under the auspices of the Academies of Science, as distinguished from the unofficial Organisation of Directors of Institutes, and secondly a separate sum is available for the development, under the same auspices, of Terrestrial Magnetism and Atmospheric Electricity, which form the subject of one of the Commissions of the Organisation of Directors. Meanwhile the Meteorological Section of the U.G.G.I. has interested itself in the publication of data for the upper air, the study of the upper air in localities which are not touched by the ordinary activities of the several institutes, in the study of solar and terrestrial radiation, in observations of atmospheric dust, in the resuscitation of the series of daily charts of the Atlantic Ocean, in the choice of units of time for meteorological purposes and their relation to the kalendar, and the use of statistical methods in meteorology. The Commissions now acting on the appointment of the International Conference at Utrecht are: Terrestrial Magnetism and Atmospheric Electricity, Solar Radiation, Investigation of the Upper Air, Synoptic Weather Information, Maritime Meteorology, Réseau Mondial and Polar Meteorology, Agricultural Meteorology, Application of Meteorology to Aerial Navigation, Investigations of the Sound of Explosions, Study of Clouds, The Formation of an International Bureau.

From the enumeration of subjects it is apparent that both associations, namely that of the Directors of Institutes on the one hand, and that of the Academies on the other, are interested in the scientific aspects of meteorology and their development. The invitations to conferences of the International Meteorological Organisation are now limited to the directors of réseaux. For the meetings of the Union there is no such limitation, the selection of representatives rests with the several National Research Committees, but although the controlling bodies are different in aim and in constitution the persons actually engaged in either are for economic reasons mainly the same. Consequently some kind of mutual understanding will become necessary as to co-operation between the two associations; and since one of them possesses funds contributed by the several countries independently of the meteorological budgets and disposable by the Union the germ of an international Bureau has already been planted. We look therefore to an international Office for the development of the science of meteorology along those lines which are not provided for by the separate institutes even when they are in co-operation.

AN INTERNATIONAL METEOROLOGICAL COLLEGE

The progress of meteorology depends upon an adequate knowledge of the circulation of the atmosphere and of the sequence of its changes, and for this purpose nothing less than the whole globe is adequate. For years a *réseau mondial* has meant such stations as may be available. With a little effort the meaning may be increased to include observations of the air over the oceans which are regularly traversed and of the sea-water itself. But that extension is not sufficient, observations must sooner or later be obtained from places chosen for their meteorological importance independently of their economic value.

Let us consider the position in the light of the history which we have sketched. There are various ways of approaching effective knowledge. First of all the Shepherd of Banbury, typical also of the weather-wisdom of the ancients—day by day he studies the sky—he sees perhaps, if he is fortunate, fifty miles in any direction; the area under observation is, on a liberal estimate, 5000 square miles of sky. He draws his conclusion from the daily panorama spread out before him, he learns what different combinations indicate, and he becomes a more or less skilled forecaster for a few hours ahead, or longer on occasions when the weather is settled. Then the map-maker, who has access to observations extending a thousand miles in either direction, whose area of observation is a million square miles; he uses conventional signs instead of the personal observations of the Shepherd but his method is to a great extent the same; day by day he sees a panorama spread out before him, he watches its changes and from his knowledge of the groupings of the observations he becomes a skilled forecaster with a range of twenty-four hours, or much more on occasions when the weather is settled. He may improve his accuracy by knowing the conditions of the upper air or in other ways, but, on the analogy of the Shepherd and his sky, all has to be referred to the sequence of events shown on the map. If he wants to go beyond that he must think out the actual physics of the weather and that means a knowledge of the atmosphere of the world, not necessarily at the moment when the forecast is made but in general. For example we may have a heavy shower of rain, a thunderstorm in summer, and if its successor is to be forecast its genesis must be known. Where did the water come from which formed the rain? and why did it choose that particular locality for a downpour? The answer to the second question requires a knowledge of the structure of the atmosphere overhead and where it came from; and an answer to the first question implies a study of the general question of the origin of rain. In the particular case it might have been part of the effect of sunshine upon the water at the mouth of the Channel, but the whole question of where the rain comes from is a world-question, and one of no little interest. There seems little doubt that the water that comes down the Amazon or the Mississippi first flows up one of those valleys—but the water of the Nile comes probably from the Atlantic Ocean over Uganda, and what feeds the Volga or the Yang-tze-Kiang is a very intricate question. But it is part of the meteorology of the world.

We may postpone our inquiry into these physical details and deal with coming weather or coming seasons by numerical methods such as the correlations between Indian rainfall and the pressure-conditions elsewhere. That as much as anything requires a complete investigation of the general circulation of the atmosphere. It is a question of finding what are called "centres of action."

Or again we may seek, as many authors have done, to detect in the chronological sequence of weather in any particular locality a combination of periodic terms which will give the past with such accuracy that it may be used to anticipate the future by continuing the same periods. For single points or even for large areas of the earth's surface that has been attempted, as for example by Douglass's analytic curve for tree-growths for 3000 years or Sir William Beveridge's curve for seventy years. Further progress must be fortuitous until the periodic terms are explained as being either of world-wide or local significance, and again the study must be taken up as a world-study.

The next step

So meteorology is now approaching, if it has not already reached, the parting of the ways where it must either be content to achieve a success, something better than the Shepherd of Banbury's, but of the same type, or advance boldly to a real knowledge of the structure of the atmosphere and its changes so that not only the whole sequence may be expressed, but the physical processes may likewise be explored. Then we may really understand weather.

This step certainly involves an international weather-office with adequate funds to initiate and maintain observations which are not part of the economic necessities of any special country. It must also co-ordinate the observations when they have been obtained. It will certainly be the most important meteorological establishment in the world and ought therefore to be staffed with the most competent of meteorologists and physicists.

Proposals for such a bureau have not been accepted. National opinion has insisted that expenditure of national resources shall be limited to national objects. Yet in the longer view the game is worth the candle. After the loss of the *Titanic* an organisation for protection against danger from ice in the Atlantic was set on foot with international funds, and now that a League of Nations looks after political interests and the weather is recognised to be of the first economic importance in every part of the world, it is possible to contemplate the effort; but it ought not to be an apanage of any national establishment. It must include not only a bureau for the collection and co-ordination of observations, and a library for the preservation of data, old and new, with any literature that may be necessary to elucidate them, but also an efficient observatory to enable the workers to appreciate the behaviour of the instruments from which the data are derived and the conditions for successful observation. That necessarily implies experimental observation or research and the facilities of a workshop, and consequently a staff capable of carrying out researches of an experimental or theoretical character. Hence we arrive at the idea not merely of an international bureau for the collection, compilation and distribution of numerical data contributed by observers elsewhere, but of a real international meteorological college where the meteorology of the globe can be studied and developed under conditions necessary for real success.

Survey of the Surface Air of the World

The extent to which meteorology has been recognised as an international science may be exhibited by reference to a list of stations which was prepared by H. N. Dickson, D.Sc., and is preserved in the Meteorological Office. Its epoch may be regarded as 1914. We give here a numerical summary of the available data arranged as stations of first, second or third order, with a supplementary heading O to indicate other stations that cannot be classified.

EUROPE

Country			I	II	III	O	Country			I	II	III	O
Albania	...	...	—	1	3	—	Italy	...	...	65	99	269	3
Austria	...	...	23	184	230	6	Light Ships	...	...	—	—	—	18
Belgium	...	...	2	4	85	2	Luxembourg	...	...	—	—	—	1
Bosnia	...	...	3	5	75	1	Malta	...	...	—	3	6	—
Bulgaria	...	...	1	16	18	—	Montenegro	...	...	—	1	—	—
Crete	...	...	—	1	1	—	Norway	...	...	5	62	19	3
Denmark	...	...	1	20	133	1	Portugal	...	...	1	13	—	—
Faroe Is.	...	...	—	1	4	—	Roumania	...	...	2	56	2	—
Finland	...	...	1	53	38	1	Russia in Europe	...	...	4	790	—	4
France	...	...	21	154	3	11	Serbia	...	...	1	20	17	2
Germany	...	...	18	312	123	2	Spain	...	...	2	81	186	2
Greece	...	...	1	27	3	2	Spitsbergen, Barents Is.						
Greenland	...	...	—	6	7	9	and Franz Josef Land			1	—	—	14
Herzegovina	...	...	1	1	30	—	Sweden	...	...	3	57	108	1
Holland	...	...	5	13	7	—	Switzerland	...	...	3	132	52	—
Hungary	...	...	6	154	66	3	Turkey	...	...	—	1	—	—
Iceland	...	...	—	6	15	1	United Kingdom		...	13	371	215	8

ASIA

Country	I	II	III	O	Country	I	II	III	O
Afghanistan	—	—	2	—	New Guinea (British)	—	—	18	—
Arabia	—	—	6	1	New Guinea (German)	—	5	1	1
Asia Minor	—	11	4	6	Persia	—	3	11	—
Assam	—	—	20	—	Philippine Is.	1	20	25	1
Bhotan	—	—	3	—	Russia (Asiatic)	1	243	—	26
British N. Borneo	—	—	6	1	Sakhalin	—	5	—	—
Burma	—	1	18	—	Siam	—	1	—	1
Ceylon	—	16	3	—	Straits Settlements	—	6	1	—
China	4	20	25	29	Syria (Palestine)	—	9	8	4
China Sea:					Tibet	—	—	2	—
Bonin Is.	—	2	—	—	Turkestan and Mongolia	—	1	2	—
Ladrone Is.	—	—	1	—					
Lu-Chu Is.	1	2	—	—	*Islands of Indian Ocean*				
Pescadores Is.	—	1	—	—	Andaman and Nicobar	—	—	4	—
Cyprus	—	6	—	—	Cocos (Keeling Is.)	—	1	—	—
Dutch E. Indies	1	22	20	6	Diego Garcia	—	—	—	1
Formosa	1	6	—	2	Kerguelen Is.	—	—	—	1
India	3	1	202	—	Laccadive Is.	—	—	1	—
India (French)	—	5	—	—	Madagascar	1	1	—	3
Indo-China (French)	1	20	3	3	Mauritius	1	4	1	—
Japan	22	60	1	4	Minicoy	—	—	1	—
Korea	—	15	—	—	Réunion	—	1	—	2
Malaya	—	—	26	3	Rodrigues	—	1	—	—
Manchuria	—	3	—	—	Seychelles	—	1	1	—
Mesopotamia	—	2	3	2					

AFRICA

Country	I	II	III	O	Country	I	II	III	O
Abyssinia	—	1	5	2	Kamerun	—	9	16	3
Algeria	2	36	36	4	Liberia	—	—	—	1
Cape Colony	2	61	20	5	Mauretania	—	2	—	—
Congo (Belgian)	—	8	39	3	Morocco	—	11	2	1
Congo (French)	—	7	3	1	Natal	1	—	20	1
Dahomey	—	1	—	1	Nigeria	—	—	25	2
E. Africa (British)	—	3	8	5	Nyasaland	—	—	1	5
E. Africa (German)	5	41	10	—	South Rhodesia	—	12	14	1
E. Africa (Portuguese)	—	—	—	7	Senegal	—	4	—	3
Egypt	1	18	10	1	Sierra Leone	—	1	9	—
Eritrea	—	1	4	7	Somaliland:				
Gambia	—	—	2	—	British	—	1	—	1
Gold Coast	—	—	13	3	French	—	1	—	—
Guinea (French)	—	1	—	2	Italian	—	—	5	—
Guinea (Portuguese)	—	—	—	1	S.W. Africa (German)	—	5	—	7
Islands:					Sudan (British)	1	15	7	—
Ascension	—	—	—	1	Sudan (French)	—	7	—	10
Azores	—	3	—	—	Swaziland	—	—	—	3
Canary Is.	—	5	2	1	Togoland	—	8	—	2
Cape Verde Is.	—	2	—	—	Transvaal	1	29	34	—
Fernando Po	—	—	—	1	Tripoli	—	3	5	—
Madeira	—	1	—	—	Tunis	—	4	1	5
St Helena	—	1	—	—	Uganda	—	2	12	2
St Thomas Is.	—	—	—	2	W. Africa (Portuguese)	—	4	1	2
Tristan Da Cunha	—	—	—	1	Zanzibar Prot.	—	—	3	—
Ivory Coast	—	1	—	1					

NORTH AMERICA

Country	I	II	III	O	Country	I	II	III	O
British North America	5	480	82	27	Newfoundland and Labrador	—	15	3	1
Mexico	1	24	80	2	United States... ...	42	186	2896	13

WEST INDIES AND CENTRAL AMERICA

Country	I	II	III	O	Country	I	II	III	O
West Indies					St Kitts	—	—	1	—
Antigua	—	—	1	—	St Lucia	—	—	2	—
Bahamas	—	6	—	1	St Thomas ...	—	1	—	—
Barbadoes	—	1	1	—	St Vincent	—	—	2	—
Bermuda	—	1	—	2	Santa Cruz	—	1	—	—
Cuba	—	3	59	2	Sombrero	—	1	—	—
Dominica	—	1	1	—	Trinidad ...	—	—	1	1
Guadeloupe ...	—	1	—	1	*Central America*				
Haiti and S. Domingo	—	3	12	—	Costa Rica	—	1	1	—
Jamaica	—	1	12	—	Guatemala	—	2	3	—
Martinique	—	2	—	1	Honduras	—	—	2	1
Montserrat Is. ...	—	1	—	—	Nicaragua	—	—	5	—
Puerto Rica	—	—	—	44	Panama	—	3	7	—
San Domingo (Repub.)	—	3	1	1	San Salvador ...	—	1	1	1

SOUTH AMERICA

Country	I	II	III	O	Country	I	II	III	O
Argentina	28	92	5	12	Guiana (British) ...	—	1	4	1
Bolivia	—	4	1	4	Guiana (Dutch) ...	—	—	1	—
Brazil	12	60	57	13	Guiana (French) ...	—	1	—	—
Chile	4	12	27	3	Paraguay	—	5	1	3
Colombia	—	1	—	7	Peru	1	2	1	16
Ecuador	3	2	2	—	Uruguay	—	21	3	3
Falkland Is.	—	1	1	—	Venezuela	—	6	—	7
S. Georgia	—	1	3	1					

AUSTRALIA AND NEW ZEALAND

Country	I	II	III	O	Country	I	II	III	O
Australia	2	85	106	4	Tasmania	—	13	7	—
New Zealand... ...	—	—	35	1					

AUSTRALASIA AND OCEANIA

Antarctic   Eleven stations associated with special expeditions.

Country	I	II	III	O	Country	I	II	III	O
Campbell Is.	—	—	—	1	New Hebrides ...	—	—	1	3
Caroline Is.	—	—	2	2	New Pomerania ...	—	1	—	1
Chatham Is.	—	1	1	—	Norfolk Is. ...	—	—	—	1
Cook Is.	—	2	2	1	Ocean Is. ...	—	1	—	—
Fanning Is.	—	—	1	—	Isla de Pascua	1	—	—	—
Fiji Is.	—	2	1	2	Paumotu Arch. ...	—	—	—	1
Gilbert Is.	—	1	—	1	Pleasant Is. ...	—	1	—	—
Hawaii	—	3	—	3	Rapa Is. ...	—	1	—	—
Lord Howe Is. ...	—	—	—	1	Samoa... ...	—	1	—	—
Louisiade Arch. ...	—	—	—	1	Society Is. ...	—	2	—	—
Malden Is.	—	1	—	—	Solomon Is. ...	—	1	—	2
Marshall Is.	—	2	—	—	Swan Is. ...	—	1	—	—
New Caledonia ...	—	5	—	—	Tonga Is.	—	2	1	1

High level stations

Range of height of stations m	EUROPE Order				ASIA Order				N. AMERICA Order			
	I	II	III	O	I	II	III	O	I	II	III	O
Above 4000	Mont Blanc				—	—	2	—	—	1	—	—
3000–4000	1	—	—	—	—	3	1	—	—	—	5	—
2000–3000	4	20	3	—	—	5	1	—	1	3	114	1
1000–2000	7	56	58	1	—	25	14	5	1	26	349	14
200–1000	2	21	19	1	—	12	6	3				

Range of height of stations m	AFRICA Order				AUSTRALASIA Order				C. AND S. AMERICA Order			
	I	II	III	O	I	II	III	O	I	II	III	O
Above 4000	—	—	—	—	—	—	—	—	—	2	—	3
3000–4000	—	—	—	—	—	—	—	—	—	—	1	3
2000–3000	—	5	5	3	—	—	—	1	3	9	2	3
1000–2000	3	60	51	12	—	1	1	—	3	21	13	5
200–1000	—	11	8	2					1	10	12	4

Observations from ships

Observations from ships according to international forms have been collected in considerable numbers since 1854 by various Meteorological Institutes, notably Denmark, France, Germany, Great Britain, Holland, Japan, and the United States. The results have been expressed in charts of mean values of the meteorological elements over the oceans issued by the various Institutes, and in other ways, notably the daily morning and evening charts of the Danish and German Offices to which reference has already been made.

We have no figures to indicate the total mass of information thus collected which is available for study. We find however that for the year 1913–14 the Meteorological Office collected 279 four-hourly logs and 2459 other forms[1]. If we regard the activity of the world in this department as four times that of the Meteorological Office there should be over a thousand log-books and ten thousand other forms out of which a monthly summary of the weather of the sea might be compiled.

[1] *Ninth Annual Report of the Meteorological Committee*, 1914, p. 31.

CHAPTER X

THE METEOROLOGICAL OBSERVATORY: THE SURFACE AIR

THE EQUIPMENT OF NORMAL CLIMATOLOGICAL STATIONS

Books of Instruction to Observers,
in the Libraries of the Meteorological Office and the Royal Meteorological Society

Africa. Hints to meteorological observers in tropical Africa, with notes on methods of recording lake-levels and a memorandum on the organisation of meteorological observations. M.O. No. 162. London, H.M. Stationery Office, 1907. *Nairobi.* Agricultural Department. Hints to meteorological observers in British East Africa with instructions for taking observations, by L. D. Carpenter. *Agric. Jour. Brit. E. Africa,* Nairobi, vol. v, 1913, p. 38.

Australia. Australian Meteorological Observer's Handbook 1925, published...under the direction of H. A. Hunt, Melbourne.

Austria. Jelinek's Anleitung zur Ausführung meteorologischer Beobachtungen nebst einer Sammlung von Hilfstafeln. 5th edition; Wien, 1905. In Zwei Teilen. Published by Zentralanstalt für Meteorologie und Geodynamik.

Belgium. Instructions pour les stations météorologiques belges préparées par A. Lancaster. 2nd edition, Observatoire Royal de Belgique, 1884.
Instructions pour les stations de troisième ordre. Observatoire Royal de Belgique, Service Climatologique. Brussels, s.a.
Observatoire Royal de Belgique. Annuaire météorologique pour 1909, p. 51. Instructions pour les Observateurs.

Brazil. Instrucções meteorologicas (Trabalho destinado aos estacionarios da rêde meteorologica do Brazil) by J. de Sampaio Ferraz. Ministerio da Agricultura, Industria e Commercio, vols. I and II. Bruxelles and Paris, 1914.

Canada. Instructions to observers connected with the Meteorological Service of the Dominion of Canada by G. T. Kingston. Toronto, 1878.
Instructions to observers at ordinary stations of the second class by G. T. Kingston, M.A. Meteorological Service, Dominion of Canada. Toronto, 1878.

Chile. Instituto Central Meteorológico y Geofísico de Chile. Santiago de Chile. Instrucciones Meteorológicas, 1919. Instrucciones para los Observadores de Estaciones Pluviométricas, 1917.

China. Instructions in the use of meteorological instruments, compiled by R. F. Marc Dechevrens, S.J. Zi-ka-wei, 1880.
Instructions for making meteorological observations prepared for use in China by W. Doberck. Hong Kong, 1883.

Egypt. Ministry of Public Works. Handbook of instructions for meteorological observers in Egypt, the Sudan, and Palestine. Cairo, 1923.

France. Instructions météorologiques par Alfred Angot. Sixième édition, revue et corrigée. Paris, 1918.
Service Météorologique, Ministère de la Guerre. Notice sur les Instruments météorologiques et sur les sondages aérologiques. Paris, 1919.

Germany. Instruction zur Führung des meteorologischen Journals (Ship's log) der Deutschen Seewarte, Hamburg, 1876.
Anleitung zur Anstellung und Berechnung der Beobachtungen an den deutschen meteorologischen Stationen. Erster Teil. Beobachtungen an Stationen II und III Ordnung. Bearbeitet von G. Lüdeling. Veröff. des Preuss. Met. Inst., No. 321. Berlin, 1924. Zweiter Teil. Besondere Beobachtungen und Instrumente (Dritte erweiterte Auflage). Veröff. des Preuss. Met. Inst., No. 268. Berlin, 1913.

Anleitung für die meteorologischen Stationen im Grossherzogtum Baden. Karlsruhe, 1908.
Instruktion für die Beobachter der württembergischen meteorologischen Stationen. Herausgegeben vom k. Statistischen Landesamt. Stuttgart, 1893.
Instruction für die Beobachter an den Meteorologischen Stationen im Königreich Bayern. Herausgegeben von der k. b. meteorologischen Central-Station. München, 1878.

Great Britain. Marine Observer's Handbook. M.O. publication No. 218, 3rd edition. London, 1922.
The Observer's Handbook. Meteorological Office publication, No 191. London, 1926.
The Observer's Primer, M.O. 266. London, 1924.
Rules for Rainfall Observers. Air Ministry, Meteorological Office, British Rainfall Organization. 1920.
Hints to Meteorological Observers, prepared under the direction of the Council of the Royal Meteorological Society by the late W. Marriott, 8th edition revised by R. Corless. London, 1924.

Holland. Kon. Ned. Met. Inst. Utrecht. No. 99. Handleiding bij het waarnemen van den Neerslag en het melden van bijzondere verschijnselen in den dampkring. 1905. No. 101. Handleiding bij de Waarnemingen op Termijnstations 1905.

India. The Indian meteorologist's Vade-Mecum. Part I, Instructions to Observers; Part II, Meteorology of India, by H. F. Blanford. Calcutta, 1877.
Instructions to observers of the India Meteorological Department by J. Eliot. 2nd edition, Calcutta, 1902.

Italy. Manuale per le osservazioni meteorologiche a mare, per il prof. L. Marini. R. Comitato Talassografico Italiano, Memoria IV. Venezia, 1912.
Istruzioni per le Osservazioni Meteorologiche e per l'Altimetria Barometrica del P. F. Denza. Associazione Meteorologica Italiana. Torino, 1882.
Istruzioni per la Misura delle Precipitazioni Meteoriche concordate coi servizi interessati nelle conferenze del febbraio 1913 presso l'Ufficio Centrale di Meteorologia e Geodinamica. Parte I, Pluviometro Comune; Parte II*, Pluviografo; Parte III*, Pluvionivometro. Ufficio idrografico del Po. Parma, 1913.
Norme ed istruzioni per il Servizio Meteorologico, Venezia, 1916. R. Magistrato Alle Acque, Ufficio Idrografico, Publ. 53, Parte I*.

Japan. The Organisation of the Meteorological Service of Japan. Department of Education, Central Meteorological Observatory of Japan, Tokyo, 1904. The list of publications set out therein includes:
No. 13. Instructions for meteorological observers.
No. 18. Additional instructions to meteorological observers; comparison of scales.
No. 22. Instructions for meteorological observers with tables.

Manila. Department of Agriculture and Natural Resources, Weather Bureau.
(1) Practical Instructions and regulations for the observers of the Weather Bureau. Manila, 1917.
(2) Instrucciones Prácticas y Breves Nociones de Meteorologia.

Norway. Vejledning til Udførelse af meteorologiske Iagttagelser ved det norske meteorologiske Instituts Stationer. Udgivet af det norske meteorologiske Institut, 1888, Christiania.
Veiledning i Meteorologiske Iagttagelser. I. Utgit av det Norske Meteorologiske Institut. Kristiania, 1915.

Poland. Sieć Meteorologiczna Warszawska. Instrukcya dla Stacyj Meteorologicznych łącznie z Obserwacyami gruntowemi i plantacyjnemi ułożona przez Biuro Meteorologiczne przy Muzeum w Warszawie wydana staraniem Związku Zawodowego Cukrowni Królestwa Polskiego. Warszawa, 1910.

Ksiazki dla Wszystkich: "Prawidła pogody oparte na spostrzezeniach meteorologicznych" by L. Gorczynski. Warszawa, 1910.

Portugal. Observatorio do Infante D. Luiz. Instrucções Meteorologicas por J. C. de Brito Capello. Lisbon, 1890.

A note of the type of instruments used at the observatory is given in the Advertencia to the Anais do Observatorio "Infante D. Luis," vide vol. XLIX, 1911. Lisboa, 1915.

Roumania. Observatorul Astronomic şi Meteorologic: Instructiuni prescurtate pentru Staţiunile Meteorologice de G. D. Elefteriu (Editiune Oficiala). Bucuresci, 1908.

Instructiuni pentru Statiunile Termo-Udometrice by S. C. Hepites. Analele Institutului Meteorologic al Romaniei,Tom. II, Part II, Anul 1886. Bucuresci, 1888.

Russia. Instruction der Kaiserlichen Akademie der Wissenschaften für Meteorologische Stationen. St Petersburg, 1887.

General Hydrographic Administration Hydrometeorological Service. Instructions for taking hydrological and meteorological observations at coast-stations and floating lightships of the Marine Department, St Petersburg, 1913. (Russian title translated.)

Spain. Instituto Geográfico y Estadístico: Instrucciones para el Servicio Meteorológico, por el Jefe José Galbis y Rodriguez. Madrid, 1913.

Sweden. Handledning vid Meteorologiska Observationers Anställande by E. Edlund. Stockholm, 1882.

Instruktion för Nederbördsobservationers Utförande vid Svenska Fyrstationer af Nautisk-Meteorologiska Byrån i Stockholm.

Instruktion för Hydrografiska Observationers Utförande vid Svenska fyr- och lots-stationer af Nautisk-Meteorologiska Byrån i Stockholm. Stockholm, 1879.

Instruktion för Meteorologiska Observationers Utförande vid Svenska-Fyrstationer af Nautisk-Meteorologiska Byrån i Stockholm. Stockholm, 1879.

Instruktion för Meteorologisk Loggboks Förande af Nautisk-Meteorologiska Byrån i Stockholm, 1879.

Instruktion för Iakttagelser öfver Vattenstånd, Is, Flottning och Vatten-temperatur, M. M. samt Insamling af Vattenprof. Utgifven af Hydrografiska Byrån, Stockholm, 1911.

Switzerland. Instruktionen für die Beobachter der meteorologischen Stationen der Schweiz. Zweite

umgearbeitete und vermehrte Auflage. Herausgegeben von der Direktion der Schweizmeteorologischen Centralanstalt. Zurich, 1893.

United States. U.S. Department of Agriculture, Weather Bureau, Washington.

Instructions for obtaining and tabulating records from recording instruments by C. F. Marvin. Circular A. Instrument Division. 3rd edition, 1913 (W.B. No. 509) with Appendix on the hygrograph and tabulating records therefrom, 1911 (W.B. No. 450).

Instructions for co-operative observers. Circulars B and C. Instrument Division. 7th edition, 1924 (W.B. No. 843).

Instructions for the installation and maintenance of wind-measuring and recording apparatus. Circular D. Instrument Division. 4th revision, 1914 (W.B. No. 530).

Measurement of precipitation. Instructions for the measurement and registration of precipitation by means of the standard instruments of the U.S. Weather Bureau. Circular E. Instrument Division. 4th edition, 1922 (W.B. No. 771). Appendix No. 2 to Circular E. Instructions for use of the Marvin float rain-gage. 3rd edition, 1915.

Barometers and the measurement of atmospheric pressure by C. F. Marvin. Circular F. Instrument Division. 4th edition, 1912 (W.B. No. 472).

Instructions for the care and management of electrical sunshine-recorders (Weather Bureau Pattern). Circular G. Instrument Division. 5th edition, 1923 (W.B. No. 802).

Instructions for erecting and using Weather Bureau nephoscope, 1919 pattern. Circular I. Instrument Division, 1920 (W.B. No. 712).

Instructions for the installation and maintenance of Marvin water-stage registers with specifications. Circular J. Instrument Division, 1921 (W.B. No. 746).

Instructions for the installation and operation of Class "A" evaporation stations by B. C. Kadel. Circular L. Instrument Division, 1915 (W.B. No. 559).

Instructions to special river and rainfall observers of the Weather Bureau by H. C. Frankenfield, 1912 (W.B. No. 469).

Instructions to storm-warning displaymen of the Weather Bureau, 1912 (W.B. No. 80).

Instructions to operators on the U.S. Weather Bureau telegraph and telephone lines, by J. H. Robinson, 1906.

Instructions to marine meteorological observers. Circular M. Marine Division. 4th edition, Washington, 1925 (W.B. No. 866).

Meteorological Instruments and Apparatus employed by the United States Weather Bureau by Roy N. Covert, Optical Society of America, 1925.

International Instructions

"Tables météorologiques internationales" publiées conformément à une décision du Congrès tenu à Rome en 1879. Paris, Gauthier-Villars et Fils, 1890.

Des principales méthodes employées pour observer et mesurer les nuages par H. Hildebrand Hildebrandsson et K. L. Hagström. Upsala, 1893.

Atlas international des nuages publié conformément aux décisions du Comité Météorologique International. H. Hildebrandsson, A. Riggenbach,

L. Teisserenc de Bort. Paris, 1896; 2nd edition, Paris, 1910.

Internationaler meteorologischer Kodex. Im Auftrage des Internationalen Meteorologischen Komitees bearbeitet von G. Hellmann und H. H. Hildebrandsson. Deutsche Ausgabe besorgt von dem Königlich Preussischen Meteorologischen Institut, Berlin, 1907. 2 Aufl. Berlin, 1911.

METEOROLOGICAL EQUIPMENT

THE general subject of the development of meteorological instruments was treated in an address before the Royal Meteorological Society by Mr R. Bentley, president, in 1905[1]. From that address and from the particulars given in the biographies of chapter VIII, a general view of the gradual progress of the development of meteorological equipment may be obtained. We do

[1] *Q. J. Roy. Meteor. Soc.*, vol. XXXI, 1905, p. 173.

not propose to deal with detailed descriptions of the instruments now in use because they are set out in the books of instructions of the various meteorological services. A list of books of this kind which are to be found in the libraries of the Meteorological Office and of the Royal Meteorological Society is given at the head of this chapter. Here we shall only notice briefly some of the salient features which should be remembered by those who wish to follow historically the progress of the science and indicate the treatment which the crude readings have to receive by correction or reduction to complete their preparation as meteorological data.

Time-keeping

Before we enter upon the consideration of the instruments which are specially meteorological we must refer to certain instruments which are in common use in all observatories or laboratories. A primary requirement of any establishment of that kind is a proper time-keeper. Accurate time-keeping is generally said to derive from the discovery by Galileo of the synchronism of the oscillations of a pendulum which found its application in the pendulum clock, though clocks were certainly in existence before the sixteenth century. Originally each clock had to be controlled by direct observations of the transit of the sun or stars, or more roughly by a sun-dial. Hence the parson in charge of any parish which possessed a clock must have retained some at least of the qualifications of the "astronomer-priest" in order to keep the parish-clock in due regulation. He must also have been the source of correct information about the orientation cross above which the wind-vane was mounted. In modern days, since the incumbent ceased to have the necessary special knowledge, the local coppersmith has in many cases been guided by the magnetic compass, and the orientation is often incorrect by the amount of the variation. In past times, even if the parish priest were not responsible for the initial setting of the gnomon of the sun-dial in line parallel with the earth's axis and the proper graduation of the horizontal or vertical surface on which the shadow fell, he must have been sufficiently instructed in astronomy to make allowance for the difference between "local apparent noon," the time of passage of the sun across the meridian as shown by the sun-dial, and "local mean noon," the time of transit of an imaginary body which completes 365 transits at equal intervals during an ordinary year.

The difference between apparent noon and mean noon is by no means negligible. As timed by a clock which keeps mean time the sun was behind the clock by 3 mins. 13 secs. in crossing the meridian on January 1, 1924, and 3 mins. 35 secs. behind on January 1, 1925, the change in the year being accounted for by 1924 having carried an extra day as leap year, when the divergence gradually diminishing through three years is brought back approximately to its correct position by the additional day. The times of apparent noon on the fifty-two Sundays during the year 1924 were as follows:

Clock Times of Apparent Noon

No. of Sunday	Time h m s	No. of Sunday	Time h m s	No. of Sunday	Time h m s	No. of Sunday	Time h m s
6 Jan. 1924 1	12 5 32	14	12 2 30	27	12 4 31	40	11 48 28
2	8 28	15	12 0 36	28	5 32	41	46 33
3	10 53	16	11 58 56	29	6 9	42	45 2
4	12 42	17	57 37	30	6 20	43	44 2
5	13 52	18	56 43	31	6 2	44	43 38
6	14 23	19	56 16	32	5 14	45	43 54
7	14 15	20	56 17	33	3 58	46	44 50
8	13 31	21	56 45	34	2 17	47	46 27
9	12 19	22	57 37	35	12 0 16	48	48 44
10	10 42	23	58 49	36	11 58 0	49	51 32
11	8 48	24	12 0 14	37	55 34	50	54 43
12	6 43	25	1 44	38	53 5	51	58 9
13	4 35	26	3 12	39	50 41	52	12 1 39 (Dec. 28)

4 Jan. 1925 12 4 59

The sunshine-recorder operating by the scorching of a card in the optical focus of a glass-sphere with the mounting designed by Sir George Stokes is an excellent sun-dial and requires all the adjustments which are necessary for a dial.

The sun-dial as a means of regulating mean-time clocks was superseded first by a telegraphic time-signal and later by a wireless signal sent out from a central observatory such as that of Paris or Greenwich, at which the regulation of a mean-time clock is a subject of special care. Signals are now sent by radiotelegraphy from many stations for the special benefit of ships at sea which are enabled thereby to regulate the ship's chronometers while they are still at sea. They are thus able to use the instruments with greater confidence for the determination of their longitude, one of the primary objects of state observatories.

The arrangements for the transmission of time-signals on an international basis were first drawn up at a "Conférence internationale de l'heure" held at Paris in 1912[1]. The times at which time-signals are sent out are:

Time of Greenwich Noon in Local standard time	Station	Time of signal [2] G.M.T.	Time of Greenwich Noon in Local standard time	Station	Time of signal [2] G.M.T.
23½ h	Wellington	2300–2305	19 h	Kien an	0215–0219
22 h	Melbourne	0200, 1400	17½ h	Calcutta	0127–0130 1327–1330
21½ h	Adelaide	0230, 1430			
21 h	Funabashi	1200–1204	14 h 1 m	Moscow	2158–2200
,,	Choshi	1200–1204	,,	Leningrad	1903–1905
20 h	Hong Kong	0156–0200 1256–1300	14 h	Lourenço Marques	0800, 1900
,,	Shanghai	0255–0259 0855–0859	,,	Cape Town	2059–2100
,,	Philippine Is. Kavite	0300, 1400	13 h	Nauen	1158–1200 2358–2400
,,	Perth	0300, 1500	12 h	Paris	0928–0930 1000
19 h 19 m	Batavia	0100–0104			1045–1049
,,	Surabaya	0210–0214			2245–2249
,,	Malabar	0058–0100			2200

[1] *Conférence internationale de l'heure* (Paris, Octobre 1912). Paris, Gauthier-Villars, 1912.
[2] The times are given to the nearest minute in civil mean time. The information is extracted from *The Admiralty List of Wireless Signals*, 1923.

Time of Greenwich Noon in Local standard time	Station	Time of signal G.M.T.	Time of Greenwich Noon in Local standard time	Station	Time of signal G.M.T.
12 h	Bordeaux	2000	7 h	Washington (Arlington)	0300, 1700
,,	Lyon	{0800 {0900–0904	,,	Annapolis	0300, 1700
9 h	Rio de Janeiro	0000, 1400	6 h	Great Lakes	1700
8 h	Halifax	1400	5 h 23 m	Mexico City	0100, 1900
,,	Buenos Aires	0156–0200	4 h	San Diego	2000
7 h 17 m	Valparaiso	0056–0100	,,	San Francisco	0600, 2000
7 h	Colon	1000, 1800	,,	Eureka	2000
,,	Balboa	1000, 1800	,,	North Head	2000
,,	New Orleans	1700	1½ h	Honolulu	0000
,,	Key West	0300, 1700			

From the commencement of 1925 *astronomical mean time* has been dated from midnight to midnight instead of noon to noon.

Before the time of railways and electric telegraphs, each clock was regulated by a local sun-dial or by some other form of astronomical observation to keep its own local mean time depending upon the time at which "mean sun" crossed the meridian. It followed that places with different longitudes kept different times, the difference being one hour for fifteen degrees of longitude. But when locomotion became more rapid and telegraphs were introduced the railway clocks were synchronised and "railway-time" was the same all over the system. That practice has now been regularised for the whole world by the system of zone-time originally adopted for use on land and in 1919 for the sea as well. The following is the scheme of times: "The world is considered as being longitudinally divided into 24 zones of 15° each; the centre zone lies between the meridian of 7½° E and that of 7½° W, and is described as Zone o; the zones lying to the eastward are numbered in sequence from 1 to 12 with a minus prefix, those to the westward being similarly numbered but with a plus prefix; the 12th zone is divided centrally by the 180th meridian (the date line), and the prefixes minus and plus are used in the eastern and western halves respectively." "The essence of the method is that the time shall always be a whole number of hours fast or slow on Greenwich. Each system of time is denoted by a positive or negative number as follows: Greenwich Mean Time is denoted by the number o or the word 'Zero.' A time fast on G.M.T. is denoted by a number with a minus prefix equal to the number of hours fast on G.M.T. A time slow on G.M.T. is denoted by a number with a plus prefix equal to the number of hours slow on G.M.T.[1]"

At the same time the practice of reckoning the hours from o midnight to 24 midnight has been growing. It was practically universal for military purposes during the great war, and is now used in many railway time-tables. A further alteration that would be very convenient would be to use decimal fractions of an hour instead of minutes and seconds in order to get rid of the awkward discontinuity between metres per second and kilometres per hour, but that is impossible now that the second has been adopted as the basis of the c.g.s. system.

[1] *System of Timekeeping at Sea by means of Time-Zones.*

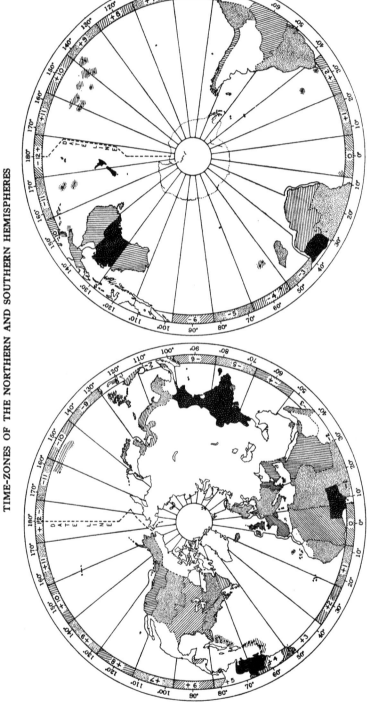

TIME-ZONES OF THE NORTHERN AND SOUTHERN HEMISPHERES

Fig. 14. Charts showing the relation of standard time of zones of sea and land to Greenwich mean time. The zone-times for the sea are indicated by figures in the margin which give the number of hours to be added (+) to or subtracted (−) from the local zone-time to give Greenwich mean time. The land areas are shaded to correspond with the adopted zone-time. Countries in which the adopted time differs from zone-time by half an hour are shaded dark. Countries that have not yet adopted the zone system are unshaded.

The following list, taken from the *Observer's Handbook* of the Meteorological Office, gives the Standard Zone Times that have been adopted for railway and other purposes, referred to the Meridian of Greenwich: the boundaries of the zones, and the "date-line," are shown in the maps of fig. 14. They are not always strictly along meridians either on land or sea.

TABLE OF ZONE TIMES

Countries which have adopted standard zone times	Time Meridian	Amount by which clock is fast or slow on Greenwich	Local Time of Greenwich Noon
Fiji	180° E	12 hours fast	24 h
New Zealand	172½° E	11½ ,, ,,	23½ ,,
Victoria, New South Wales, Queensland, Tasmania	150° E	10 ,, ,,	22 ,,
South Australia	142½° E	9½ ,, ,,	21½ ,,
Japan, Corea	135° E	9 ,, ,,	21 ,,
Western Australia, Hong Kong, China Coast, Philippine Islands, British North Borneo, Labuan, Macao, Portuguese Timor	120° E	8 ,, ,,	20 ,,
Straits Settlements, Indo-China	105° E	7 ,, ,,	19 ,,
Burma	97½° E	6½ ,, ,,	18½ ,,
India	82½° E	5½ ,, ,,	17½ ,,
Mauritius, Seychelles	60° E	4 ,, ,,	16 ,,
Italian Somaliland	45° E	3 ,, ,,	15 ,,
(East Europe.) Bulgaria, Roumania, Eastern Turkey, South Africa, Egypt, Portuguese East Africa	30° E	2 ,, ,,	14 ,,
(Mid Europe.) Germany, Luxemburg, Austria, Hungary, Denmark, Sweden, Norway, Switzerland, Italy, Bosnia, Serbia, Malta, Portuguese West Africa	15° E	1 ,, ,,	13 ,,
(Greenwich.) British Isles, France, Belgium, Spain, Portugal, Gibraltar, Algeria, Faröe Islands	0°	—	—
Iceland, Madeira, Portuguese Guinea, Sierra Leone	15° W	1 hour slow	11 ,,
Azores and Cape Verde Islands ...	30° W	2 hours ,,	10 ,,
(*Atlantic.) Maritime Provinces of Canada, Leeward Islands ...	60° W	4 ,, ,,	8 ,,
(Eastern.) Jamaica, Western Labrador, Quebec, Ontario to 80° 30′ W, New Brunswick, Eastern Zone of the United States, Panama, Peru, Chile ...	75° W	5 ,, ,,	7 ,,
(Central.) Central Zones of Canada and United States	90° W	6 ,, ,,	6 ,,
(Mountain.) Mountain Zones of Canada and United States	105° W	7 ,, ,,	5 ,,
(Pacific.) British Columbia and Pacific Zone of United States	120° W	8 ,, ,,	4 ,,
Yukon, Alaska	135° W	9 ,, ,,	3 ,,
Sandwich Islands	157½° W	10½ ,, ,,	1½ ,,
Samoa	172½° W	11½ ,, ,,	0½ ,,

* Argentine. Uruguay from April 30, 1920.

A further proposal[1] is made for the numeration of meridians from 0° to 360° so that the meridian of Greenwich should be 180° and the sun should be on meridian 180° at 12 h and on meridian 0 at 0 h and 24 h.

[1] *The Times*, January 8, 1925.

The proposal is certainly a rational step in the simplification of reckoning. If it should be adopted a day of particular date would begin from midnight at Fiji, long. o°, o h at Greenwich would be 12 h (noon) of the same kalendar date at long. o° (180° E), it would be noon along long. 15° W at 1h G.M.T. and so on, at 23 h G.M.T. in long. 345°, and finish at midnight in long. 360°, when at Greenwich it would be noon of the next day. Thus the same kalendar day would be counted as ending at midnight for long. o° and beginning at the same hour for long. 360°; two kalendar dates are always running at the same time.

Surveying instruments

Next in importance to the means of measurement of time may be ranged the instruments designed for measuring angles, of which we may mention the azimuth-compass, prismatic compass, clinometer, theodolite, sextant and theodolite-sextant. The simplest are the azimuth-compass for measuring angles in the horizontal plane and the clinometer for measuring the angle between a sighted line and the vertical as indicated by a plummet. Concerning these we need only mention that in the prismatic compass, a reflecting prism brings an image of the point of the compass-needle and part of the scale beneath it into the line of sight upon the mark of which the direction is sought.

The object of the compass-needle is to obtain the North and South line. This requires a knowledge of the angle between the line of equilibrium of the needle and the true North and South line at the particular place and at the particular time. That angle is called the magnetic declination by writers on terrestrial magnetism, but for official purposes of navigation it is called the "variation of the compass." It is subject to diurnal variation which on the average amounts to 8·03' on "quiet" days in this country (varying from 11·01' in August to 3·34' in December) and to 12·00' on disturbed days, as well as to occasional fluctuations, which when they are of considerable magnitude are attributed to magnetic storms. It is also subject to a secular variation.

The normal variation of the compass is set out in charts issued by the Royal Observatory at Greenwich. They are constructed for a definite epoch which is marked on the charts, and the rate of the secular change in variation is marked upon them. For the transient changes caused by magnetic storms recourse must be had to the curves of a self-recording magnetograph or to the tabulations of the curves which are published by the magnetic observatories[1].

In order to avoid the necessity for determining the North and South line on every occasion of the determination of an azimuth it is desirable for each observing station to have a distant fixed mark of which the azimuth from the true North and South line has been determined once for all.

The next instrument in common use at meteorological observatories is the theodolite which is provided with a means of levelling, and with vertical and

[1] *Studies in Terrestrial Magnetism*, by C. Chree, Macmillan and Co., Ltd., London, 1912. A transcript of the charts for the epoch 1922 will be found in *Manual of Meteorology*, vol. II.

horizontal circles, for determining the altitude and azimuth of any visible object. Many special forms have been designed within the last twenty years in order to observe the altitude and azimuth of pilot-balloons, small balloons which are filled with hydrogen. After being liberated, they rise in consequence of the buoyancy and are carried along by the air-currents which the balloon traverses.

The common characteristic of the theodolites now used for pilot-balloons is a prism, which deflects the beam of light from the object sighted along the axis of rotation of the altitude circle, and which thus keeps the eyepiece of the instrument always at the same level. This modification of the ordinary theodolite of the surveyor was introduced by de Quervain in about 1905[1] and has added very largely to the facility of observation. Many minor modifications and improvements have been introduced subsequently by various makers of scientific instruments.

At sea the sextant becomes an essential instrument for meteorological observations. Its primary use is to determine the angle between the horizon and an object above it. It is so arranged that the horizon and the object are brought into coincidence, in the field of view of the object-glass of a telescope, by rotating a mirror. The angle of rotation of the mirror from its zero is one-half of the angle of which the measure is sought. The instrument can be used in like manner for measuring the angle subtended by the distance between any two visible objects by bringing the two objects into coincidence in the eyepiece of the telescope.

For watching pilot-balloons at sea a combination of the theodolite and the sextant has been introduced with success by Wegener and Kuhlbrodt[2]. The theodolite is hung from a stand so that its plane may remain horizontal in spite of the motion of the vessel. The addition of the sextant is of great assistance because when the horizon and the balloon are brought into coincidence in the eyepiece they remain in coincidence notwithstanding the slight motion of the whole instrument. A deviation of the balloon from the image of the horizon results from the motion of the balloon itself and that motion can be followed by the adjustment of the sextant, while the azimuth with reference to the ship's head can be read on the horizontal circle of the theodolite.

The photographic camera

Another instrument of common use which should find a place in every meteorological observatory is the photographic camera, though its use is not nearly so frequent as it deserves to be. The forms of cloud are the objects upon which the instrument is most frequently used, but they are only one among many desirable uses. The technical manipulation of the camera is a matter for photographic dealers, here we need only mention that for cloud

[1] *Über die Bestimmung atmosphärischer Strömungen durch Registrier- und Pilotballons*, by A. de Quervain, *Met. Zeitschr.*, vol. XXIII, 1906, p. 149.
[2] *Pilotballonaufstiege auf einer Fahrt nach Mexiko März bis Juni* 1922, von Alfred Wegener und Erich Kuhlbrodt. *Aus dem Archiv der Deutschen Seewarte*, vol. XL, 1922, Hamburg.

forms various coloured screens are desirable to obtain a good contrast in the picture. The instructions which were issued for the photography of clouds with a view to an international cloud-week in 1923, are as follows:

INSTRUCTIONS FOR TAKING PHOTOGRAPHS OF CLOUDS

Extract from a memorandum drawn up by G. A. Clarke for the *International Photographic Survey of the Sky*, September 24–30, 1923.

1. *For those who are not accustomed to Panchromatic plates.*

(a) The worker should use "Process" or "Slow Ordinary" plates, of speeds between 25 and 70 H and D, or about Wynne F 20 to F 40, *without* a filter.

Exposure-times at f 8 should be about $\frac{1}{25}$ or $\frac{1}{80}$ sec.

Develop with Hydroquinone at a temperature of 60° F or 16° C.

(b) If the worker prefers to use Orthochromatic or Isochromatic plates, it will be better to select those of the lower speeds, about 150 to 250 H and D, or Wynne F 60 to F 80. With these plates a yellow filter doubling the exposure should be employed, and a shutter-speed of $\frac{1}{50}$ sec. at f 8. If a yellow filter increasing exposure five times is used, the corresponding shutter-speed would be $\frac{1}{20}$ sec.

2. *For those who are accustomed to Panchromatic plates.*

Much depends upon the particular brand of plate employed, and the filters used with them. The filter screens used may be yellow, orange or red, the last named being the best for faint cirrus or cirro-stratus. Exposures will depend upon the plates and filters used.

In the case of Ilford Panchromatic plates, which are highly recommended for the purpose, the approximate details are as follows:

1. *Rapid Process Panchromatic.*

(a) Using Ilford Tricolour Red filter: exposure $\frac{1}{8}$ sec. at f 8.

(b) Using Wratten K2 filter (yellow): exposure $\frac{1}{18}$ sec. at f 8.

Develop in darkness for two minutes in the Pyro-Soda developer recommended by the makers of the plate, at a temperature of 65° F or 18° C.

2. *Special Rapid Panchromatic.*

(a) Using Ilford Tricolour Red filter: exposure $\frac{1}{20}$ sec. at f 8.

(b) Using Wratten K2 filter (yellow): exposure $\frac{1}{80}$ sec. at f 8.

Develop in darkness for six minutes in the Pyro-Soda developer recommended by the makers of the plate, at a temperature of 65° F or 18° C.

(The K2 filter is mentioned particularly, because many amateur workers already possess the filter.)

Note. The Pyro-Soda developer recommended above is mentioned merely because it is suggested by the plate-makers. Photographers who use other developers should modify the time of development accordingly.

INSTRUMENTS FOR THE EXPLORATION OF THE ELEMENTS OF ATMOSPHERIC STRUCTURE AND THEIR CHANGES

We may now refer to some points of special interest in connexion with the development of the equipment of observatories and stations which are in the more technical sense meteorological. We take first into consideration the instruments by which we obtain the information necessary to form correct ideas about the structure of the atmosphere, the general circulation and its changes. And here we may remark that the ideal instruments for this purpose must be self-recording, and with these we may include the photographic camera as a self-recording instrument for visual phenomena. We have seen that the results of a fully equipped observatory or station of the first order, according to the international code, are either autographic records or hourly observations by eye. The full use of stations of the second order, at which observations are taken at two, three or four fixed hours each day, called "term-hours," assumes the possibility of interpolating values between those at the term-hours, when the structure of the atmosphere in the interval has to be visualised. The ultimate impression of the atmosphere which we desire to obtain includes the continuous sequence throughout the day; and, though the interpolation between the fixed hours can be accepted with sufficient confidence for most meteorological purposes, it is extremely useful to have continuous records in reserve for the purposes of reference in case any discontinuity is suspected as forming part of the phenomena.

Wind

The chief element in the structure of the atmosphere is the wind, the motion of the air. It is also the most difficult to visualise from fixed instruments, designed for the express purpose, under the general name of anemometers[1], because all such instruments have to be attached to some structure. Within the limit of height of all solid structures the motion of the air is complicated by the eddies due to what is called the friction of the ground, partly the actual resistance of fixed obstacles, hills, buildings, trees, etc., or moving obstacles, waves of water, etc., and partly to the molecular viscosity of the air which would still produce some effect even if the surface of the land or sea were perfectly level. In Part IV we treat more fully of the light which has been thrown on this part of the subject by the discussion of actual observations; we may content ourselves for the time being by pointing out that the layer of the atmosphere near the surface, where instruments for measuring the direction and velocity of the wind are of necessity installed, is most unsatisfactory for the purpose, and this opinion is amply justified by the extraordinary complication which the records of any anemometer disclose. Some are more complicated than others but all show some degree of complication, and the problem of obtaining an idea of the changes of the general atmospheric structure from the records of an anemometer is a very awkward one.

The number of anemometers designed for measuring or recording the direction and velocity of the wind is very large. One of the earliest kind, and perhaps the most common, is shown in fig. 12. Hand-instruments of that type were largely used during the war. Meteorological practice has in the course of time concentrated its attention on two forms of anemometer, the Robinson anemometer which consists of horizontal cross-arms with hemispherical cups at their ends, the rate of rotation of which is very nearly proportional to the "mean" velocity of the wind, and the Dines anemograph which, like the Pitot tube, depends upon the pressure exerted by the wind upon the opening of a tube. It is measured by some form of pressure-gauge

[1] A detailed description of the instruments is given in vol. III of the *Dictionary of Applied Physics*, Macmillan and Co., Ltd., London, 1923, s.v. Meteorological Instruments. Specimens of records are given in Part IV of this Manual.

always with the proviso that since a pressure-gauge has two openings, one for each side of the gauge, the instrument must make due allowance for the effect of the wind upon the second limb of the gauge. Another common form of anemometer that is sometimes arranged to record automatically is the wheel with inclined vanes, on the same principle as an ordinary wind-mill.

In each of the typical instruments the direction of the wind is separately recorded. The Robinson anemometer was put into recording form by Beckley at Kew Observatory (Richmond) in 1856 and a direction-recorder, actuated by double fan-wheels, on an axis which is in equilibrium when it lies across the wind, brought the instrument within the full definition of an anemograph. A full description of the original instrument is given in a paper by T. R. Robinson, 'Description of an improved anemometer for registering the direction of the wind and the space which it traverses in given intervals of time,' published in the *Transactions of the Royal Irish Academy*, vol. XXII, 1850, pp. 155–178. Its most notable characteristic is its consistent working over a long period of years. Its chief feature from the meteorological point of view is that in virtue of the momentum of the cross and cups it does not respond to the variations of the wind which are due to the eddies of local origin, a feature of surface-winds which in this book is referred to as gustiness; it gives a measure of the wind from which the transient variations have been smoothed. The actual record is made on a revolving drum by silver spirals which are turned by the wind.

A modification of the Robinson anemograph is in general use at the meteorological stations of the United States and Canada, in which the record is obtained by dots made at intervals by electrical contacts. The record is in consequence not actually continuous. Directions are shown by operating four levers, one for each cardinal point, the intermediate direction is indicated when two of the levers operate simultaneously.

The anemograph of W. H. Dines on the other hand, being dependent upon the pressure of the wind upon the opening of a tube transmitted to a cistern in which it can operate a float carrying a recording pen, shows with remarkable fidelity the eddy-variations of the wind; as many as twenty changes of wind velocity may be indicated in the course of a minute[1]. The use of the instrument was indeed discountenanced for some years because of the detailed nature of its record. There was no simple method of obtaining an indication which would correspond with the mean wind of the Robinson anemograph. For the study of atmospheric changes however these details of wind-variation are of interest and importance.

The record of direction is obtained in this instrument by recording the motion of the vane which brings the opening of the tube to face the wind. In 1917 objection was taken to the shape of the vane used for this purpose as having no relation to stream-line shape and being in consequence itself liable to affect the trace of the instrument by eddies of its own causing.

[1] See Part IV, chap. V, p. 41.

A new form of head was therefore substituted provided with a "stream-line" appendage in order to reduce these instrumental irregularities.

A fourth method of recording the force and direction of the wind is by means of a plate which is turned by a vane to face the wind and which records the force by the compression of a spring. This plan is used in Osler's anemometer which has been in continuous operation at the Royal Observatory, Greenwich, since January 13, 1841. The records are published in the reports of the Observatory.

They ought to show a relation with the records of the Robinson anemograph not differing much from:

$$F = \cdot 003 V^2,$$

where F is the pressure in lbs. per sq. ft. and V the velocity in miles per hour. They do not do so. The differences may perhaps be accounted for by the fact that the plate would adjust its record to the extremes of a transient gust and the anemograph shows only the smoothed value. But that is only a partial explanation, the differences are too great. One of the curiosities of meteorological work upon wind is that differences of the kind here referred to are tolerated for years without anyone feeling it necessary to explore the subject to the point of actual conviction.

The secret of that really intolerable toleration is the basic difficulty of all anemographic records—the exposure. The reading of any anemograph is a function not only of the instrument but of the site, and of the shape and orientation of the structure upon which the instrument is mounted. Any flat vertical surface exposed to the wind produces a localised eddy analogous to what is treated elsewhere as a cliff-eddy, and a few degrees of difference in the orientation of the wind may have a considerable effect on the record. The conclusion arrived at in the Meteorological Office was that nothing short of a separate structure, a tower of open ironwork, on a very open space of level ground was really efficient and even in that case, as may be seen in Part IV, distant geographical features may have a paramount influence upon the record of wind. The standard mounting of a tube-anemometer is with the vane 40 feet above the ground. When a site well away from trees or buildings cannot be secured the vane should be at least 20 feet above them.

Hence it has come about that unless the local opportunity for exposure was exceptionally good it was not thought desirable to insist upon, or even to advise, the erection of an elaborate instrument for recording the wind. Wind did not really lend itself to recording, except in a specially local sense, local as to building as well as site and general locality. It was thought better to get the general impression of the wind which is expressed by the adaptation of the Beaufort scale to observations on land than to obtain a more precise numerical value which had no meteorological significance of the same order of accuracy. No structure of meteorological reasoning can be raised without a tolerance of at least 20 per cent. in the assigned values of surface-winds.

Pressure

From the consideration of the surface-wind, the most intractable of all the meteorological elements in respect of accurate representation, we pass to pressure, the most delicately accurate of meteorological measurements. We enter a new field in which we want an accuracy of one part in ten thousand instead of one part in five, and yet we are so sure that in a strictly meteorological sense there is a very close relation between the wind and the distribution of pressure that we have devoted a separate part of the book, Part IV, to its consideration. It took however some 250 years to arrive at what we will call finality in the representation of pressures. People are apt to think when they keep a barometer in their hall, and read it as they tap it, that they understand what a meteorologist means by atmospheric pressure. That is far from the truth and on that account, if on no other, atmospheric pressure, in the sense in which a meteorologist uses the word, deserves a special unit with a special name—millibar. We have already considered the history of the barometer. Very soon after its invention it was recognised as being closely associated with changes in the weather. Pascal found that the length of the column of mercury varied with elevation. Hooke provided the mercury with an easily legible index in the shape of a wheel-barometer, and somebody attached legends to the position of the index which are perhaps more appropriate to London than anywhere else but which are certainly not appropriate to any place that is not very near sea-level in the temperate latitudes of Europe.

The effect of placing the familiar legends on the dial is to suggest that the weather is dependent directly upon the height of the barometer, or, in modern phraseology, that the weather is the same all along a line of equal pressure, not indeed merely along an isobar as drawn on a modern map but all over an isobaric surface. That is one of the ideas which has to be unlearned by a beginner in the study of meteorology. Precision was introduced into the measurement of the length of the column and when temperature also had become capable of exact measurement it was learned that the height of the mercury-column required correction for temperature. Laplace provided a formula for the variation of the pressure with height, and when maps came to be made it was found that the variation was so marked that no regularity was indicated unless the pressures at the various points of observation were referred to the same horizontal surface or the pressure was "reduced to sea-level." Finally it was agreed that the hydrostatic pressure of the air, not the height of the barometer, was the quantity which lent itself to meteorological computation, and the correction for latitude was introduced into synoptic charts but only in the twentieth century (see *Codex of Resolutions adopted at International Meteorological Meetings*, 1872–1907, M.O. No. 200, p. 11). When all this transformation of the readings of the "argento vivo" was agreed upon and accomplished it was found that to draw the isobaric line with sufficient regularity the readings as corrected and reduced must be accurate within the tenth of a millimetre or millibar, that a difference of pressure on the map of one millibar in forty kilometres was as a rule significant of a gale in the latitude of London. Nearer the equator the importance of pressure-difference becomes enhanced until in the neighbourhood of the equator we cannot account for persistent difference of local pressure except by supposing that the isobar is suitably curved.

Besides other things reduction to sea-level requires an accurate knowledge of the height of the station at which observations are made or, more strictly speaking, of the cistern of the barometer. The height can only be correctly ascertained by levelling. In new countries where levelling has not been carried out it is possible to plot and work with difference from mean values at the stations instead of the computed pressure at sea-level.

The maps would be quite similar if the pressure were normally uniform at the sea-level; that is not, strictly speaking, the case, but the difference is not great enough to obscure the main outlines of notable distributions. Differences from mean values were used for Europe by Brandes in 1820 and for the Transvaal by Innes in 1906.

The historical development of the mercury-barometer led to its being graduated in such a way that its readings required a large correction for temperature. Barometers graduated in millimetres gave correct readings of pressure in standard millimetres in latitude 45° when the barometer was in that latitude, and the mercury and scale at the freezing-point of water. The Kew-pattern barometer, which had a small cistern and a scale adjusted to allow for the change in the level of the mercury in the cistern, was specially designed for use at sea. Its readings could be certified as giving correct readings in true inches of standard mercury when the barometer was in latitude 45°, the mercury at 32° F and the scale at 62° F. That was a complication which no seaman was expected to understand. It was therefore ordered that the barometric readings and the temperature of the attached thermometer should be entered and the readings reduced by a staff of clerical experts on shore. Something less complicated was desirable and in 1914, when barometers graduated in a scale of millibars were introduced into British practice, the procedure was simplified by certifying a temperature called the "fiducial temperature" at which the readings were correct in a given latitude. Simple instructions for correction in that case were substituted for the previous complication.

Another notable attempt to provide for satisfactory observations at sea is to be found in the sympiezometer which was in fact an "air-thermometer" with an open tube and consequently affected by variations of temperature as well as those of pressure. An auxiliary mercury-thermometer and a scale adjusted thereto gave the means of correcting the reading for variations of temperature. But the sympiezometer went out of use after the development of the Kew-pattern barometer.

Towards the middle of the nineteenth century (1843) the aneroid barometer was introduced, the invention of Lucien Vidie. It consists of a shallow

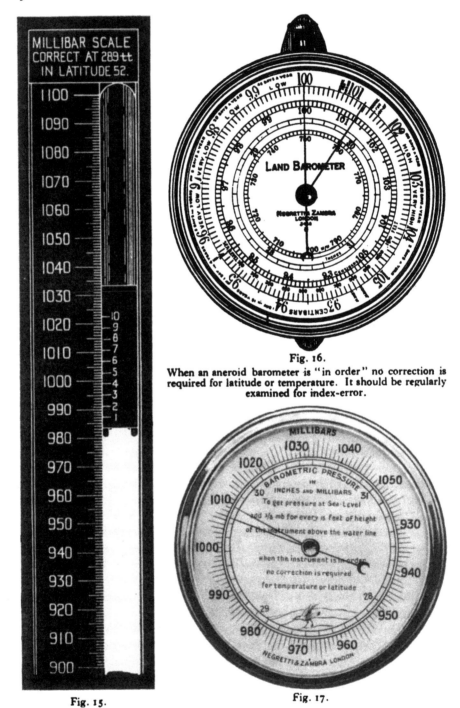

Fig. 16.

When an aneroid barometer is "in order" no correction is required for latitude or temperature. It should be regularly examined for index-error.

Fig. 15.

Fig. 17.

Fig. 15. Mercury barometer, Fortin pattern, adjusted to read correctly in millibars at 289tt in latitude 52°. The reading requires correction for other latitudes and temperatures, and reduction to sea-level.

Fig. 16. Barometric pressure in millibars, "mercury millimetres" and "mercury inches." Aneroid barometer with adjustable scale for automatic reduction to sea-level.

Fig. 17. Barometric pressure in millibars. Aneroid barometer for use at sea.

capsule of thin metal very nearly exhausted of air. The opposite faces of the capsule are kept apart by a metal spring. The residual air can be so adjusted as to give an automatic correction for temperature, and the displacements of the face of the capsule under the variations of barometric pressure are indicated on a dial by a train of mechanism. It can be made extremely sensitive and as a portable instrument is quite unrivalled. It has the further advantage that its indication depends on the pressure of the atmosphere alone without being affected by variation in the force of gravity. It is however subject to vexatious errors on account of irregularities in the elasticity of the metals of which it is constructed. On that account it is not admitted to the position of a standard instrument for meteorological observations. Nevertheless with the customary legends added it is the most popular instrument for showing variations of barometric pressure.

Self-recording barometers were attempted in very early times. The regular recording with the accuracy necessary for meteorological work was provided at Kew Observatory (Richmond) in 1862 and instruments of that type were installed in many parts of the world.

For the Meteorological Office portable self-recording instruments using an aneroid were obtained from the firm of Richard Frères, Paris, in about 1889.

Part of the provision for accuracy was the regular observation of standard barometer-readings at fixed hours as check-readings, and these are even more important with the reading of portable barographs than with the photographic record; but in fact they are indispensable for both.

In 1903 however W. H. Dines, at the request of the present writer, designed an air-thermometer the container of which could be protected by casing against rapid changes of temperature. The variations in volume in consequence of changes of external pressure were recorded on a revolving drum; the instrument is still used as a microbarograph. The movement of the pen is effected through the lifting or sinking of a cylindrical bell which rests in a bath of mercury. It magnifies the rapid changes of pressure twenty-fold; at the same time a small orifice in the tube between the container and the floating bell forms a very slow leak which relieves the index from the effect of slow changes of pressure and temperature. Subsequently Mr Dines introduced a well-designed form of the wheel-barometer which gives sufficiently accurate readings for meteorological purposes and a sufficiently large scale to make further mechanical magnification unnecessary.

The recording mercury-barometer which is in most frequent use on the continent is Sprung's weight barograph[1].

Temperature. Screens

The measurement of temperature, whether of the dry or wet bulb, is susceptible of ample accuracy so far as the instrument is concerned whether in the form of the mercury-thermometer, or the bimetallic spiral or Bourdon tube used in portable self-recording instruments, but as a meteorological element the temperature of the air is a very indefinite quantity. It varies

[1] 'Eine neue Form des Wagebarographen,' Rep. für Exp. Physik, vol. XIV, 1878, p. 46.

considerably in neighbouring positions according to the degree of shade, sunshine or other form of radiation. The temperature indicated on the instrument depends not only upon the temperature of the air which it is intended to measure but also on the loss or gain of heat by radiation to or from surrounding objects. It is an accepted principle that the true temperature of the air will be obtained by sufficiently rapid motion of the bulb of the instrument through the air, and on that account provision is made for drawing air mechanically over the bulb of the thermometer as in the Assmann psychrometer, or by causing the thermometer to make rapid motion through the air by whirling it. A thermometer used in this way is called a "sling thermometer" and various contrivances for whirling or slinging have been used in the United States[1] or in the Antarctic[2]. There is of course a limit to the speed of relative motion if the bulb is to register the true temperature of the air. With slow motions the wave of compression of the air in front of the moving bulb is transmitted with the velocity of sound and is negligible so long as the relative motion is small, but with increase of velocity a state of things is arrived at when the air immediately in contact with the moving body is appreciably warmed by compression. With high velocities the warming may be very considerable. Thus a rifle-bullet is always entering warm air of its own creation and the luminosity of meteors can be explained in a similar way and indeed in no other[3]. We have no definite information as to the limit of speed which is permissible with a sling thermometer, but it would not be difficult to calculate it.

Meanwhile in order to secure the most favourable conditions for getting the temperature of the air at an ordinary meteorological station general agreement has been arrived at to keep the thermometers encased in a "screen," a box to be set on a stand in the open, all four sides of the box composed of "double louvres" and free passage of air allowed through the bottom and under the roof. A screen of this form was designed by Thomas Stevenson, C.E.[4], engineer to the Scottish Board of Northern Lights, perhaps better known now as the father of Robert Louis Stevenson. After examination a standard form was elaborated by a Committee of the Royal Meteorological Society and this seems now to be adopted by many countries under the name of the "English screen," "Englische Hütte." Other larger screens more or less on the same principle are adopted for special purposes such as the housing of self-recording thermometers too large for the Stevenson screen. The screen should stand on four legs so that its base is about 3 feet 6 inches above the level of the ground, and there should be a space of at least 3 inches between the bulbs of the thermometers and the top, bottom, or sides of the screen.

[1] *The Aims and Methods of Meteorological Work*, by Cleveland Abbé, Maryland Weather Service, The Johns Hopkins Press, 1899.
[2] *British Antarctic Expedition* 1910–13, *Meteorology*, by G. C. Simpson, vol. 1, Discussion, Calcutta, 1919, p. 18.
[3] 'A theory of meteors and the density and temperature of the outer atmosphere to which it leads,' by F. A. Lindemann and G. M. B. Dobson, *Proc. Roy. Soc.* A, vol. CII, 1922–3, p. 411.
[4] *Journ. Scott. Meteor. Soc.*, vol. I, 1866, p. 122.

The original principal observatories of the British Meteorological Service repeat the model adopted at Richmond (Kew Observatory) in 1866. This consists of a large single-louvred screen attached to the north wall of the observatory through which the long tubes of the thermometers pass to the contrivance for photographic registration.

Another form of exposure which has been used at the Royal Observatory, Greenwich, and elsewhere, for many years is the Glaisher screen, a vertical board upon which the thermometers are hung. The board has its back to the sun and has to be turned in the long days to prevent the sun from shining on the instruments.

Another form of screen much smaller than the standard size is used for exposing the dry and wet bulb thermometers on ships at sea. One of the most striking ironies of meteorological work is the difficulty of getting a proper approximation to the temperature of the air at sea, still greater is the difficulty of getting accurate simultaneous readings of the dry and wet bulb.

The meteorological service of India was accustomed to use an open-sided hut with a thatched roof as a shelter in which to hang a wire-cage containing thermometers, and the same form of exposure has been recommended to other tropical countries by the Meteorological Office. It is however now understood that the Stevenson screen is preferred in India to the hut[1].

The real uncertainty of the meaning of the word temperature as used in meteorology is eloquently expressed by the comparisons of the readings of the maxima and minima with various exposures in the same locality which are published from time to time[2]. We give an example from a recent paper[3]:

Comparison of temperatures in Glaisher and Stevenson screens
35-year Averages, British Rainfall Organization, London, 1881–1915

	Max. Temp. °F		Min. Temp. °F		Mean Temp. °F		
	Ste.	Gla.	Ste.	Gla.	Ste.	Gla.	Diff.
January	43·9	43·5	34·7	34·0	39·30	38·75	0·55
February	45·8	45·4	35·3	34·4	40·55	39·90	0·65
March	50·2	50·1	36·5	35·6	43·35	42·85	0·50
April	57·0	57·4	40·4	39·4	48·70	48·40	0·30
May	64·2	64·9	46·2	45·2	55·20	55·05	0·15
June	69·8	71·0	51·9	51·0	60·85	61·00	−0·15
July	72·8	74·1	55·4	54·4	64·10	64·25	−0·15
August	71·5	72·6	54·7	53·6	63·10	63·10	0·00
September	66·7	67·4	50·9	49·8	58·80	58·60	0·20
October	57·4	57·5	44·9	43·9	51·15	50·70	0·45
November	50·0	49·7	39·7	38·9	44·85	44·30	0·55
December	45·5	45·1	36·5	35·7	41·00	40·40	0·60
Year	57·9	58·2	43·9	43·0	50·90	50·60	0·30

[1] 'On Exposures of Thermometers in India,' by J. H. Field, *Memoirs of the Indian Meteorological Department*, vol. XXIV, part III, Calcutta, 1922.

[2] G. M. Whipple, 'Report on Experiments made at Kew Observatory with Thermometer Screens of different patterns during 1879, 1880 and 1881,' *Quarterly Weather Report of the Meteorological Office*, 1880, Appendix II, pp. [13]–[16]. R. T. Omond, 'Temperatures in thermograph and Stevenson screens,' Edinburgh, *Jour. Scott. Meteor. Soc.*, 3rd ser. vol. XIV, 1906, p. 15.

[3] I. D. Margary, 'Glaisher stand versus Stevenson screen,' *Q. J. Roy. Meteor. Soc.*, vol. L, 1924, pp. 209–26.

As regards the extreme values of the difference of the daily maxima after elimination of all apparent errors of reading, etc., there are several genuine cases of 4·4°, but 3·5° may be taken as the usual upper limit. Similar negative values occur, the highest being − 4·3°, while several exceed − 3·5°.

Extreme values of the difference of the daily minima are rather smaller than for the maxima values. The highest value is 3·7° and many exceed 3°. Only a few days have negative values over − 1°, and none over − 1·5°.

Further difficulties in respect of measurements of temperature are concerned with the estimation of the "mean temperature of the day" from observations at fixed hours alone, or from the maximum and minimum readings alone, or from a combination of the two. Too much variety is shown in the formulae used in different parts of the world. A résumé is given by C. E. P. Brooks in the volumes of the *Réseau Mondial*[1]. It is a subject to which J. Hann had devoted a great amount of attention[2]; but even now there is much diversity of practice that is not likely to be removed until a Meteorological Office for the world undertakes the compilation of the data from the world point of view. In the meantime too much stress must not be laid upon the details of the distribution of temperature, a good deal of variation may come within the rather wide limits of probable errors of the measurement, which arise not from defects in the instruments but from local differences in the elements to be measured.

With this long preface as to the inherent uncertainties of measurements of temperature which are really not much less than those of wind we can only enumerate the instruments which after very great variety of experiments commend themselves in practice.

For the dry and wet bulb, ordinary mercury-thermometers graduated in Fahrenheit or Centigrade, or in the Tercentesimal scale, or in McAdie's Kilograd scale[3], with absolute zero and the freezing-point of water 1000.

For maximum temperature, the constricted tube of Messrs Negretti and Zambra.

For minimum temperature an alcohol-thermometer with an index.

An Assmann psychrometer with dry and wet bulbs in separate tubes of highly polished metal with means for pulling a current of air past them.

A frame for whirling wet thermometers or both, on the lines of the sling thermometer.

A self-recording thermometer of Richard Frères with Bourdon tube, or of Short and Mason, Ltd., with bimetallic spiral, provided with an auxiliary mercury-thermometer to check the accuracy of the scale of the paper and its setting.

The Callendar self-recording electrical thermometer.

[1] *British Meteorological and Magnetic Year Book*, 1916, Part v, Réseau Mondial, M.O. No. 227 g.

[2] J. Hann, *Der tägliche Gang der Temperatur in der äusseren Tropenzone*, 2 vols., Wien, 1907. *Der tägliche Gang der Temperatur in der inneren Tropenzone*, Wien, 1909.

[3] As I was writing this sentence (January 12, 1925) a parcel was brought to me from Messrs Negretti and Zambra containing a specimen of McAdie's multiple scale thermometer which, by means of additional scales hinged at the top of the metal frame on which the thermometer was mounted, made it easy and even convenient to read the temperature in any one of the four scales and, in addition, two other scales similarly hinged gave at sight the saturation density and pressure of water-vapour.

Humidity

The next meteorological element to be considered is the water-vapour in the atmosphere which may be represented by the number of grammes of water-vapour per cubic metre at the time of observation, or the partial pressure of the vapour, that part of the atmospheric pressure which is supported by the vapour, or the relative humidity, the percentage ratio of the density or pressure of water-vapour to the maximum possible corresponding with the temperature at the time. The same figure expresses the ratio of the vapour-pressure at the time of observation to the pressure of saturation at the observed temperature of the air. Questions of humidity are of such frequent occurrence that it may be well here to repeat the calculations which will be considered in relation to another context further on, when we will call attention to a suggestion of S. Skinner that water in the spheroidal state in a red-hot crucible, and therefore equally Carnelley's "hot-ice" in an atmosphere at very low pressure, may be regarded as exaggerated examples of wet bulb temperatures, that is to say of temperatures kept low by evaporation in an enclosure of higher temperature.

The density of water-vapour under all circumstances is a fixed fraction ·622 of the density of dry air at the same temperature and pressure, and this latter can be calculated from the observed pressure of water-vapour e and the temperature t at the time. Thus if d is the density of the water-vapour, D the saturation density at the temperature of observation t, Δ the standard density of dry air = ·001201 grammes per c.c. at 1000 mb and 290 t, E the saturation vapour-pressure at the observed temperature and H the relative humidity,

$$d = 0.622 \times \frac{e}{1000} \times \frac{290}{t} \Delta,$$

$$D = 0.622 \times \frac{E}{1000} \times \frac{290}{t} \Delta.$$

Hence $\qquad e/E = d/D = H.$

The instruments in use for determining humidity are chiefly the wet and dry bulb thermometers which require a formula of reduction. According to experimental work carried on from the time of Regnault to the introduction of the tables for the Assmann psychrometer, different tables are required for calm air, and for a strong wind, with an intermediate formula for a light wind. Tables in this form were introduced by C. Jelinek[1] of the Austrian Meteorological Service. We quote from the *Computer's Handbook*.

The formulae used by J. M. Pernter (a successor of Jelinek's) in a revised edition of the tables are, when the cover of the wet bulb is not frozen:

Calm (wind-velocity in the screen 0 to 0·5 m/s)
$$e'' = e' - ·0012\, B\, (t - t')\, (1 + t'/610).$$
Light wind (wind-velocity in the screen 1 to 1·5 m/s)
$$e'' = e' - ·0008\, B\, (t - t')\, (1 + t'/610).$$

[1] *Psychrometer-Tafeln für das Hunderttheilige Thermometer nach Dr H. Wild's Tafeln*, bearbeitet von Dr C. Jelinek, Zweite Auflage, Wien, 1876.

13-2

Strong wind (wind-velocity in the screen greater than 2·5 m/s)

$$e'' = e' - \cdot 000656 \, B \, (t - t') \, (1 + t'/610),$$

where e'' and e' are the vapour-pressures at the dew-point and at the temperature of the wet bulb respectively; t and t' are the readings of the dry and wet bulbs in degrees Centigrade and B is the pressure of the air measured in the same units as e' and e''.

When the water in the cover of the bulb is frozen the formulae are modified by reducing the numerical coefficients to ·001060 for calms, ·000706 for light wind, and ·000579 for strong wind, and by writing 689 for 610 in the last factor.

For the tables prepared for *Assmann's ventilated psychrometer* the numerical co-efficient is taken as ·5/755 = ·000662, and agrees very closely with Pernter's formula in the case of strong wind.

The tables used in India compiled originally by Blanford and recently revised are based upon the formula:

$$e'' = e' - \cdot 000437 \, B \, (T - T') \, \{1 + (T' - 32)/1098\},$$

where T and T' are the temperatures in degrees Fahrenheit. When converted to the Centigrade scale the formula becomes

$$e'' = e' - \cdot 00079 \, B \, (t - t') \, (1 + t'/610),$$

and is almost identical with Pernter's formula for light winds.

The tables used in the United States of America are those in the Fahrenheit scale compiled by Marvin from Ferrel's formula

$$e'' = e' - \cdot 000367 \, B \, (T - T') \, \{1 + (T - T')/1571\}.$$

They are intended for use with sling thermometers which are whirled in the air before taking the reading: they are therefore comparable with Pernter's table for strong wind.

The formula includes a term which involves $(T - T')^2$; the reader will note that the importance of that term increases very rapidly with increasing dryness of the air. Details are given in the *Computer's Handbook*, Section 1, p. 15, London, H.M.S.O. 1916.

This really leaves the measurement of humidity in an unsatisfactory position. Certainly in a closed room without special ventilation of the wet bulb the "calm" formula is required and with the Assmann psychrometer or some other means of securing a vigorous air-current over the bulb the "strong wind" formula is needed, in other circumstances the general opinion as expressed by common practice is that the light wind formula will do.

British observations of dry and wet bulbs have, up to recent years, been reduced by tables compiled by J. Glaisher from many thousands of observations of comparison between the readings of the dry and wet bulbs and of the dew-point by Daniell's hygrometer. The method of the tables was based upon an empirical table of Greenwich factors by which the "depression of the wet bulb" must be multiplied to give the depression of the dew-point. One of the strongest points in favour of Glaisher's tables is that they never give manifestly absurd results, which may happen in very dry circumstances to the adjusted formula of Regnault. We must postpone the physical aspects of this interesting question to Vol. III and in the meantime in the meteorological world opinion appears to be gravitating towards the light wind formula as a workable compromise for a wet bulb with ordinary exposure. We give on p. 197 a brief table for the reduction of readings of dry and wet bulbs based upon it.

TABLE FOR OBTAINING THE RELATIVE HUMIDITY FROM READINGS OF THE WET AND DRY BULB THERMOMETERS

Barometer 755 mm = 1007 mb
Light Wind

Saturation vapour-pressure	Dry bulb	Relative humidity								
		90	80	70	60	50	40	30	20	10
mb	t	Readings of the wet-bulb on tercentesimal scale								
62·2	310					301·3	299·3	296·9	294·5	
58·9	309					300·5	298·3	296·3	293·9	
55·7	308				301·5	299·7	297·5	295·5	293·2	
52·7	307				300·7	298·9	296·9	294·7	292·5	
49·8	306				299·9	298·0	296·1	294·1	291·7	
47·1	305			300·7	299·0	297·1	295·3	293·4	291·3	
44·5	304			299·7	298·1	296·5	294·5	292·7	290·7	
42·0	303			298·9	297·2	295·5	293·9	291·9	289·9	
39·7	302		299·3	297·9	296·3	294·7	293·1	291·3	289·3	
37·4	301	299·7	298·4	297·0	295·5	293·9	292·1	290·5	288·5	
35·3	300	298·7	297·5	296·1	294·5	293·1	291·4	289·7	287·9	
33·3	299	297·7	296·5	295·1	293·7	292·3	290·6	288·9	287·3	
31·4	298	296·8	295·5	294·3	292·8	291·4	289·9	288·3	286·7	
29·5	297	295·8	294·6	293·3	291·9	290·5	289·1	287·6	285·9	
27·8	296	294·8	293·7·	292·5	291·1	289·8	288·4	286·8	285·3	
26·2	295	293·9	292·7	291·5	290·3	288·9	287·6	286·1	284·7	283·1
24·6	294	292·9	291·8	290·6	289·4	288·1	286·8	285·5	284·0	
23·1	293	291·9	290·9	289·7	288·6	287·3	286·0	284·7	283·3	
21·8	292	291·0	289·9	288·8	287·7	286·5	285·3	284·0	282·6	281·3
20·4	291	290·0	289·0	287·9	286·8	285·7	284·5	283·3	282·0	
19·2	290	289·0	288·0	287·0	285·9	284·8	283·7	282·5	281·3	280·0
18·0	289	288·1	287·1	286·1	285·1	284·0	282·9	281·7	280·6	
16·9	288	287·1	286·1	285·2	284·2	283·1	282·1	281·0	279·9	
15·8	287	286·0	285·2	284·3	283·3	282·3	281·3	280·2	279·2	278·0
14·9	286	285·1	284·2	283·4	282·5	281·5	280·5	279·5	278·5	
13·9	285	284·1	283·3	282·5	281·5	280·7	279·7	278·7	277·7	
13·0	284	283·2	282·4	281·6	280·7	279·8	278·9	278·0	277·1	
12·2	283	282·3	281·5	280·7	279·8	279·0	278·1	277·2	276·3	
11·4	282	281·3	280·5	279·7	278·9	278·1	277·3	276·5		
10·7	281	280·3	279·5	278·9	278·1	277·2	276·5	275·7		
10·0	280	279·3	278·7	277·9	277·1	276·5	275·7	274·9	274·1	
9·3	279	278·3	277·7	277·0	276·3	275·6	274·9	274·1		
8·7	278	277·4	276·7	276·1	275·4	274·7	274·1	273·3		
8·1	277	276·4	275·8	275·2	274·5	273·9	273·3			
7·5	276	275·4	274·9	274·3	273·7	273·0				
7·0	275	274·5	273·9	273·3						
6·5	274	273·5								
6·1	273									

Add 10 per cent. to relative humidity for 3000 metres of height for wet bulb readings on right of ladder.

Add 5 per cent. to relative humidity for 3000 metres of height for wet bulb readings between the two ladders.

We give also in fig. 18 a series of graphs for the determination of the relative humidity from the readings of the dry and wet bulbs according to Jelinek's formulae for calm and strong wind.

To use the graphs, start from the dry bulb temperature as indicated on the right hand edge; travel from that point parallel to a full line until the temperature has descended to that of the wet bulb. Read off the relative humidity indicated by the horizontal distance from the left as marked at the bottom. That gives the value for strong wind. For calm travel along a dotted graph instead.

We must repeat however that the measurement of humidity is by no means disposed of by the acceptance of a common formula for the reduction of observations with the wet and dry bulbs. We quote from the *Computer's Handbook*[1] the results of some observations of the behaviour of a bulb surrounded by a jacket of wet cotton-wool in a current of air dried by liquefaction.

Depression of the Wet Bulb in perfectly dry air

(From the laboratory of the Royal Institution)

Number of experiment	Flow of air*	Temperature of dry bulb	Temperature of water-bulb†	Temperature of ice-bulb	"Depression of wet"
n	li/sec	tt	tt	tt	tt
5	·22	272·9	<272·4	269·3	3·6
2	·067	273	—	270·2	2·8
4	·094	273	<271·2	270·1	2·9
6	·16	273	<269·3	270·2	2·8
3	·134	273·1	<269·9	—	>3·2
15	·22	278·2 278·7	272·4	—	6·3
16	·24	278·6 278·1	272·2 272·6	—	5·8
9	·22	281·1	<274·1	—	7·0
10	·47	281·2	274·1	—	7·1

The symbol < is inserted when the shape of the curve suggests that the temperature had not reached a steady state when freezing took place or the experiment terminated.

The two temperatures of the dry bulb in experiments 15 and 16 are the extremes between which the water-bath varied during the experiment.

* The cross-section of the channel was of the order 1 cm² so that the velocity of the stream of air passing the wet bulb may be determined very roughly in metres per second by multiplying the numbers given in this line by 10.

† The figures in this column show the extent of the super-cooling.

We draw the conclusion that freezing may or may not occur at any temperature over a considerable range, and that as the supply of moisture to the wet bulb by capillary action is fundamental for the instrument it ceases to behave as a scientific instrument when the temperature of the wet bulb is below the freezing-point of water.

Consequently some other form of hygrometer is not only desirable but even necessary. The hair-hygrometer originally elaborated by De Saussure was condemned by Regnault, but has been re-introduced into practice by the Norwegian Meteorological Service; and the change in the tension of a hair or a bundle of hairs has been made a means of recording the variations of humidity by Richard Frères in their recording-hygrometer and by many designers of meteorographs for the upper air.

[1] *Computer's Handbook*, M.O. 223, Section 1.

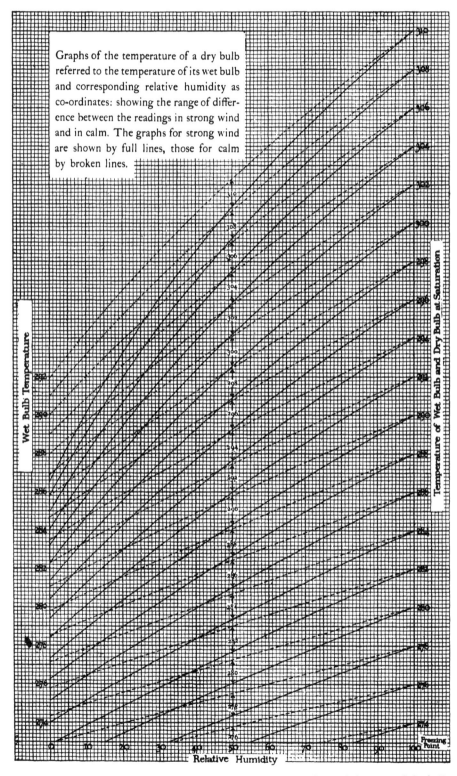

Fig. 18. Diagram for obtaining the relative humidity from readings of the wet and dry bulb thermometers in strong wind and calm.

Note on the density of atmospheric air

The density of the air is naturally an element of great importance in the dynamics of the atmosphere but there is no separate instrument for its determination. The values are obtained when required from the simultaneous observations of pressure p, temperature t and pressure of aqueous vapour e. The formula by which the density d is calculated is

$$d = \frac{p - (1 - a)\,e}{1000} \times \frac{290}{t} \times \Delta,$$

where a is the specific gravity of aqueous vapour referred to dry air at the same temperature and pressure, and Δ is the standard density of dry air containing 0·04 per cent. of carbon dioxide. The value of a is the same for all pressures and temperatures within the range of meteorological observation and is numerically equal to 0·622 which does not differ materially from $\frac{5}{8}$, and Δ in grammes per cubic metre is 1201; 290t is a convenient temperature at which to take the standard density because it is a temperature which frequently occurs in practice and differs in fact little from the comfortable temperature of a living room. It corresponds with 62·6° F or 17° C.

Rainfall

The measurement of rainfall is one of the oldest of meteorological measurements. An ancient Corean rain-gauge dating from the fifteenth century is figured in many books[1]. The earliest known measurements of rainfall in Great Britain are said to be those made at Townley near Burnley, Lancashire, dating from 1677. We need not enter into the details of the development of the modern rain-gauge which differs from the most simple form of instrument in the care taken to secure a true circular edge, to prevent the loss of water from the receiving vessel by evaporation and to guard against the loss of precipitation in the form of snow in consequence of the eddies due to the wind passing over the gauge.

The effect of wind on the amount of precipitation collected in a rain-gauge is indeed a question of some importance in exposed situations and not without difficulty. In order to reduce the effect a shield has been proposed by Professor F. E. Nipher of St Louis and is used occasionally. It consists of a conical jacket surrounding the body of the gauge. Its diameter varies from that of the gauge itself at the bottom to four or five times that diameter at the top. The top of the shield is level with the receiving surface. The shield may be made either of solid metal or of fine wire-netting, in some cases the inner surface of the jacket is covered with wire-netting. The netting is found to be almost as effective as the solid metal in reducing the eddying and is not so liable to cause insplashing of rebounding drops of water.

In the absence of any such protection the amount of rainfall is found to be less in elevated or exposed situations than in situations at ground-level with the ordinary protection of surrounding fences, trees or buildings, and

[1] *Q. J. Roy. Meteor. Soc.*, vol. XXXVII, 1911, p. 83, and vol. L, 1924, p. 26.

therefore it is necessary to note a difference of practice in British and continental stations. The normal height for the rim of the rain-gauge on the continent is a metre whereas in Britain it is a foot. No Nipher shield is in general used to make things equal. The difference of practice arises doubtless from the experience of the different countries as regards snow. In the British Isles snow is infrequent and hardly ever so deep as to submerge a rain-gauge, but on the continent snow-storms are a more important source of precipitation and have to be considered more seriously.

Rain-gauges in Britain are of 5 inches or 8 inches in diameter, rarely 3 inches, a diameter which at one time was not uncommon in Scotland. Special forms of gauge are sometimes used for collecting the rainfall in sites which are only visited at long intervals, such as the catchment areas of water-reservoirs.

Self-recording rain-gauges are very numerous and of varied patterns. The instruments in use now for nearly sixty years at the official observatories of the Meteorological Office are those constructed to the original design of Beckley of Richmond (Kew Observatory) in 1869. It depends upon the prompt action of a siphon to empty the receiving vessel when it gets full of water. An improvement in that respect was introduced as the Stonyhurst discharger. Objection used to be taken in official quarters to the gauges which depend upon a "tipping bucket," one of the earliest forms of recorder, because an uncertain amount of water is left in the bucket and evaporates after it is tilted. Less expensive forms of gauge which are free from that objection are the Halliwell gauge and the hyetograph of Messrs Negretti and Zambra. Gauges have also been designed by J. Baxendell and W. H. Dines. On the continent a well-known form of rain-gauge, suitable also for collecting and measuring snowfall, is that of Dr G. Hellmann.

Evaporation

The complementary instrument to the rain-gauge is the evaporation gauge. In some countries evaporation exceeds the rainfall and in all tropical countries the loss of water by evaporation is a matter of great economic importance. It has not yet reached its proper position in the discussion of the physical processes of the sequence of weather, partly perhaps because these processes have not yet been quantitatively explored and partly also because the greater amount of evaporation takes place in an entirely uncontrolled manner from seas, lakes and rivers on so vast a scale as to make insignificant anything that can be measured in a gauge. Moreover the actual measurements themselves are by no means easy or adequately controllable[1]. Temperature of the air and water, wind and sun, all affect the evaporation in ways which are not easily brought into account and moreover the state of the evaporating surface

[1] 'Report on Evaporimeters,' *Quarterly Weather Report of the Meteorological Office*, 1877, Appendix III. 'Studies of the phenomena of the evaporation of water over lakes and reservoirs,' Nos. 1–4, by F. H. Bigelow, Washington, D.C., *U.S. Dept. of Agriculture Monthly Weather Review*, vol. xxxv, 1907, p. 311; vol. xxxvi, 1908, pp. 24, 30, 437.

is a factor which has not yet been dealt with. A film of oil will inhibit evaporation from the surface affected and it becomes difficult to draw any general inference from what happens to the water in any one form of gauge. The subject deserves more consideration than it has yet received.

The gauges which are most in use at present are those known by the names of their originators, Piche and Wild, and the evaporation tank which has been supplied to a number of stations in England by Negretti and Zambra. Two gauges by De la Rue and by Lamont, which have been tried, do not appear to have been continued in use. The Piche gauge, a very handy instrument, consists of a long graduated glass tube about 20 cm long and 1 cm² in area; it is sealed at one end where a finial in the form of a glass ring affords a convenient means of suspension. The open bottom is covered by a disc of blotting-paper with a radius three times that of the tube, kept in place by a metal plate of the diameter of the tube carried by a bent wire leading to a metallic clasp sliding on the tube. The tube is filled with water and inverted, the end being covered by the paper disc. Evaporation takes place from the wet paper and small bubbles creep into the tube to replace the loss of water by evaporation. It can be exposed anywhere and the loss of water is very easily read.

The instrument designed by Wild is a balance or steel-yard. A short arm carries a pan about fifteen centimetres in diameter and two centimetres deep containing water, the other arm is a weighted pointer which travels over a graduated arc and indicates by its movement the weight of water lost by evaporation.

The evaporation-tank is a large tank 6 feet square containing the water to be evaporated. In a separate side-chamber is a counterpoised float the level of which is shown by the motion of a pointer.

We know of no self-recording gauge for evaporation. That the whole subject deserves further investigation, and in a sense requires it, will be evident from the map of the results obtained in different parts of the world shown in the second volume. We see there that the total depth of water evaporated at the gauge may exceed the total depth of rainfall in the locality, from which it might be inferred that the locality is drying up at an alarming rate. But it is obvious that actual evaporation depends upon the supply of water to be evaporated. At the evaporation-gauge a continuous supply of water to be evaporated is provided, but in the surrounding country only the surfaces of water, moist soil or trees are available to supply water for evaporation, and the figure for the measured evaporation may be applicable only to a very small part of the area over which the fall of rain is distributed.

Reduction to sea-level

We have noted that it is common practice for measures of pressure and temperature to be reduced to sea-level, the first by Laplace's formula, using some convention as to the value to be assigned to the temperature of the air in the formula, and the second by an empirical formula which is surprisingly accurate on the average but which has some striking exceptions. The absolute humidity or vapour-pressure is also reduced to sea-level for the maps of Vol. II in accordance with an exponential formula which will be quoted later on. No sort of reduction to sea-level is possible for relative humidity; saturation or extreme dryness may occur at any level. Moreover it is not usual to make any reduction in the case of wind or rainfall, not because there is no variation of wind or rainfall with height but because the variations are too irregular to be represented by a general formula. We have dealt with the changes in wind with height near the ground in the volume which has been published as Part IV. In Vol. II we shall call attention to some meteorological sections round the globe along different circles of latitude which will make it abundantly clear that rainfall is strikingly dependent upon orographic features, but the variation is dependent upon wind-direction as well as height and the variation with height is controlled by local circumstances which have not so far yielded to the efforts for generalisation.

Even in the case of pressure, the reduction of which to sea-level is the subject of one of the most celebrated of meteorological formulae, the results may be misleading when heights above a few hundred metres are concerned unless precautions are taken which are beyond the practice of the ordinary observer. In such cases as those of Greenland and the Antarctic continent, for example, or still more notably the high land of Central Asia, reduction to sea-level gives values which are purely conventional and the wise course is to stop the isobars where they cut high land and leave the figures for high levels without reduction or reduce them to some uniform high level.

The duration of sunshine, and of clear sky at night

The duration of sunshine as indicating the freedom of the atmosphere from cloud in the particular direction of the sun during the day is obtained exclusively by self-recording instruments, which are called heliographs or sunshine-recorders. Three different types of recorder are in more or less common use. The first is the glass ball with a suitably mounted card. It acts as a recorder of the scorching power of the sun's rays focussed upon cardboard specially prepared so that it will char but not take fire; it is mounted in a frame so that it will carry the image of the sun if it is shining and so indicate the duration of sunshine by the length of the trace. This instrument was developed by Sir G. Stokes from the original instrument devised by J. F. Campbell of Islay, which was a hemispherical bowl of wood, mahogany or cedar, in the centre of which a glass sphere was mounted. It was left exposed to the sun for the six months from solstice to solstice, each day left a score by the charring of the wood along a new line and the total amount of sunshine for the six months was indicated by the amount of charring. There was however no easy means of determining that amount and Stokes's development first made it possible to measure the duration with some approach to accuracy. At first sight the instrument seems to be as nearly absolute as any instrument could be, in view of the charred trace due to the sun itself without any interference; yet, in practice, the most scrupulous care is necessary; we must specify the size of the glass sphere, the material of which it is composed, the dimensions and adjustments of the bowl, the material of which the card is to be made and the graduation of the card in due relation to the frame that is to carry it. Even with all precautions the measurement is not very precise on account of the spread of the scorching beyond the actual image of the sun and its variation according to the intensity of the sunshine. But the instrument is in itself most attractive on account of its astronomical associations, and the gradual transformations of the cutting and adjustment of the belt from that suitable for the invariable twelve-hour record of the equator to the twenty-four hour record of the pole, equally invariable for six months of the year, afford very instructive exercises.

Similar in fundamental characteristics are the photographic recorders of sunshine with which the names of their inventors, T. B. and J. B. Jordan, are associated. That has to be set up so that the beam of sunshine passing through a linear aperture falls on paper which is sufficiently sensitive to direct sunlight, though not appreciably sensitive to the diffused daylight that gets through the aperture. The trace thus obtained can be fixed by washing in water and the duration estimated by the graduation of the paper. Here instead of the uncertainty due to the spreading of the trace in strong sunshine, or its failure in a weak sun, we have the uncertainties of photographic sensitiveness of paper, superposed upon the difference between the thermal intensity of the sun's rays and their photographic intensity. The difference may or may not be of importance to the comprehension of the general circulation and its changes. The mistake in the attitude of meteorological science to the subject seems to be that its exponents have addressed themselves to the question whether a scorching recorder or a photographic recorder could be regarded as interchangeable, and if not which of the two should be approved as the standard. The opportunity of using the two instruments side by side as a means of investigating the variations of the photographic effect in relation to the thermal effect is still open to those who are interested in such matters.

Meanwhile the power of the sun's heat and light is competent to set in motion many kinds of mechanism other than mere scorching or photographic recording, and we are thereby led to the third type of sunshine-recorder depending upon a tracing pen or recording beam that is operated when the solar intensity exceeds some small limit dependent upon the mechanism. Such a recorder is that proposed by W. H. Dines[1] which depends upon the differential heating by the sun of a glass tube one end of which is blackened. The tube contains a small quantity of mercury and some ether, but is completely exhausted of air. The result of the sun shining upon the tube is that the blackened end is warmed more than the other, the pressure of the ether-vapour in the blackened end is therefore greater than in the other and the mercury is driven away from the blackened end, the metal holder in which the tube is mounted falls over and is stopped by an electric contact-piece. When the sun ceases to shine the mercury returns to its lowest position and the holder also returns to its normal position.

[1] *Q. J. Roy. Meteor. Soc.*, vol. xxvi, 1900, p. 243.

A somewhat similar instrument devised by Dr C. F. Marvin is used in the stations of the United States Weather Bureau. The principle employed is similar to that of Leslie's differential air-thermometer, but the expansion of the air under the sun's radiation causes the mercury index to make an electrical contact and the duration of the contact is recorded automatically[1].

These secondary effects of sunshine are capable of unlimited exploitation as the powers of concentration are developed. As we shall see in Volume III full sunshine amounts to 100 kilowatts per square dekametre, a very substantial power, quite sufficient to insure an unmistakable record, and perhaps meteorology ought to turn its attention more strenuously to the development of some simple method of recording the intensity of the solar radiation and not be content merely with the record of bright sunshine as a climatic element without regard to its physical significance.

As a matter of fact the measurement of the duration of sunshine does not form part of the international scheme of observations for a climatological station of the second order, although in British stations a sunshine-recorder is generally included in the equipment. The necessary knowledge of the state of the sky is provided for by estimates at each of the term-hours of the number of tenths of the whole area of the vault of heaven which is covered by clouds, with or without a specification of the form of the clouds and the fraction of the sky which each type occupies. Such estimates are from the nature of the case not susceptible of any great accuracy but they are sufficient for the purpose. Instructions for procedure are to be found in the books for observers.

Pole-star recorder

At night when there is not sufficient light to see the clouds the estimation of the amount of cloud is specially uncertain and on that account it is remarkable how few observatories and stations take advantage of the opportunity which the photographic camera affords of getting over the difficulty by exposing a photographic plate to the sky in the neighbourhood of the pole during the night. It has been done with a special camera at Blue Hill Observatory, Massachusetts, for many years, and an ordinary camera has been used now for some six years at the Royal Observatory, Greenwich, for the same purpose[2]. A photograph of a trace thus obtained is reproduced in fig. 19.

An interesting point is raised as to the relation of the visibility of the sun or a particular star to the amount of cloud in the sky. Circumstances arise in which a particular part of the sky, as seen from some spot on the earth, becomes, for some hours perhaps, a locality of convergence or divergence as the case may be and remains cloudy or clear in spite of the fact that other clouds that are to be seen in the sky are obviously travelling. These instances of concentration or dissipation may be associated with the lenticular clouds

[1] 'Instructions for the care and management of electrical sunshine-recorders,' *U.S. Department of Agriculture, Weather Bureau*, Circular G, Instrument Division, 5th ed. 1923 (W.B. No. 802).

[2] 'Description of the Night-sky Recorder, recently brought into use at the Royal Observatory, Greenwich,' *Q. J. Roy. Meteor. Soc.*, vol. XLVI, 1920, p. 243.

which are to be seen at times in all districts within the range of influence of hills even at a considerable distance.

RECORD OF THE NIGHT SKY

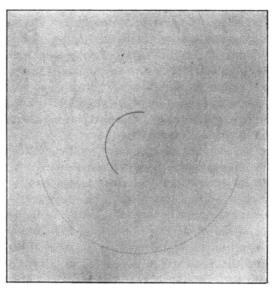

Fig. 19. Photographic record of the sky at the Royal Observatory, Greenwich, on the night of September 23 to 24, 1924, from a negative lent by the Astronomer Royal.

The strongest trace is that of *Polaris* representing 9 h 58 m duration, 100 per cent. of the time of exposure—the next is that of *δ Ursæ minoris* representing 9 h 47 m, 98 per cent. of the time of exposure.

The details of exposure are: commenced 19 h 0 m, concluded 5 h 0 m, the sun being approximately 10° below the horizon in each case. The small breaks due to passing cloud occurred at 21 h 26 m to 27 m, 22 h 26 m to 33 m, 35 m to 39 m and 4 h 55 m to 56 m.

Certificates of examination of common meteorological instruments

We have now gone through the instruments which form the usual equipment of a climatological station, with some additions that are not generally included but might be so with advantage. We have only now to add that nearly all the instruments, barometers, thermometers, rain-gauges, etc., require careful comparison with standards in order to avoid errors due to imperfect construction or graduation. That service is provided in this country by the National Physical Laboratory at Teddington. Its certificates are a recognised part of official practice.

We have written about the ordinary instruments in a very general manner without going into the details of their use or of the necessary notebook and forms, because, in any case, a book of instructions in the technique will be required and a suitable one can be selected from the list that we have given.

The non-instrumental observations which supplement the routine of instrumental measurements have already been considered in chapter II.

CHAPTER XI

THE METEOROLOGICAL OBSERVATORY: THE UPPER AIR

1749	Dr Alexander Wilson of Glasgow raised thermometers by kites.
1752	Franklin's kite-ascents.
1784	Dr John Jeffries and the aeronaut Blanchard made the first balloon ascents for meteorological purposes; subsequent ascents were made by Robertson in 1803–4, by Biot and Gay-Lussac in 1804 and by Barral and Bixio in 1850.
1809	Pilot-balloons: Thomas Forster watched the movement of small balloons filled with inflammable gas.
1822–3	Rev. George Fisher and Captain Sir Edward Parry raised self-registering thermometers by means of kites in the Arctic regions.
1852	Four balloon-ascents by John Welsh at Vauxhall for the Committee of Kew Observatory reaching 22,930 ft., 6988 m, giving pressure, temperature and humidity.
1862	September 5. Glaisher reached an estimated height of 11,200 m in a balloon.
1869	Glaisher made the first meteorological observations by means of a captive balloon.
1874	Pilot-balloons: the use of small free balloons suggested by Le Verrier.
c. 1882	First use of wire for kites by Archibald.
1891	M. Bonvallet sent up paper balloons with post-cards attached.
c. 1893	Lawrence Hargrave invented the box-kite.
1893	March 21. First registering balloon-ascent by Hermite and Besançon.
1894	The first continuously recording instrument was raised by kite at Blue Hill Observatory.
1894–5	Ascents at Berlin with free balloon of silk ("Cirrus") to 18,500 m.
1896	November 14. First international upper air ascents.
1899–1902	Discovery and demonstration of the stratosphere by Teisserenc de Bort and Assmann.
1901	Kites first used at sea by Lawrence Rotch.
1901	Berson and Süring reached 10,800 m in a balloon.
1902	Dines flew kites from a steamship.
1902–3	Teisserenc de Bort arranged for kites to be flown day and night, when possible, for nine months at Hald, Jutland.
1904	Rotch made the first registering balloon ascents in America.
1907	First ascents with a Dines meteorograph.
1909	First pilot-balloon ascent at Blue Hill in the United States.

The dates contained in the heading to this chapter have been extracted from the following papers: (1) Gold and Harwood, 'The Present State of our Knowledge of the Upper Atmosphere,' Report to the British Association, 1909; (2) *The Principles of Aerography*, by A. McAdie: Rand McNally and Co. 1917; (3) 'Upper Air Research,' by C. J. P. Cave, *Q. J. Roy. Meteor. Soc.*, vol. XL, 1914, p. 97.

THE VARYING ATMOSPHERE

WE now proceed to give some account of the equipment which has been devised and brought into operation for extending our knowledge of the structure of the atmosphere above the surface-layer, the region which is generally indicated by the title of the "free air."

There are many ways in which observers at the surface can appreciate the variations in the state of the atmosphere above them. Astronomers who are engaged in the study of the sun, planets and stars are always conscious of effects which are attributable to the atmosphere; the twinkling of the stars is one of the most familiar instances. There are many others; the observers of the planet Mars, for example, have to depend on periods of clearness in the atmosphere for the best definition of the planet in their telescopes and on this account visual observations give more information than photographs[1]. The edges of an image of the sun projected upon a screen present a flickering appearance which is called the "boiling of the limb." Meteorologists have been invited to regard it as furnishing a means of forecasting weather.

Such interferences with "good seeing," however, cannot at present be classed as atmospheric measurements. By astronomers they are regarded mostly as things to be avoided or reduced to the smallest possible limits by choosing special sites for their observatories and special times of the day or night for their observations. Observers of solar radiation however make a practice of forming a numerical estimate of the absorption of the atmosphere by separate observations with the special object of making the correction; and since the navigation of the air has become the special care of meteorologists, observations of visibility of distant objects through the free atmosphere have become part of the recognised duty of many meteorological stations.

We cannot however discuss here the bearing of such observations upon our knowledge of the structure of the atmosphere further than to mention that a "dust-horizon" (fig. 24), the appearance of a horizontal surface extending to the boundary of vision, is easily recognised by aviators in favourable circumstances. It marks the top of the layer of air, generally but not always the surface-layer, which contains sufficient dust to differentiate it from the layer above by a well-marked boundary. The conditions are favourable when the surface-layers are comparatively free from eddies. The evening and early morning are favourable times.

CLOUD-FORMS

Among observations which are more definitely within the domain of meteorology in relation to the structure of the atmosphere we must refer to clouds as being one of the most obvious sources of information about the free air. We have already mentioned the amount of cloud as one of the regular observations of meteorological stations. In that respect, as also in respect of the observatories at high levels, the surface-survey and the survey of the upper air overlap. We have now to consider the development of the study of clouds which is concerned with their form, height and motion.

The forms of cloud must have been an object of interest from earliest times and the classical languages have a number of names νέφος, νεφέλη, nubes, nebula, nimbus, imber, translated by the word *cloud* or *mist* without any specific definition. We have noted on p. 101 the important references in Theophrastus to "streaks of cloud" in a wind from the south and "clouds like fleeces of wool"; the classification thereby indicated was not developed. The credit of classifying the forms of cloud in a scientific manner belongs to Luke Howard[2]. He named the main classes, cirrus, stratus, cumulus, and gave a definite meaning to nimbus.

[1] Sir Frank Dyson, *Nineteenth Century*, February, 1925. Rev. T. E. R. Phillips, 'Observations of Mars in 1924,' *Monthly Notices R.A.S.*, vol. LXXXV, December, 1924, p. 179.
[2] Luke Howard, *On the Modifications of Clouds*, London, 1803.

PLATE I

THE FREE AIR

Fig. 20. Snow cloud on Mt Everest.

Fig. 21. Smoke cloud in London.

Fig. 22. Dust storm in Iraq.

Fig. 23. Dust whirlwinds moving swiftly.

Fig. 24. Dust-horizon at 1250 m. seen from 450 m. above, with clouds beneath it.

PLATE II
HIGH CLOUD-FORMS: CIRRUS AND FALSE CIRRUS

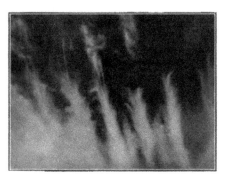

Fig. 25. Cirrus feathers.

Fig. 26. Detached cirrus.

Fig. 27. Cirrus with vortical whorl.

Fig. 28. Cirrus uncinus or caudatus.

Fig. 29. Fibrous cirrus.

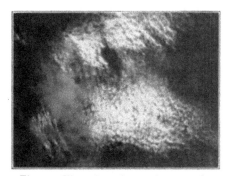

Fig. 30. Flocculent cirrus, cirro-macul

Fig. 31. False cirrus above cumulo-nimbus.

Fig. 32. False cirrus leaving parent cu.-nb.

PLATE III

LOW CLOUD-FORMS: STRATO-CUMULUS TO FOG

Fig. 33. Typical strato-cumulus.

Fig. 34. Low cumulus dissolving.

Fig. 35. Stratus cumuliformis.

Fig. 36. Stratus cloud in an Alpine valley.

Fig. 37. Ground-fog filling a valley.

Fig. 38. Ground-fog in sunshine.

Fig. 39. Fog cascades in California.

PLATE IV

LENTICULAR CLOUDS, OROGRAPHIC EDDIES

Fig. 40. Lenticular clouds.

Fig. 41. Group of lenticular clouds.

Fig. 42. Lenticular alto-cumulus.

Fig. 43. Distant lenticular alto-cumulus.

Fig. 44. Foreground of lenticular cloud.

Fig. 45. Lenticular cloud (Föhnwolken). Greenland.

Fig. 46. "Contessa del Vento."

Fig. 47. Mountain cloud-cap. Java.

PLATE V
LENTICULAR CLOUDS, TURBULENCE LOCALLY ACCENTUATED

Fig. 48. Pico with its Baleia (whale), thin cloud-cap, and a sea-gull.

Fig. 49. Nuvem da Prainha, seen from Horta, looking North.

Fig. 50. Horizontal bank of local turbulence caused by Pico.

Fig. 51. A "water-spout" over Prainha.

PLATE VI

CLOUD-FORMS: TOWERING CUMULUS

Fig. 52. Eruption cloud. Mount Asama.

Fig. 54. Cumulo-nimbus. Buitenzorg.

Fig. 55. Small cumulus reaching 2350 m.

Fig. 53. Cumulo-nimbus. Stoner Hill.

Fig. 56. Cumulus of Fig. 54 after 15 minutes

PLATE VII

CLOUD-FORMS: CUMULUS AND THUNDER-CLOUDS

Fig. 57. Detached cumulus.

Fig. 58. Cumulus with rounded base (globosus).

Fig. 59. Alto-cumulus castellatus.

Fig. 60. Alto-cumulus castellatus.

Fig. 61. Cumulo-nimbus at sunrise.

Fig. 62. Cumulo-nimbus at 100 km. distance.

Fig. 63. Vertical eddies in a small line-squall.

Fig. 64. Mammato-cumulus.

PLATE VIII

CLOUD-GROUPS: DISORDERLY GROUPING

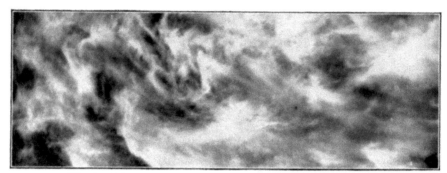

Fig. 65. Disorderly cirrus in the centre of a summer anticyclone.

Fig. 66. Disorderly strato-cumulus with a head of cumulus. Colombo.

Fig. 67. Irregular stratus.

Fig. 68. Irregular distribution of cumulus. W. Java.

PLATE IX

CLOUD-GROUPS: LINEAR GROUPING

Fig. 69. Alto-stratus with a linear boundary.

Fig. 70. Linear fracto-nimbus or scud below heavy nimbus cumuliformis.

Fig. 71. Fracto-cumulus linealis with thin alto-stratus behind, beneath corrugated alto-stratus.

Fig. 72. Coast-line cloud. Strato-cumulus along the coast for at least 16 km.

PLATE X

CLOUD-GROUPS: CORRUGATION

Fig. 73. Corrugation in cirrus.

Fig. 74. Corrugation in cirro-cumulus.

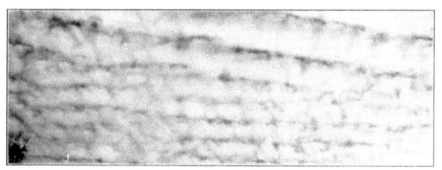

Fig. 75. Corrugation in alto-cumulus.

Fig. 76. Corrugation in strato-cumulus. Roll-cumulus.

PLATE XI

CLOUD-GROUPS: TESSELLATION

Fig. 77. Cirro-cumulus.

Fig. 78. Cirro-cumulus.

Fig. 79. Alto-cumulus.

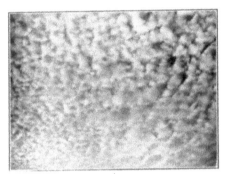

Fig. 80. Alto-cumulus.

Fig. 81. Alto-cumulus.

Fig. 82. Alto-cumulus in large patches.

Fig. 83. Strato-cumulus advancing from seaward.

Fig. 84. Tessellated cumulus.

PLATE XII

CONTINUITY

Fig. 85. Continuous cirro-stratus.

Fig. 86. Alto-stratus with small patches of strato-cumulus below.

Fig. 87. Continuous sheet of strato-cumulus at about 500 m.

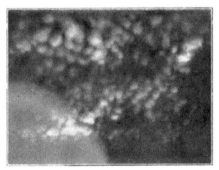 

Fig. 88. Cirro-cumulus. Under surface.   Fig. 89. Alto-cumulus. Upper surface.

PLATE XIII

NEBULA

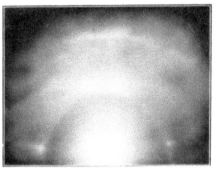

Figs. 90 and 91. Cirro-nebula with haloes. Aberdeen and Potsdam.

Fig. 92. Cirro-nebula without halo, or alto-nebula.

Fig. 93. Alto-nebula visible beyond the edges of small cumulo-castella.

Fig. 94. Surface nebula; indefinite fog on the ground.

PLATE XIV

THE WHOLE SKY: STEREOSCOPIC VIEW

Fig. 95. Stereoscopic pair of photographs of the whole sky at Cambridge, 1924. vii. 29, 1447 by the hemispherical lens and camera designed by Robin Hill. The two photographs are from the ends of a base line 384 metres long from NNE (on the left) to SSW (on the right). The clouds shown are a lower layer with castellated cumulus, cumulo-nimbus, and above a plume of false cirrus reaching from the sun's position nearly to the zenith.

The sun's position is indicated by the images of the three inch circular screens used as sunshades.

INDEX OF CLOUD-FORMS AND CLOUD-GROUPS
WITH ACKNOWLEDGMENTS
CLOUD-FORMS

Clouds of solid particles.

20. Snow cloud on Everest. Mount Everest Committee. *The Air and its Ways,* p. 153.
21. Smoke cloud in London. Dr J. S. Owens.
22. Dust storm on the plains of Iraq. R.A.F. (Flight Lieut. J. F. Lawson).
23. Dust whirlwinds moving swiftly over sandy soil, India. Baddeley. London, 1860.
24. Dust-horizon at 1250 m. seen from 450 m. above, with cumulus clouds beneath it. 1920. x. 11. Air Ministry. A. W. Judge.

High Clouds.

25. Cirrus feathers. 1923. vii. 30, 0910. South Downs, overhead. C. J. P. Cave.
26. Detached cirrus. 1922. v. 6, 1415. South Downs, overhead. C. J. P. Cave.
27. Cirrus with vortical whorl. 1923. iii. 6, 1200. Overhead in the Alps. C. J. P. Cave.
28. Cirrus uncinus or caudatus. 1924. iii. 9, 1035. South Downs to NW. C. J. P. Cave.
29. Fibrous cirrus. 1924. i. 20, 1200. South Downs, overhead. C. J. P. Cave.
30. Flocculent cirrus, cirro-macula. 1921. x. 7, 1551. Aberdeen. G. A. Clarke.
31. False cirrus above cumulo-nimbus. 1919. ix. 26, 1312. Aberdeen. G. A. Clarke.
32. False cirrus separating from cumulo-nimbus, which is passing off from NNW, view ESE. 1920. iv. 29, 1610. Aberdeen. G. A. Clarke.

Low Clouds.

33. Typical strato-cumulus. 1916. xi. South Downs to S. C. J. P. Cave.
34. Low cumulus dissolving. 1923. v. 13, 1630. South Downs to S. C. J. P. Cave.
35. Stratus cumuliformis under dark alto-stratus, looking West. 1920. ix. 16, 1830. Aberdeen. G. A. Clarke.
36. Stratus cloud in an Alpine valley. 1923. iii. Montana to SW. C. J. P. Cave.
37. Ground-fog filling the Petersfield valley. 1923. xi. 25, 1530. From Stoner Hill to E. C. J. P. Cave.
38. Wreaths of ground-fog in sunshine on snow-covered aerodrome. 1917. iv. 3, 1145. South Farnborough to S. C. J. P. Cave.
39. Fog cascades in California. A. McAdie, *The Clouds and Fogs of San Francisco,* 1912.

Lenticular Clouds. Orographic eddies (turbulence locally accentuated).

40. Lenticular clouds to E over the South Downs. 1924. iv. 21, 0800. C. J. P. Cave.
41. Group of lenticular clouds to SE, South Downs. 1923. viii. 25, 0900. C. J. P. Cave.
42. Lenticular alto-cumulus. 1921. i. 23, 1250. Aberdeen. G. A. Clarke.
43. Distant lenticular alto-cumulus just over the horizon, telephotographic view. 1921. v. 13, 0650. Aberdeen. G. A. Clarke.
44. Lenticular cloud in front of alto-cumulus castellatus. A. de Quervain, *Neujahrsblatt,* 1912.
45. Lenticular cloud (Föhnwolken) seen from SSW. 1908. vi. 25. A. de Quervain, *Danmark-Ekspeditionen til Grønlands Nordøstkyst,* Bd. XLII, Nr. 4, Tafel XVII.
46. Turbulence locally accentuated. "Contessa del Vento." Etna. 1897. iv. 21. Catania. A. Mascari. Reproduced from *Le Nubi* by L. Taffara. Rome, 1917, Fig. 24.
47. Mountain cloud-cap on Semeroe. Java. Dr Kemmerling. *Het Klimaat van Nederlandsch-Indië,* vol. 1, part 4, by C. Braak. *K. Mag. en Met. Obs. te Batavia,* 1923, Fig. 10.

Lenticular Clouds. Turbulence locally accentuated. Azores. Stereoscopic views from Horta towards Pico.

48. View due East; Pico (2320 m.) with its Baleia (whale), on the South side a fine weather cloud, thin cloud-cap and a sea-gull. Col. F. A. Chaves.
49. Nuvem da Prainha, seen from Horta, looking North. Col. F. A. Chaves.
50. Horizontal bank of local turbulence caused by Pico. Col. F. A. Chaves.
51. A "water-spout" over Prainha. Col. F. A. Chaves.

Towering Cumulus: volcanic and atmospheric.

52. Eruption cloud. Mount Asama. Reproduced from a post-card sent by T. Kobayasi.
53. Towering cumulus or cumulo-nimbus. 1923. vii. Stoner Hill to E. C. J. P. Cave.
54. Towering cumulus, cumulo-nimbus developing in the evening in the direction of Buitenzorg, as seen from the Observatory at Batavia. The height is estimated at 15 km. 1921. iv. 22, 1730. (See Fig. 47.)
55. Small cumulus of penetrative convection breaking through a sheet of cloud at 460 m., reaching 2350 m. 1918. ix. 27, 0700. France. C. K. M. Douglas.
56. The towering cumulus of Fig. 54 after an interval of 15 minutes.

Cumulus and Thunder-clouds, in prospect and in being.

57. Detached cumulus. Aberdeen. G. A. Clarke.
58. Cumulus with rounded base (globosus) in process of degradation from flat-bottomed cumulus. From the summits pieces are becoming detached and floating away. 1920. viii. 3, 1150. Aberdeen. G. A. Clarke.
59. Alto-cumulus castellatus. South Downs to E. 1925. v. 17, 1850. C. J. P. Cave.
60. Alto-cumulus castellatus. Aberdeen looking NW. 1920. v. 26, 1200. G. A. Clarke.

INDEX OF CLOUD-FORMS AND CLOUD-GROUPS

Classification of Clouds

During the first thirty years of the existence of the International Meteorological Organisation much attention was given to the development of Howard's classification. The principal contributors were Riggenbach, Abercromby, Clement Ley, Rotch, Mohn, Hildebrandsson and Teisserenc de Bort. The effort reached a definite stage in the *International Cloud Atlas* published in 1896; a further stage was arrived at in the second edition of the *Atlas*, entrusted to Hildebrandsson and Teisserenc de Bort, and published in 1910[1].

For the purposes of reference we reproduce, slightly modified, the definitions which accompanied the second edition of the *Atlas*. Illustrations of cloud-forms and cloud-groups are separately inset.

Cirrus (Ci.). Detached clouds of delicate appearance, fibrous (thread-like) structure and feather-like form, generally white in colour.

Cirrus clouds take the most varied shapes, such as isolated tufts of hair, i.e. thin filaments on a blue sky, branched filaments in feathery form, straight or curved filaments ending in tufts (called Cirrus uncinus), and others. Occasionally cirrus clouds are arranged in bands, which traverse part of the sky as arcs of great circles, and as an effect of perspective appear to converge at a point on the horizon, and at the opposite point also if they are sufficiently extended. Cirro-stratus and cirrocumulus also are sometimes similarly arranged in long bands.

It may be noted that the outline of the sun is visible, and his rays cast a shadow in spite of the presence of clouds of this type, unless the clouds and the sun are both low down on the horizon.

Cumulus (Cu.), Woolpack or Cauliflower Cloud. Thick cloud of which the upper surface is dome-shaped and exhibits protuberances while the base is generally horizontal.

These clouds appear to be formed by ascensional movement of air in the day-time which is almost always observable. When the cloud and the sun are on opposite sides of the observer, the surfaces facing the observer are more brilliant than the margins of the protuberances. When, on the contrary, it is on the same side of the observer as the sun it appears dark with bright edges. When the light falls sideways, as is usually the case, cumulus clouds show deep shadows.

[1] The *Meteorological Magazine*, 1897, reviewing a volume of the cloud observations of Blue Hill Observatory, Cambridge, U.S.A., 1896, gives the following list of authorities on cloud-forms quoted by H. H. Clayton in the volume:

1801–04	Lamarck	1880–94	Hildebrandsson	1887–88	Abercromby
1803	Howard	1881–92	Möller	1887–89	Clayton
1815	Forster	1881	Weilbach	1889	Maze
1828	Clos	1882	Vettin	,,	von Helmholtz
1831	Kaemtz	,,	Klein	1890	Neumayer
1841–68	Loomis	1883–90	Köppen	,,	van Bebber
1846	Fritsch	1883	Russell, Rollo	1891	Abbé, C.
1857	Jevons	,,	Mohn	,,	Singer
1860	Schmid	1884	Tissandier	,,	Vincent
1861	Herschel	1885	Scott	1892	Kassner
1863	FitzRoy	,,	Sprung	1893	Gaster
1863–79	Poëy	,,	Capello	,,	von Bezold
1874	Muhry	,,	Wilson Barker	1894	Davis, W. M.
1875	Blasius	1886	Jesse	,,	Riggenbach
1879–94	Clement Ley	,,	Toynbee	,,	Teisserenc de Bor

True cumulus has well-defined upper and lower margins; but one may sometimes see ragged clouds—like cumulus torn by strong wind—of which the detached portions are continually changing; to this form of cloud the name Fracto-cumulus may be given.

Nimbus (Nb.). A dense layer of dark, shapeless cloud with ragged edges from which steady rain or snow usually falls. If there are openings in the cloud an upper layer of cirro-stratus or alto-stratus may almost invariably be seen through them.

If a layer of nimbus separates in strong wind into ragged cloud, or if small detached clouds are seen drifting underneath a large nimbus (the "Scud" of sailors), either may be specified as Fracto-nimbus (Fr. Nb.).

Alto-cumulus-castellatus. "Little miniature cumulus rising in many heads from a more or less compact layer of alto-cumulus."

Not a very common cloud in these latitudes but sometimes seen in summer, and when coming from a Westerly or South Westerly point is almost always a sign of the approach of shallow depressions which bring thunderstorms[1].

Cumulo-nimbus (Cu. Nb.), the Thunder Cloud; Shower Cloud. Great masses of cloud rising in the form of mountains or towers or anvils, generally having a veil or screen of fibrous texture (false cirrus) at the top and at its base a cloud-mass similar to nimbus.

From the base local showers of rain or of snow, occasionally of hail or soft hail, usually fall. Sometimes the upper margins have the compact shape of cumulus or form massive heaps round which floats delicate false cirrus. At other times the margins themselves are fringed with filaments similar to cirrus clouds. This last form is particularly common with spring showers. The front of a thunderstorm of wide extent is frequently in the form of a large low arch above a region of uniformly lighter sky.

Lenticular Cloud-banks. Banks of cloud of an almond or airship shape, with sharp general outlines, but showing, on close examination, fretted edges, formed of an ordered structure of cloudlets similar to alto-cumulus or cirro-cumulus which is also seen in the bank itself when the illumination is favourable.

Sometimes the body of the cloud-bank is dense, and the almond shape is complete, fore and aft, but sometimes the bank thins away from the forward edge of clear sky within, so that the bank presents the appearance of a horse-shoe seen in perspective from below at a great distance. The bank appears nearly or quite stationary, while the cloudlets move rapidly into it at one side and away from it at the other.

Stratus (St.). A uniform layer of cloud, like fog, but not lying on the ground.

The cloud-layer of stratus is always very low. If it is divided into ragged masses in a wind or by mountain tops, it may be called Fracto-stratus. The complete absence of detail of structure differentiates stratus from other aggregated forms of cloud.

[1] Captain C. J. P. Cave, R.E., 'The Form of Clouds,' *Q. J. Roy. Meteor. Soc.*, vol. XLIII, 1917, p. 68.

Strato-cumulus (St. Cu.). Large lumpy masses or rolls of dull grey cloud, frequently covering the whole sky, especially in winter.

Generally strato-cumulus presents the appearance of a grey layer broken up into irregular masses and having on the margin smaller masses grouped in flocks like alto-cumulus. Sometimes this cloud-form has the characteristic appearance of great rolls of cloud arranged in parallel lines close together. (Roll-cumulus in England, Wulst-cumulus in Germany.) The rolls themselves are dense and dark, but in the intervening spaces the cloud is much lighter and blue sky may sometimes be seen through them. Strato-cumulus may be distinguished from Nimbus by its lumpy or rolling appearance, and by the fact that it does not generally tend to bring rain.

Cirro-cumulus[1] (Ci. Cu.) (Mackerel Sky). Small rounded masses or white flakes without shadows, or showing very slight shadow: arranged in groups and often in lines.

French, *Moutons*; German, *Schäfchen-wolken*.

Alto-cumulus (A. Cu.). Larger rounded masses, white or greyish, partially shaded, arranged in groups or lines, and often so crowded together in the middle region that the cloudlets join.

The separate masses are generally larger and more compact (resembling strato-cumulus) in the middle region of the group, but the denseness of the layer varies and sometimes is so attenuated that the individual masses assume the appearance of sheets or thin flakes of considerable extent with hardly any shading. At the margin of the group they form smaller cloudlets resembling those of cirro-cumulus. The cloudlets often group themselves in parallel lines, arranged in one or more directions.

Cirro-stratus[1] (Ci. St.). A thin sheet of whitish cloud; sometimes covering the sky completely and merely giving it a milky appearance; it is then called cirro-nebula or cirrus haze; at other times presenting more or less distinctly a fibrous structure like a tangled web.

This sheet often produces halos round the sun or moon.

Alto-stratus (A. St.). A dense sheet of a grey or bluish colour, sometimes forming a compact mass of dull grey colour and fibrous structure.

At other times the sheet is thin like the denser forms of cirro-stratus, and through it the sun and the moon may be seen dimly gleaming as through ground glass. This form exhibits all stages of transition between alto-stratus and cirro-stratus, but according to the measurements its normal altitude is about one-half of that of cirro-stratus.

Pallio-nimbus[2]. A pall of cloud covering the whole sky which may be described as of the same type as alto-stratus but from which rain falls.

Fog[3]. A cloud, devoid of structure, formed on land in the layers of air which, though nearly stationary, really move slowly over the ground.

The same kind of surface-cloud may be found at sea in nearly calm weather or with wind of considerable force. Viewed from a distance fog has a definite boundary whereas some other forms of obscurity have not.

---

[1] The outline of the sun is visible and his rays cast a shadow in spite of the presence of clouds of these types, unless the clouds and the sun are both low down on the horizon.

[2] J. Vincent, 'Atlas des Nuages,' *Annales de l'Obs. Royal de Belgique*, tome xx, Bruxelles, 1909.

[3] *Forecasting Weather*, 2nd edition, p. 45.

The following remarks are added in the international atlas as instructions to observers:

(*a*) In the day-time in summer all the lower clouds assume, as a rule, special forms more or less resembling cumulus. In such cases the observer may enter in his notes "Stratus- or Nimbus-cumuliformis."

(*b*) Sometimes a cloud will show a mammillated surface and the appearance should be noted under the name Mammato-cumulus.

(*c*) The form taken by certain clouds particularly on days of sirocco, mistral, föhn, etc., which show an ovoid form with clean outlines and sometimes irisation, will be indicated by the name lenticular, for example: cumulus lenticularis, stratus lenticularis (Cu. lent., St. lent.).

(*d*) Notice should always be taken when the clouds seem motionless or if they move with very great velocity.

The classification is perhaps not yet in its final form. The following remarks are taken from a circular on the subject addressed in the year 1923 to the International Commission for the Study of Clouds.

The pictures of cloud-forms which are included in the inset between pp. 208 and 209 have been selected to illustrate these remarks.

THE FORMS OF CLOUDS AND THEIR NOMENCLATURE

The International Cloud Atlas has ten principal forms of cloud: Cirrus, Cirro-stratus, Cirro-cumulus, Alto-cumulus, Alto-stratus, Strato-cumulus, Nimbus, Cumulus, Cumulo-nimbus, Stratus. Fog is regarded as stratus on the ground.

These names and no others are included in the code for reporting cloud-forms in telegraphic reports. Other forms mentioned in the international classification but not in the code are: Cirrus uncinus, Cirro-nebula, False cirrus, Roll-cumulus, Scud (Fracto-nimbus), Fracto-cumulus, Fracto-stratus, Stratus-cumuliformis, Nimbus-cumuliformis, Mammato-cumulus, Cumulus lenticularis, Stratus lenticularis.

Since the issue of the cloud atlas the following atlases or collections of photographs of cloud-forms have come under my notice:

Clayden, A. W. Cloud Studies. London, 1905. 2nd edition, 1925.

De Quervain, A. Beiträge zur Wolkenkunde. Met. Zeitschr., xxv, 1908, pp. 433–453.

Vincent, J. Atlas des Nuages. Annales de l'Observatoire Royal de Belgique. Nouvelle Série, Annales Météorologiques, tome xx. Bruxelles, 1909.

Loisel, Julien. Atlas photographique des nuages. Paris, 1911.

Wegener, A. Danmark-ekspeditionen til Grønlands Nordøstkyst 1906–1908, Bd. xlii, No. 4. Copenhagen. (Contains photographs of clouds in Greenland.)

Neuhaus, E. Die Wolken in Form, Färbung und Lage als lokale Wetterprognose. Zurich, 1914.

Taffara, L. Le nubi, parte I testo, parte II atlante. R. Ufficio Centrale di Meteorologia e Geodinamica. Rome, 1917.

Kusnetzov, B. B. Cloud Atlas. Nicholas Central Physical Observatory, Petrograd, 1917.

Cave, C. J. P. The forms of clouds. London, Q. J. R. Meteor. Soc., 1917.

Douglas, C. K. M. The lapse-line and its relation to cloud-formation. With thirteen photographs of clouds. Edin. J. Scott. Meteor. Soc., vol. xvii, 1917, pp. 133–147.

London, Admiralty, Hydrographic Department. Naval meteorological service cloud atlas. London, 1918 (later edition, London, 1920).

Douglas, C. K. M. Clouds as seen from an aeroplane. London, Q. J. R. Meteor. Soc., vol. xlvi, 1920, pp. 233–242.

London, Meteorological Office. Cloud forms according to the international classification. London, 1918 (2nd edition, London, 1921).

Clarke, G. A. Clouds. A descriptive illustrated guide-book to the observation and classification of clouds, London, 1920.

Matteuzzi, T. Colonnello L. Le Nubi Fondamentali. Rome, 1921.
Washington, Department of Agriculture. Weather Bureau. Description of cloud forms. New edition, 1921.
Besson, L. La classification détaillée des nuages en usage à l'Observatoire de Montsouris. Ann. des Services Techniques d'Hygiène de la ville de Paris, tome I, 1921.
McAdie, A. A Cloud Atlas. Rand McNally, 1923.
Besson, L. Aperçu historique sur la Classification des Nuages. Mémorial de l'Office National Météorologique de France, Iʳᵉ Année, No. 2. Paris, 1923.
Fontseré, E. Atlas elemental de núvols. Barcelona, 1925.

It is understood that the classification of cloud-forms is based on their appearance to the observer for the purpose of identification. It has been suggested that a new classification should be based upon certain views of the mode of formation; adjectives have been assigned to express those views; but there is at present no general international agreement as to the physical processes which are represented by the several cloud-forms. I should find it very difficult to reconcile the adjectives *ventosus* or *fractus* for example with the views of the physical processes which seem to me to be established; in the case of ventosus because it suggests a relation between the form of cloud and variations of wind of which we have no evidence, and in the case of fractus because it suggests, equally without evidence, that the separated clouds were originally united.

The advantage of classification for the science of meteorology is not so much to express accepted or suggested opinions, as to group together phenomena which have common aspects with a view to evolving an idea of the physical processes which they represent. Hence the classification ought not to depend upon any controversial opinions.

Remarks on the International Classification

1. *The general principles of selection of forms*

I note first that of the ten primary cloud-forms five are named on account of the recognisable shape or texture of the individual cloud-mass and may be taken as defining cloud-units, these are: cirrus, cumulus, nimbus, cumulo-nimbus and stratus. The remainder, cirro-stratus, cirro-cumulus, alto-cumulus, alto-stratus, strato-cumulus refer to groups of clouds that appear to be in horizontal sheets or layers.

This observation suggests a systematic classification of cloud-forms according to (1) the cloud-unit of which the form is composed, and (2) the manner in which the units are grouped. And here we may notice that in the supplementary forms included in the International Atlas there is clearly shown a tendency of clouds to arrange themselves in lines, often in a series of parallel lines presenting the general appearance of a corrugated surface. For example fracto-cumulus generally gives a distinct impression of arrangement in line, along the course of the wind. Matteuzzi notes the analogy with the smoke of a chimney. It is probably the arrangement in line along the wind which has led observers to use the adjective *fractus* in that connexion. The natural development of *fracto-cumulus* is to form a cloud-bank such as that which Matteuzzi calls *cumuli a diga*. The series of cumulus cloud along a coast-line is another familiar case of arrangement in line which may in time give a clue to the causes of linear arrangement.

Looking at the various forms of cloud from the point of view of their arrangement in line we cannot fail to note the frequency of grouping in parallel lines such as the form of strato-cumulus called roll-cumulus. The lines may be along the wind or across the wind; and, whatever may be the cause, the corrugated appearance of the sky is very characteristic of certain forms of cloud as seen from an aeroplane.

There are many varieties of such corrugations, from the continuous cloud-mass with corrugations showing only as alternating lines of thin cloud and thick cloud to the parallel lines of cloud with blue sky visible between them as in certain cases of roll-cumulus.

Tessellated layers. We next notice that cloud-units may be grouped into lines in more than one direction at the same time and so form a "pattern," the arrangement is most characteristic of *cirro-cumulus* and *alto-cumulus*; even when the individual units are no longer detached the pattern may still be evident as a cross-corrugation, with the appearance of ripple-marks.

Finally we reach the continuous cloud-layer in which the continuity is more conspicuous than the corrugation. Such continuous masses have received the name of stratus. The forms in order of height are cirro-stratus, alto-stratus and strato-cumulus, whereas the simple name stratus is applied only to the lowest form of cloud-sheet. There is here an obvious incompleteness in the nomenclature. If stratus can be used in cirro-stratus the lowest example ought to be distinguished also for example as humilo-stratus, or since hill-clouds are examples of low stratus the best name might be colli-stratus. But, details apart, it seems possible to form a rational classification of cloud-forms on the general principles of the appearance of the unit-cloud on the one hand and the grouping of the units on the other.

The clouds illustrated in the International Cloud Atlas would arrange themselves according to the suggested classification in the following manner:

Unit-clouds. Cirrus (detached), False cirrus, Cumulus (detached), Nimbus (detached or cumuliformis), Cumulo-nimbus and Stratus (detached), also Nubes lenticularis.

Groups of unit-clouds, without orderly arrangement. Some forms of strato-cumulus.

Arrangement in line. Fracto-cumulus, Fracto-stratus, Fracto-nimbus.

Arrangement in parallel lines, corrugation. Cirrus, Strato-cumulus, Roll-cumulus.

Arrangement in cross-parallels, double corrugation. Alto-cumulus, Cirro-cumulus.

Continuous cloud-sheets. Cirro-stratus, Alto-stratus, Stratus (humilis).

This leaves out of account for the moment cirrus uncinus, cirro-nebula, stratus cumuliformis, mammato-cumulus, cumulus lenticularis, stratus lenticularis, but before dealing with them I wish to inquire whether the list of unit-clouds and of cloud-layers which has been enumerated, modified if necessary by qualifying adjectives, is sufficient for practical purposes.

2. *The question of the adequacy of the existing international classification*

Upon this point I note with reference to the first group (the group of individual clouds):

(1) *Cumulus.* I assume that the name cumulus should be limited to clouds which have a flat base; otherwise we should require another class for clouds with a flat base because the flatness marks the locus of condensation. But there is a lack of name for a detached cloud-mass which is truly globular—cumulus globosus seems to be a contradiction. Globus or moles or nubes globosus would be free from that objection.

(2) *Lenticularis.* A cloud of the lenticular type has an apparently smooth outline which is not characteristic of cirrus, detached cumulus, nimbus or stratus. Such clouds may be due to some physical process which is somewhat different from those which cause other clouds, and certainly *the form of the cloud is distinctive*. It may occur at various levels and for the present the adjective *lenticularis* may be used to identify the particular form of unit. It may be applied to clouds of various types; some are already included in the classification, and as a general expression of this kind of cloud we may use the term nubes lenticularis.

(3) *Nimbus.* The word nimbus is used to intimate that cloud is in sight overhead or at a distance from which rain is falling. Without more precise definition it is a description of rainy weather rather than of cloud.

If we give the name *nimbus* to a cloud of limited extent from which rain is falling, has been falling or is on the point of falling, of which a considerable part of the edge is visible, though overhead the sky is covered, we may perhaps fairly include nimbus or nimbus cumuliformis with the names of other individual clouds as we have done, and the identification of that condition taken together with the rate at which rain falls may be helpful as contributing some evidence about the physical process which causes the rain: but often when rain is falling the whole sky is covered and no limit of the cloud is visible in the horizontal direction. It is of no assistance to us to regard a vast layer of that kind as a single or individual cloud. Within the area covered by the cloud the rain may be very variable. We require a name to describe conditions of continuous and wide-spread rain from an overcast sky. Pallio-nimbus as proposed by J. Vincent is therefore rightly added to the list of sheet-clouds.

(4) *Alto-cumulus-castellatus.* The cloud called *alto-cumulus-castellatus* is distinct in its apparent structure from the other forms of cloud. It is an individual cloud or bank of clouds like cumulo-nimbus with castellated extensions upwards. It cannot fairly be called cirrus, cumulus, stratus or nimbus. It is recognised only if its vertical structure is visible and consequently it must be at a comparatively low altitude as seen by the observer. In that respect it is different from other clouds. No one could be expected for example to estimate what fraction of the sky is covered by alto-cumulus-castellatus. It should find a place in the list of individual clouds. The prefix alto is of little help and the name alto-cumulus for this kind of cloud is inappropriate if we are agreed that alto-cumulus is a doubly corrugated cloud-layer; "cumulus-castles" (*cumulo-castella*) is probably a sufficient description. It may be noted that turrets or *castella* of cloud, quite similar in appearance to those by which alto-cumulus-castellatus is recognised, have been seen projecting upwards from cloud-sheets by observers in balloons and aeroplanes. They have been sketched and photographed and are reported to be associated with thunder. We may therefore regard cumulus-castles as a view from the ground at great distance of the turret-clouds or *castella* which are the most characteristic projection upward from sheets of strato-cumulus.

(5) *False cirrus.* The observation of what is known as false cirrus has been much used in recent times. It is a cloud of thread-like form which veils the top of cumulo-nimbus but also occurs as detached cloud at various levels much lower than what is usually classed as cirrus and it appears as tufts with fibrous extensions. It is for purposes of scientific inquiry a separate form of cloud. It might perhaps be called "low cirrus," as the reason for calling it false is not apparent. (See Clarke, G. A., *Clouds*, pp. 26–28; Douglas, C. K. M., *Edin. J. Scott. Meteor. Soc.*, vol. XVII, 1917, p. 139.) C. J. P. Cave (*Nature*, Feb. 6, 1926) gives measurements made on Jan. 24 of a sheet of alto-cumulus at a height from 10,000 to 10,500 feet with cirrus uncinus below it appearing dark against the clouds above.

(6) *Names of parts of unit-cloud.* The international atlas includes a picture of mammato-cumulus which presents an undoubtedly characteristic appearance and is indeed a favourite subject for photographers of cloud. The opinion of the sub-commission which was reported to the International Meteorological Committee was that "the name does not define a type of cloud but a detail found with several types." My experience of it is that it is eminently characteristic of a certain stage in the life-history of cumulo-nimbus. It is easily identified but is not at present of any meteorological importance. (D. Brunt has recently given an experimental illustration of the physical processes underlying the formation of mammato-cumulus in an article on 'Convective Circulations in the Atmosphere' in the *Meteorological Magazine* for February, 1925.) Another example of a name being given to a separate part of a cloud is that of "velum" or "veil of false cirrus" for the anvil of cumulo-nimbus. Thus we must add a separate class for parts of unit-clouds.

(7) *Squall-cloud.* A kind of Cu.Nb. which is quite easily recognised as a separate form (see p. 210) is the arched cloud which marks the onset of a line-squall (Clarke, Plate 36 A and B). It is in fact the prelude to heavy rain accompanied in part of its extent by hail and thunder. The name *cumulo-nimbus linealis* would distinguish it from other forms of cumulo-nimbus which do not show an arrangement in line.

With regard to the stratified or layer-clouds we have:

cirro-cumulus	8–9 km	cirro-stratus	9–10 km
alto-cumulus	3–4 km	alto-stratus	4–5 km
strato-cumulus	1–2 km	pallio-nimbus	5–1 km
		stratus	less than ·5 km

The clouds in the same column differ in the amount of water available for condensation and that is largely a function of height. In this sense they may be called the same kind of cloud at different levels. They can be differentiated by the extent to which they obscure the sun and moon. Through cirro-stratus the sun may be clearly seen, in alto-stratus it shows as a bright patch, in strato-cumulus the form of the sun is completely obscured.

There can be little doubt that the cloud-forms on the same line in the two columns would merge into one another by sufficient aggregation. Cirro-cumulus becomes cirro-stratus when or where the cloudlets join to form a continuous cloud-bank over a large area; and cirro-stratus, dissolving, shows cirro-cumulus as one stage of the process; that is equally the case with alto-cumulus and alto-stratus; but the cirro-stratus that is represented by the massing together of cirro-cumulus differs in appearance from cirro-nebula which is much thinner, and the conditions represented by cirro-nebula require separate definite mention. Moreover the sky is often in a nebular condition at lower levels. For example, General Delcambre spoke at Rome of a misty condition being formed below cloud on certain occasions[1].

The clouds which have been mentioned previously had a definite outline, but now we are concerned with a cloudiness or milkiness which has no definite boundary and therefore no shape. A separate group is required for these nebulous clouds.

The names of some groups of clouds are perhaps not very appropriate. Cirro-cumulus is not tessellated cirrus but tessellated cirro-stratus; it is like cirrus only as regards the height at which it is formed, and it is like cumulus only in so far as it appears in detached masses. The lines in which these clouds are arranged are a much more characteristic feature, cirro-stratus tessellatus would be more reasonable; but the names are in common use and the clouds are easily recognisable. They therefore serve their purpose; the objection to them is that they may lead the tiro to think that the association with cirrus or cumulus means more than it ought. There is a similar objection to alto-cumulus which is more accurately described as alto-stratus-tessellatus; and strato-cumulus which is generally a group of linear clouds. They are not easily associated with cumulus.

On some occasions we find in the telegraphic reports entries such as Cu.7 or Cu.8, meaning that the greater part of the sky is covered by detached cumulus. These also may be arranged in a horizontal layer and the meteorological condition is notably different from that which is represented by Cu.1 or Cu.2 or Cu.3; we ought perhaps therefore to provide a name for the layer of cloud of detached cumulus, representing Cu.4 to 10, when the sky becomes tessellated with cumulus. Cumulus tessellatus would meet the case. In England the condition of extensive cumulus often passes to Pallio-nimbus.

Summing up these comments I find my own requirements for cloud-forms to be as set out in the following table and I have endeavoured to illustrate them in Plates I to XIV.

Class A. Individual or unit clouds. Cirrus (detached), False cirrus, Cumulus (detached), Nimbus (detached or cumuliformis), Cumulo-castella, Cumulo-nimbus, Cumulo-nimbus linealis (line-squall cloud), Nubes lenticularis, (Colli-)stratus, Detached fog.

Class B. Parts of unit-clouds. Mammato-X, Anvil or veil of false cirrus.

---

[1] 'Les Systèmes Nuageux,' by Ph. Schereschewsky and Ph. Wehrlé. *Mémorial de l'Office National Météorologique de France*, Année 1, No. 1, Paris, 1923.

Class C. Assemblies of unit-clouds without orderly grouping. Groups of cirrus without orderly arrangement. Some forms of strato-cumulus.

Class D. Linear cloud-groups. Fracto-cumulus (Cumulus linealis), Fracto-stratus, Fracto-nimbus.

Class E. Clouds arranged in parallel lines: Corrugation. Strato-cumulus (corrugatus), Roll-cumulus.

Class F. Tessellation, double corrugation: detached masses arranged in ranks in two directions forming a tessellated cloud-layer. Cirro-cumulus, Alto-cumulus, Strato-cumulus (some examples), Cumulus 4–10. Coelum tessellatum.

Class G. Continuous layer. Stratified or layer clouds. Cirro-stratus, Alto-stratus, Pallio-nimbus, Humilo-stratus, General fog.

Class H. General cloudiness or milkiness without boundaries. Cirro-nebula. Other nebular conditions. Alto-nebula. Mist.

The illustrations of cloud-forms in an atlas should certainly now include photographs of typical conditions as seen by aviators from above the cloud-layers.

In order to avoid the use of the same name for different types the nomenclature could be revised.

At the meeting of the International Meteorological Committee in 1921 it was reported that clouds were to be seen in Japan and at the Azores which were not included in the international atlas. Colonel Chaves has been good enough to supply a number of photographs of clouds in the Azores taken with the verascope of Messrs Richard Frères. They are reproduced in Plate V.

The stress which is here laid upon stratification is justified by our increasing knowledge of the stratification of the atmosphere by its potential temperature, and the association of clouds with different levels is borne out by the table of heights of different cloud-forms which is given by Hildebrandsson in *Les Bases de la Météorologie Dynamique*, livraison 8, chap. VII, and forms part of our representation of the atmospheric structure in Vol. II.

The physical processes incidental to the formation of clouds and the changes to which they are subject will be treated in Vol. III.

Among the cloud-atlases of which the titles are quoted, those by Clayden, Vincent and Besson are distinguished by the number of varieties for which they think separate names to be desirable. Clayden names nine varieties of cirrus, Vincent thirteen, Besson thirteen, and so on. Clayden adds the general terms: "nimbus," "fracto"— for clouds which are supposed to be torn to pieces by wind,—and "undatus." Vincent suggests the addition of the qualifying terms "undulatus" applying to upper clouds; "undulatus," "striatus" and "mammatus" applying to intermediate clouds; "striatus" and "mammatus" applying to low clouds.

I have tried to identify corresponding forms in Mr Clayden's and M. Vincent's types by an examination of the illustrations which those authors have given and the accompanying text. I cannot say that the success which I have achieved is entirely satisfactory to me.

Of the remaining atlases given on p. 212, Neuhaus, London Admiralty Hydrographic Department, London Meteorological Office, Washington Department of Agriculture Weather Bureau, Clarke and Matteuzzi follow the international classification, and Loisel and Taffara adopt that proposed by Vincent.

We close the question of classification for the present with a reference to a paper on the classification of cirrus by H. H. Hildebrandsson, *Geografiska Annaler*, 1922.

Measurement of the height of clouds

The determination of the height of the different forms of cloud is the next subject that claims our attention. For that purpose the most effective method is the beam of a searchlight set vertical which illuminates a patch of the cloud above it. With a sighting scale at a distance from the searchlight it is easy to read off the height of the illuminated patch. But in its ordinary form this method is only available at night-time; whether it will be possible to adapt it for use in the day-time is not yet settled. In any case a searchlight is n'ot an ordinary part of the equipment of an observatory and in the meantime the instrumental arrangements for determining the heights of clouds and the direction of their motion were reviewed in a pamphlet[1] of instructions for the observations of clouds as an international enterprise during the year 1896–97, which is commonly known in meteorological circles as the "cloud-year," "année des nuages."

A naval or military range-finder is an obvious suggestion as an instrument wherewith to read the distance and height of a cloud. It has been used for the purpose but not apparently with satisfaction. Its use becomes indeed less satisfactory when the clouds are at great heights, and such situations are of higher meteorological value.

Two special theodolites by Olsen of Oslo, designed by Mohn, having sighting tubes without lenses, are recommended, one to be at each end of a measured base 1000 metres long for low clouds and 2500 metres long for high clouds; or equally well the pair of Darwin-Hill mirrors, used in precisely the same manner as for the observation of pilot-balloons referred to in a later section. The other forms of instrument for this purpose are the photogrammeters which are adaptations of two cameras to attain the same object by photographs of the same cloud from two separate points. An arrangement of this kind is installed as part of the permanent equipment of the Meteorological Observatory at Potsdam. Special arrangements for the same purpose were set up at Kew Observatory, Richmond, by the Meteorological Council with the assistance of Sir W. Abney in the year 1885[2]. Many pairs of photographs were obtained but the results have never been fully discussed in consequence of the difficulty of identifying corresponding points of the two photographs. The base in this case was half a mile and is perhaps too large for easy identification. A. W. Clayden obtained paired pictures of clouds from the ends of a base 200 yards long by choosing occasions when the sun was within the field of view and consequently marked on each plate the position of an object at infinite distance. An endeavour has been made to use a base of ten metres on the roof of the Meteorological Office, South Kensington,

---

[1] *Des principales méthodes employées pour observer et mesurer des nuages*, par H. Hildebrand Hildebrandsson et K. L. Hagström.

[2] *Reports of the Kew Committee*, 1885–93. 'Cloud Photography conducted under the Meteorological Council at the Kew Observatory,' by R. Strachey and G. M. Whipple, *Proc. Roy. Soc.*, vol. XLIX, 1891, p. 467.

by combining on a single screen two images formed by paired lenses with vertical axes, the one directly over the screen, the other focussing an image after reflexion of the beam from a large plate-glass mirror ten metres distant and a second reflexion from a total reflexion prism. The combination of apparatus forms what may be called a *stereo-camera obscura*; when the apparatus is in perfect adjustment the separation of the two images on the screen is inversely proportional to the height of the cloud, the images being coincident for the stars or other objects "at infinity."

Measurement of the motion of clouds

The common name for an instrument to measure the direction of motion of a detached cloud or a recognisable portion of a cloud-mass is a nepheloscope or nephoscope. Most of the instruments are based upon the interesting geometrical principle that if a horizontal line is drawn and a cloud as sighted from a fixed point travels along the line, the direction of the line is parallel to the direction of motion of the cloud. No correction for projection or foreshortening is necessary. This principle is most simply illustrated in the Besson nephoscope, "herse nephoscopique Besson," a comb with eleven vertical teeth, 10 centimetres apart on a horizontal cross-rod, carried on a vertical rod about 250 centimetres high. The direction of the comb can be adjusted from a few metres distance by a bridle until the cloud under observation appears to travel along the comb. Since however the height of the cloud is unknown at the time that the observation of motion is made, it is customary to express the travel across the nephoscope in terms of the travel across a plane at 1000 metres above the surface. The formula $\Omega = 1000a/bt$ gives the angular velocity of motion at the zenith in milliradians per second, where a is the distance between the spikes, b the distance from the upper cross-piece to the marked point on the rod which has been adjusted to the level of the observer's eye, and t the observed time which the cloud takes to pass from spike to spike. a and b must be measured in the same units.

A slightly more elaborate expression of the same principle is to use lines drawn from the centre to the circumference of a horizontal mirror of black glass as in the nephoscope of Fineman. A vertical rod fixed to the circumference of the mirror provides a means of "fixing" the observer's eye, a provision which is not thought to be necessary on the larger scale of the Besson nephoscope.

All nephoscopes are practically modifications of this same principle except the *camera obscura* which is the most expressive of all. To adapt it to determine the motion of clouds all that is required is that the screen upon which the image is formed should have orientation lines drawn on it from the image of the zenith, and equidistant circles representing equal space intervals at an assumed level of 1000 metres.

The estimate of the velocity of travel of the cloud under observation is obtained in that way. Equal distances, as sighted in the nephoscope or focussed in the camera obscura, correspond with equal distances in the horizontal plane in which the cloud is supposed to move.

METEOROLOGICAL OPTICS

Closely associated with the study of clouds is the study of the optical phenomena which are seen in the free air. These include the blue colour of the sky, the red colours of the sunset and other phenomena which are apparent without any special instrument. But the polarisation of the light from the sky is a matter of some importance as it is a physical consequence of the process of scattering which produces the blue colour. For that a Jamin or Babinet compensator is quite frequently employed.

E. C. Pickering (1885) devised a special instrument for measuring the polarisation of the sky. The physical analysis of the different colours of the sky was the object of extensive researches by Abney[1].

A spectroscope is indeed a desirable adjunct to the resources of a well-equipped observatory and a short direct-vision instrument is the most convenient. From time to time a good deal of attention has been paid to the absorption bands shown in the spectroscope, particular groups of which were attributed to water-vapour in the atmosphere and were recommended as a valuable addition to the material for forecasting[2]. The suggestion has not however found any further development.

The most striking phenomena coming within the range of meteorological optics are the rainbow, iridescent clouds, haloes and coronae. These require no special instrument, though haloes and coronae are often seen more clearly as reflexions in a black glass-mirror such as that which forms the table of the Fineman nephoscope than as viewed directly. Moreover it is convenient to include in the equipment of an observatory an instrument which enables an observer to get very quickly a measurement of the angular radius of a rainbow, halo or corona. This can be done by sighting the ring to be observed from the axis of a circle so mounted as to be turned at will in any direction. A very convenient instrument of this kind is in use at the Meteorological Institute, De Bilt.

THE WINDS OF THE FREE AIR BY PILOT-BALLOONS

We now revert to the measurement of wind in order to consider it especially from the point of view of the free air. We must keep in mind not only the difficulties attaching to the meteorological significance of the measures of wind obtained by anemometers at the surface, but also the conclusion drawn from experience in the use of the barometer to the effect that the distribution in a horizontal surface is required in order to bring pressure into

---

[1] Bakerian Lecture, 1886, and 'Colour Photometry,' Parts II and III, by Capt. W. de W. Abney and Maj.-Gen. E. R. Festing, *Phil. Trans.* A, vol. CLXXIX, 1889, pp. 547–70; CLXXXIII, 1893, pp. 531–65.

[2] *A Plea for the Rainband and the Rainband vindicated*, by J. Rand Capron, London [1886]. Piazzi Smyth, *Nature*, vol. XII, pp. 231, 252; vol. XXII, p. 194; vol. XXVI, pp. 551–4; vol. XXIX, p. 525; *Journal of the Scott. Meteor. Soc.*, vol. V, 1880; and *Madeira Spectroscopic*, Edinburgh, 1882.

relation with wind. The horizontal surface most frequently employed is that of sea-level, but there is nothing in the original idea of reduction to a common level that limits the consideration to sea-level. It would be quite possible to use some other level as that to which the pressure should be reduced for the purpose of mapping. W. Köppen indeed suggested a height of 106 metres above sea-level as being more serviceable than sea-level because, among other reasons, the mean pressure at that level is 1000 millibars. We are not at the moment, however, concerned with the question of selecting a level for maps but with the representation of the structure of the atmosphere. The laws of the relation between wind and the distribution of pressure which are set at naught, or gravely impaired, by the effect of eddies in the surface-wind are operative without that specific disadvantage in the upper air, and although we may not at the moment see our way to maps of the distribution of pressure on any specific occasion at various levels, we can regard the observations of the motion of air at different levels as an important contribution to the representation of the atmospheric structure.

Here perhaps we may call attention to the meaning which ought rightly to be attached in meteorological literature to the word *horizontal*. A horizontal surface is by strict definition a surface which is everywhere at right angles to the direction of the force of gravity, that is to the plumb-line. It is on that account that "sea-level" or the surface of still water is horizontal. When we transfer our ideas from the surface to the upper air we must still understand the word *horizontal* in the same sense, that is to say as at right angles to the force of gravity. It follows immediately from this conception that horizontal surfaces in the free air are "equi-potential surfaces" because no energy is spent or required by a heavy body moving along them without friction. The vertical distance therefore between two level or horizontal surfaces represents the same difference of potential all over the surface. It is not quite the same thing as the geometrical difference of height: the surfaces approach one another more closely in the geometrical sense where gravity is greater. A horizontal surface in the upper air is to be found by tracing points above sea-level where the integral $\int g\,dh$ is the same, not where the geometrical integral $\int dh$ is the same. It is on that account that for indicating heights in meteorology the geopotential $\int g\,dh$ is to be preferred to the geometrical value $\int dh$. The difference is not very large and the transition from geometrical height to dynamical height can be made by means of tables[1].

For a long time observations of wind and other information about the free air have been regarded as an important part of meteorological enterprise. In a memorandum drawn up in 1842 in favour of the establishment of Kew Observatory (Richmond) as an observatory for the study of geophysical subjects the following paragraph occurs[2]:

[1] *Dynamic Meteorology and Hydrography*, by V. Bjerknes and different collaborators. Washington, D.C., published by the Carnegie Institution of Washington, 1910. Meteorological Tables and Appendix.

[2] Memorandum on the establishment of a Physical Observatory at Kew [Richmond] under the auspices of the British Association for the Advancement of Science, B.A. meeting, Manchester, 1842. *The History of Kew Observatory*, by R. H. Scott, p. 15.

Among instruments which have been proposed, and which will probably not be constructed and brought into use without the assistance which an institution like this alone can afford, may be mentioned: a universal meteorograph, which will accurately record half-hourly indications of various meteorological instruments, dispensing entirely with the attendance of an observer; an apparatus for recording the direction and intensity of the wind simultaneously at various heights above the earth's surface; an apparatus for telegraphing the indications of meteorological instruments carried up in balloons or by kites, to an observer at the earth's surface.

We have no information as to the particular form of apparatus that was contemplated in this memorandum; one of its signatories was Sir Charles Wheatstone, and no doubt the possibilities of electrical transmission were at that time very fully in view; but, so far as we know, the actual achievements of the Observatory in that direction were an apparatus for kite-observations in 1847, four ascents in a manned balloon by Welsh in 1852, and subsequent observations of temperature at the Pagoda in Kew Gardens. Since that time observations of the currents of the upper air by means of pilot-balloons have become very general.

In more recent days kites of various forms have been employed and kite-balloons, which are a combination of gas-bags lifted by hydrogen, but tethered. They operate in a wind, like a huge kite, and have been much used for getting information from the upper air.

Kites were employed by E. Douglas Archibald for estimating the variation of wind with height; they are now seldom used exclusively for wind but to carry meteorographs.

The most common apparatus for the measurement of wind in the free air is a small balloon filled with hydrogen, the special sizes in daily use are known as 48 inch, 70 inch and 90 inch, which means that they can be blown out to that circumference without bursting. Larger balloons can be used with advantage but the expense increases with the size. Any other visible object that floats in the air can be used in like manner. During the war the smoke of a shell-burst was observed by special forms of apparatus, the Darwin-Hill mirrors were largely used, and on comparison with the ordinary observation with a pilot-balloon a very satisfactory accordance was found.

The progress of the balloon horizontally and vertically is determined by observations with two theodolites by which the altitude and azimuth of the balloon, as viewed from each end of a base-line, can be determined; from these data the height and distance of the balloon at any instant of simultaneous observation from both stations can be calculated.

Quite early in the development of an inflated pilot-balloon made of india-rubber as a method of observing the currents of air at successive levels it was found that the rate of ascent of the balloon was approximately constant and, with the knowledge of the rate of ascent, the height could be determined by simply noting the lapse of time from the start. The original formula is due to Hergesell and after long experience the formula in use in the observations of the British Office is:

$$V = q.L^{\frac{1}{3}}/(L + W)^{\frac{1}{2}},$$

where V is the upward velocity, W is the weight, and L is the free lift, that is the excess of buoyancy over the weight.

The value of q near the earth's surface is about 275 if L and W are expressed in grammes and V in feet per minute, but it depends to some extent on the size of the balloon. If V is expressed in metres per minute the value of q is 83·8[1]. [The modern practice of the "tail" method is explained in vol. IV, p. 170.]

PRESSURE, TEMPERATURE AND HUMIDITY IN THE FREE AIR

We read in Forbes's report to the British Association that a few observations with balloons had been published by Lord Minto in the *Edinburgh Journal of Science* and that in the winter of 1822–23 Captain Parry with the Rev. George Fisher by means of a kite had found in latitude 69° 21′ that a thermometer indicated no diminution of temperature at a height of 400 feet where it stood at − 24° F.

We have seen that as early as 1842 the possibility of obtaining indications of meteorological instruments carried up in balloons and by kites was one of the reasons why certain scientific men of that time were desirous that the British Association for the Advancement of Science should initiate a physical observatory. Sir J. Herschel, Professor Wheatstone and Lieut.-Col. Sabine were most prominent in the enterprise. The records obtained by John Welsh, superintendent of the observatory, in four ascents in balloons in August, October and November 1852, show excellent observations of the temperature of the air, tension of vapour, and relative humidity, up to about 20,000 feet. On each occasion there was an inversion of lapse-rate of temperature below ten thousand feet.

Subsequently James Glaisher, of the Royal Observatory, Greenwich, carried out meteorological observations in a series of historic balloon ascents for the British Association between 1862 and 1866. On these occasions the precautions for preventing the direct influence of solar radiation upon the thermometers were apparently less effective than Welsh's. The curves obtained have been thought to require correction on that account. The Meteorological Council, appointed in 1877 to direct the Meteorological Office, proposed to continue the investigation of the upper air by balloons and to use shell-bursts for the study of the wind-currents. The balloon enterprise came to an untimely end with the loss on December 10, 1881, of the balloon "Saladin," and of Walter Powell, M.P., who was carried out to sea after an attempt had been made to secure a landing on the shore at Bridport.

The method of shell-bursts was tried by Sir Andrew Noble at the suggestion of Sir Francis Galton[2]. A height of 9500 feet was reached with 6 lb. shells fired at 75°, but the investigation was not followed up.

The next stage in this branch of the science is indicated by the publication of three volumes of results of observations in balloons under the auspices of

---

[1] A comparison of the formulae used in different countries is given by Dr Molchanoff in the Report of the Meeting of the International Commission for the Upper Air, London, 1925.
[2] *Report of the Meteorological Council to the Royal Society, London*, 1882, p. 23.

the Aeronautical Society of Berlin[1]. Then in 1896 Abbott Lawrence Rotch, the founder and Director of the Observatory of Blue Hill, Hyde Park, Mass., U.S.A., commenced a regular series of observations with self-recording instruments carried on kites[2], and Léon Teisserenc de Bort founded the Observatory at Trappes for the study of the upper air with kites and sounding balloons, that is to say balloons which carry self-recording instruments but no observer; the use of balloons of this kind had been suggested by Hermite and Besançon. Almost simultaneously a special section of the Royal Prussian Meteorological Institute, in charge of Dr R. Assmann, was specially devoted to the study of the upper air and adapted the kite-balloon, from its original military purposes, to carry self-recording instruments; ultimately an observatory for the upper air was established by the German Emperor as a separate institute at Lindenberg.

In 1896 also, a Commission for Scientific Aeronautics was appointed by the International Meteorological Organisation under the presidency of Hugo Hergesell, head of the Meteorological Institute at Strassburg, who, with the advantage of the friendship of the Prince of Monaco and of his yacht the *Princesse Alice* extended the inquiry to include observations from ships at sea. Such methods were also pursued by Teisserenc de Bort and Rotch who, after a number of preliminary trials, equipped an expedition to the intertropical regions of the Atlantic. These efforts were followed by expeditions to Victoria Nyanza by Berson from Lindenberg, *S.M.S. Planet* to the West Indies and Italian ships in the Indian Ocean. An expedition to the Amazon was organised by Patrick Y. Alexander for Berson, but had not reported when the war broke out. Mr Alexander had previously initiated observations with ballons-sondes in England from his aeronautical establishment at Bath.

So extensively had the investigation developed before the war that the pages of the international monthly issues in which the results were published, contained contributions from the following countries: Austria (Vienna and Pola), Argentina (Pilar), Belgium (Uccle), Denmark (Copenhagen), Dutch East Indies (Batavia), Egypt (Helwan), Finland (Helsingfors), France (Trappes), Germany (Aachen, Cöln, Frankfurt, Friedrichshafen, Göttingen, Hamburg, Lindenberg, München, Strassburg, Stuttgart), Great Britain (Pyrton Hill, Manchester, Limerick), Holland (De Bilt, Prinsenberg, Soesterberg), Italy (Ferrara, Firenze, Livorno, Mileto, Modena, Montecassino, Moncalieri, Milano, Pavia, Trieste, Vigna di Valle, Verona), Norway (Bergen, Kristiania), Russia (Bakou, Black Sea, Ekaterinbourg, Fort Kouchka, Kovno, Nijni-Oltchedaeff, Pavlovsk, Sebastopol, Tiflis, Vladivostok), Samoa (Apia), Spain (Tenerife), Switzerland (Zurich), United States (Blue Hill, Mount Weather). In addition observations were included from upwards of twenty mountain stations, and observations of clouds from sixteen countries.

The method of the ballon-sonde or registering balloon is to obtain a balloon filled with hydrogen, or two balloons in tandem, of sufficient capacity

---

[1] *Wissenschaftliche Luftfahrten*, 3 vols., Braunschweig, 1899–1900.
[2] *Annals of the Astronomical Observatory of Harvard College.*

taken together to carry to a height of from 15 to 20 kilometres a case containing a meteorograph, or instrument for recording pressure, temperature, and generally also humidity. By some contrivance, generally now the bursting of a balloon, the lifting power is automatically diminished to about one-half and the remaining mass falls at about the same rate as it had previously risen. Upward and downward journey each takes about one hour, or upward and downward velocities are about five metres per second. The meteorograph is fixed in some sort of cage or basket and is attached to the balloon by a long cord. For balloons that are to fall on land only a label offering a reward for the return of the instrument is required, for those to be used at sea a float is necessary to prevent the instrument dropping into the water.

In the case of soundings at sea moreover the train must not be deprived

EXPLORATION OF THE UPPER AIR

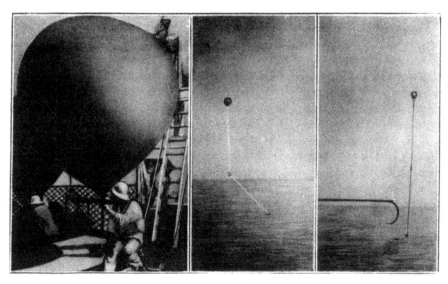

Fig. 96. The launch and recovery of sounding balloons at sea. (Compiled from illustrations in *Travaux Scientifiques de l'Observatoire de Trappes*, vol. IV.)

entirely of its hydrogen, so two balloons must be used; some sort of sea-anchor is required to prevent the balloon being drawn along by the wind faster than the observer's vessel can steam. In that case the meteorograph with its trace would be carried along out of reach after its return to sea-level. In order to give the reader a general idea of the process of sounding the air by ballons-sondes we give three pictures of the various steps of the process as employed by Teisserenc de Bort and Rotch.

Troposphere and Stratosphere

The use of these balloons has resulted in the most surprising discovery in the whole history of meteorology. Contrary to all expectation the thermal structure of the atmosphere in the upper regions is found to be such that isothermal surfaces are vertical surfaces succeeding one another with diminishing temperature from the pole outwards towards the equator, while underneath this remarkable structure, which shows the lowest temperatures of the atmosphere to be at very high levels over the equator, lies the structure with which we are ordinarily familiar consisting of approximately horizontal layers warmer in the lower latitudes and colder in the higher and, with some interesting exceptions, diminishing in temperature with height until the upper layer of vertical columns is suddenly reached. The lower region of the atmosphere in which temperature is arranged in horizontal layers was called by Teisserenc de Bort the **troposphere** and the upper region of so-called "isothermal columns" the **stratosphere**. The name **tropopause** has been coined to indicate the level at which the troposphere terminates. The conditions are effectively represented in fig. 104.

METEOROGRAPHS FOR BALLOONS AND KITES

Though the meteorographs for the British ascents carried out by Dr Varley for P. Y. Alexander were from Richard Frères, Paris, instruments for the several investigations were, for the most part, designed and made in the workshops of the respective observatories. Instruments of the type used by Hergesell were obtainable from the firm of Bosch, Strassburg; otherwise those who had not adequate workshop-facilities, obtained instruments from the several institutes. The differences are generally in detail; but there is a fundamental difference between the instruments designed by W. H. Dines and the rest, inasmuch as Mr Dines's instrument for kites used for the record a large paper disk which is rotated by a clock about its centre on a flat board instead of a drum rotating about its axis; and his instrument for sounding balloons has no clock; it merely draws a trace in which the variation of temperature or humidity is shown in relation to the variation of pressure[1]. The graph is made by a writing stilus engraving a scratch on copper coated with electro-plated silver. The extent of the record moreover is very small, the whole not being larger than a common postage-stamp; the power of the stilus to make its scratch is correspondingly great. On the other hand the instruments used in the Continent of Europe and in the United States of America are small models of a combined barograph, thermograph and hygrograph, they have the long magnifying levers which have become usual in recording instruments for meteorology and the trace is worked generally by a light point on a smoked copper or silver sheet on a drum which is revolved by clockwork within. Thus the curves drawn by the recording pens show the variation of pressure, temperature and humidity respectively in relation

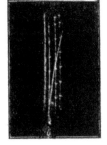

Fig. 97.
Trace of Dines
meteorograph[2].

[1] An instrument of the same character designed by Assmann is referred to in 'Ergebnisse der Arbeiten am Aëronautischen Observatorium,' *Veröff. K. Preuss. Met. Inst. Berlin*, 1902, p. 87.
[2] *Q. J. Roy. Meteor. Soc.*, vol. XL, 1914, p. 102, fig. 7.

to time. Upon the recovery of the instrument the trace is fixed by varnish. A reproduction of an original record of this character is shown in fig. 98. The clockwork instruments are many times as weighty and expensive as Dines's instruments. The only counterbalancing advantage is that the times for any intervals of height can be deciphered and an opinion formed as to whether the motion was sufficiently rapid to protect the record from the direct influence of the sun's rays. In the earlier stages of the investigation when the results obtained had to face a good deal of obstinate incredulity, that question was of primary importance, but it is less so now when the general features of the distribution of temperature in relation to pressure are recognised and the curve drawn during the descent of the balloon can be utilised as a check on the curve of ascent. Instruments on the same principle as that of Dines but even lighter and more delicate have been designed and

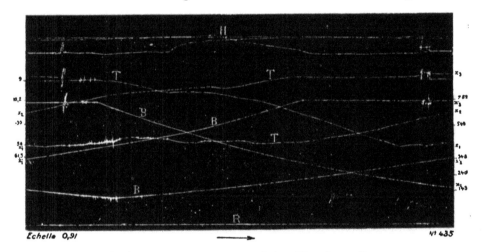

Fig. 98. A record of pressure, temperature and humidity from 36 m to 14,050 m on September 16, 1922, on smoked metal-sheet; from a meteorograph of the type used at Trappes, Office National Météorologique de France.

(The scale of the reproduction is ·56 of the original record.)

used by J. H. Field in India[1]. Each instrument has to be calibrated in a laboratory for the pressures and temperatures to which it is to be exposed. The range may be taken as from 100 millibars to 1000 millibars and from 180tt to 300tt. For this purpose special apparatus is required. The arrangements for deciphering and tabulating the records are also different. The results of observations expressed as graphs of temperature in relation to height are shown in the polygraph of fig. 104. The height aimed at by a sounding with these instruments is from fifteen to twenty kilometres: the "record" for height by this method is thirty-seven kilometres by a balloon from Pavia, Italy, in 1912.

[1] 'The free atmosphere in India.' Introduction by J. H. Field. *Memoirs of the Indian Meteorological Department*, vol. XXIV, part 5, 1924.

BALLOONS

One of the most important contributions of Assmann to this part of the investigation was the introduction of the closed balloon of india-rubber in place of an open-mouthed balloon of varnished paper which Teisserenc de Bort had employed up to that time. Two india-rubber balloons could be arranged in tandem, one of which was so filled as to burst when it had reached the desired height, leaving the gear to be brought downward at a reasonable rate with the remaining balloon taking part of the weight. In W. H. Dines's arrangement only one india-rubber balloon is employed and when that bursts the fabric acts as a parachute with sufficient effect to prevent any damage to the instrument or record.

India-rubber balloons have been found serviceable in intertropical regions, by Wagner in Tenerife, van Bemmelen in Java and by Berson in Central Africa, but they were found to deteriorate so rapidly in British India by Field that their use was abandoned and gutta-percha balloons have been used there instead. They are practically similar in behaviour to the varnished paper balloons of Teisserenc de Bort. They are apparently not so serviceable for great heights as those of india-rubber, for the stratosphere over India is not yet explored.

It is not only the measurement of pressure, temperature and humidity for which sounding balloons are serviceable. By an ingenious apparatus which opened an exhausted glass bulb by breaking a drawn-out point when pressure reached a prescribed level, and immediately closed it again automatically by fusing the broken point, Teisserenc de Bort collected samples of air from a height of 14,000 metres and, with the assistance of Sir William Ramsay, had them carefully analysed for the amount of rare gases in the samples. No difference in composition was detected and we may therefore infer that the mechanical mixing of the gaseous components of the atmosphere is effective even beyond the limit of the troposphere for the time being.

Although the result will unavoidably be imperfect we may attempt a tabular summary of the equipment that has been employed in the exploration of the free air by ballons-sondes and pilot-balloons. As some guide to the cost incurred in such investigations we shall assign prices where they are known, although they are liable to change beyond recognition even within short periods. The summary will be found on pp. 232, 233.

KITES

The kites that have been employed in the investigation of the free air are box-kites invented by Hargrave of Sydney; that is to say the structure of the kite is a cylindrical framework of thin wooden rods, the cross-section of the frame may be either rectangular, as used by Hargrave, and adopted generally elsewhere, or semi-circular as used in Russia, or lozenge-shaped with diagonal struts as used by W. H. Dines. A broad band of fabric—silk, cotton or linen, surrounds either end of the frame to the extent of about one-quarter of the length, the middle part being left open. The wire which holds the kite in the wind is steel piano-wire, so chosen as to sustain the pull unless the wind proves to be unexpectedly strong. For the details of the bridle the several papers may be referred to.

Great heights are reached by putting supplementary kites on the line to carry the weight of the lower portion. As many as seven kites have been used. The real difficulty about the continued use of kites is that on occasions the

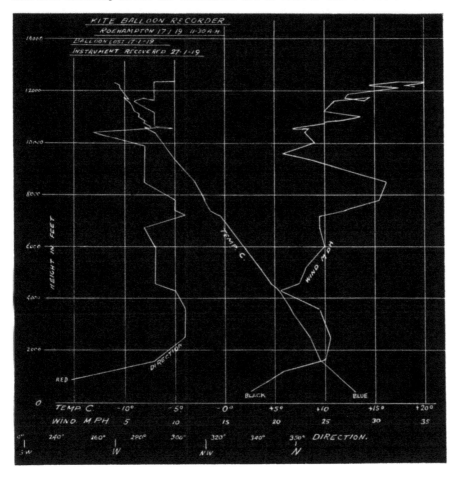

Fig. 99. Graphs of temperature and of direction and velocity of wind at heights up to 12,000 feet on January 17, 1919, obtained from the records of Dobson's meteorograph for kite-balloons.

stress in the wire is too great and the kites and wire are carried away. The length of wire is very great, sometimes several miles, and a trailing wire of that length and strength cannot safely be left about in populated districts. Teisserenc de Bort has some thrilling experiences about the effect of loose steel-wire upon locomotives, steamboats and rowing-boats in the neighbourhood of Paris.

It may therefore be expected that in course of time the aeroplane will replace the kite for the exploration of the free air up to 25,000 feet (8 kilometres), though at present the expense of aeroplanes is found to be prohibitive of any extensive use of them for this purpose.

KITE-BALLOONS, DRACHEN-BALLONS, BALLONS-CAQUOT

The use of kite-balloons for various military purposes was very widespread during the war, and the technical equipment for raising the balloon including balloon, gas-gear, wire, winch, etc., was standardised and easily procurable. In peace-time however it does not seem to be at all popular partly, at any rate, for the reason that a kite-balloon to reach a height of 5 kilometres has to be of considerable dimensions. It therefore requires a special house and a larger crew to manage it than is available at a meteorological observatory. One of its advantages is that the weight of the meteorograph is of little importance compared with the weight of the balloon itself. A special meteorograph was designed in 1919 by G. M. B. Dobson and tried at Roehampton on January 17 of that year. On that occasion the balloon was lost but recovered ten days later. Its records were intact and from them the graphs represented in fig. 99 have been obtained.

Some Books of Instruction in the Technique of Upper Air Soundings

Ballons-sondes—Pilot-balloons

Germany: (1) Anleitung zur Auswertung von Registrierballonaufstiegen nach den neuen Bestimmungen der Intern. Kommission für wiss. Luftfahrt. J. Reger, Beitr. z. Phys. d. fr. Atm., Bd. v, 1913.

(2) Die Drachentechnik am Aeron. Obs. Lindenberg v. O. Tetens. Jahrbuch, Bd. IX, 1913.

(3) Das königl. Preuss. Aeron. Obs. Lindenberg v. Richard Assmann, 1915. (Braunschweig, bei Fried. Vieweg u. Sohn.)

Great Britain: Computer's Handbook, Section II, Subsections I and II. (London, H.M. Stationery Office.)

The free atmosphere in the region of the British Isles. First Report by W. H. Dines, F.R.S., M.O. publication No. 202, 1909.

India: Memoirs of the Indian Meteorological Department, vol. XXIV, part 5.

Italy: Il "Meteorografo Gamba" per palloni-sonda. Estratto dagli Atti del Congresso della Società Meteorologica Italiana, Torino, 1921.

I Lanci di Palloni-Sonda eseguiti nel R. Osservatorio Geofisico di Pavia nell' anno 1906 dal Dott. Pericle Gamba. Ann. dell' Ufficio Cent. Met. e Geo. Ital., vol. XXVIII, parte 1, 1906. Rome, 1909.

Il R. Osservatorio Geofisico di Pavia dal Prof. Pericle Gamba, Direttore. Pavia, 1925.

United States: U.S. Department of Agriculture. Weather Bureau. Instructions for Aerological Observers by W. R. Gregg. Washington, 1921. W.B. 740.

For observations at sea

Ballons-sondes and pilot-balloons.

Exposé technique de l'étude de l'atmosphère marine par sondages aériens. Extrait du mémoire de MM. L. Teisserenc de Bort et L. Rotch (Travaux Scientifiques de l'Observatoire de Trappes, vol. IV, 1909, pp. 39–51). Reprint by U.G.G.I. 1924.

Sur les lancements de ballons-sondes et de ballons-pilotes au-dessus des océans, Note de S.A.S. le Prince de Monaco. Comptes Rendus de l'Académie des Sciences, vol. CXLI, 1905, pp. 492–3.

Pilotballonaufstiege auf einer Fahrt nach Mexiko März bis Juni 1922, von Alfred Wegener und Erich Kuhlbrodt. Aus dem Archiv der Deutschen Seewarte, 40 Jahrg. 1922, nr. 4.

H. Hergesell: 'Instructions for the use of ballons-sondes on board ships at sea,' The free atmosphere in the region of the British Isles, M.O. 202, Appendix 1, p. 52, London, 1909.

Kites

E. Douglas Archibald: 'On the use of kites for meteorological observation,' Q. J. Roy. Meteor. Soc., vol. IX, 1882–83, p. 62.

W. H. Dines: 'The method of kite-flying from a steam-vessel and meteorological observations obtained thereby off the west coast of Scotland,' Q. J. Roy. Meteor. Soc., vol. XXIX, 1903, p. 65.

Lawrence Hargrave: 'Flying-machine motors and cellular kites,' Journ. of the Roy. Soc., N.S. Wales, vol. XXVII, 1893.

H. Hergesell: 'Drachenaufstiege auf dem Bodensee,' Beitr. Phys. freien Atmosph., Strassburg, Bd. 1, 1904–05, p. 1.

A. L. Rotch: 'The exploration of the free air by means of kites at Blue Hill Observatory,' Smithsonian Report, 1897, p. 317, Washington, 1898; Q. J. Roy. Meteor. Soc., vol. XXIV, 1898, p. 250.

—— 'Meteorclogical observations with kites at sea,' Science, vol. XIV, 1901, p. 896, and vol. XVIII, 1903, p. 113.

—— 'The exploration of the atmosphere at sea by means of kites,' Q. J. Roy. Meteor. Soc., vol. XXVIII, 1902.

Kurt Wegener: Die Technik der Drachen- und Ballonaufstiege im Winterquartier 1912/13 zu Ebeltofthafen (Spitzbergen). Braunschweig, 1916.

[In connexion with the proposal for an international association with the title *Aeroarctic* under the leadership of the late Fridtjof Nansen to make use of an airship for initiating and maintaining observing stations in the polar regions, suggestion has been made by P. Moltchanoff of Sloutzk in Russia and at Lindenberg in Germany to introduce a system of automatic signalling of the readings of meteorological elements from instruments fixed in selected positions on land, water or ice, and also from those carried by sounding-balloons. See 'Erforschung der höheren Atmosphärenschichten mit Hilfe eines Radiometeorographen' (Vorläufige Mitteilung), Sloutzk Observatorium.]

Particulars of balloons used in the several meteorological observatories for soundings of the upper air

	France	Germany	Great Britain		Holland	Italy — Pavia	Italy — Rome	Russia		Spain
Manufacturer	Hutchinson, Paris	Continental, Hanover	Paturel (Paris), Pirelli (London) and Continental Rubber Co. (London)		Continental, Hanover	Pirelli, Milano	Pirelli, Milano	"Tréougolnik," Leningrad		Hutchinson (Paris) and Delasson (Paris)
Material of the balloon	—	Rubber	Rubber		Rubber	—	—	Rubber		—
Weight in grammes	1140	1430–1650	400	800	1000–1100	700–720	700–750 feuille de para	80	470	450*
Thickness of the original rubber-sheet in mm	0·6	0·2	0·4	0·4	0·3	0·4	0·4	0·4	0·4	0·6
Diameter of the circle occupied by the material lying flat on the table (cm)	80	160	70	100	100	110	100	20	50	60
Circumference when filled with gas (m)	1·75	6·0	3·3	4·15	5·40	Pas toujours sphérique	4·4†	240	‡	4·4
Rate of ascent (m/min.)	200	250	250	300	—	—	180	200 (sans enregistreur)	—	—
Volume of gas at standard pressure required for filling (m³)	3	3·6	0·6	1·2	2·58	1·500	1·4	—	—	1·3
Free lift (g)	950§	760–1280	250–340	560–840	1000–1100	600	1100	195	—	1000
Price	90 fr.	50–60 Mark	10/- to 20/-	20/- to 40/-	15 fl.	75 It. lire	86 It. lire	1 r. 40 k. (about 3/-)	25 r. (£5)	1400 pts for 30 balloons (avec les droits de la Douane)

* On emploie dans chaque lancement trois ballons, dont deux portent 1 kg force utile et le troisième 0·8 kg. Ce dernier en substitution du parachute.

† En effet on use au lieu d'un ballon unique No. 3 ballons gonflés respectivement à 750, 500 et 450 g (force ascensionnelle libre).

‡ Les données pour les grands ballons, lancés jusqu'à 1914, sont imprimées dans les "Publications de la Comm. Intern. pour l'Aéro. Scient." A présent des grands ballons ne sont lancés que très rarement et les coefficients correspondants ne sont pas encore déterminés.

§ Force ascensionnelle 1100 g.

In Canada balloons of 250 g to 330 g are employed, they are made in the country. Smaller balloons are used in all countries as pilots for observations of wind. In Norway for example balloons of 33 g to 35 g made by Macintosh and Co., Manchester, are employed.

Particulars of meteorographs used in the several meteorological observatories for soundings of the upper air

	France		Germany	Great Britain	Holland	Italy		Russia	Spain†	United States‡
	Strasbourg*	Trappes				Pavia	Rome			
Manufacturer	J. and A. Boech	Observatoire de Trappes	Boech and Boech, Hechingen	Short and Mason, Negretti and Zambra (Dines)	Bunge, Berlin	L. Fascianelli, Rome	Laboratorio R. Stazione Aerologica Principale di Vigna di Valle	Observatoire Aérologique	Boech and Boech-Hergesell	
Weight of instrument in grammes	980–1420 (including basket)	340	620	70 (including case)	260	250	437	180–200	—	180
Weight of parachute in grammes	140–250	270	360	—	210–230	(200)	—	—	—	50
Weight of basket (grammes) ...	—	180	300	43	70–80	80	158	—	—	45
Weight of "Parasoleil" ...	—	120	—	—	—	—	—	—	—	—
Total weight (grammes) ...	1120–1670	910	1280	127	540–560	420	595	180–200	800–1250	275
Cost of instrument ...	—	440 fr.	250 Mark	5£	75 fl.	800 It. l.	1271 It. l.	2·5£	250 pts.	$100
Cost of accessories ...	—	50 fr.	—	Negligible	15 fl.	15 It. l.	45 It. l.	Negligible		$3·50

* Annuaire de l'Institut de Physique du Globe, 1922, Strasbourg.
† The substitution of the Dines equipment for the heavier model is contemplated.
‡ The data for the United States are taken from a paper by S. P. Ferguson entitled 'New Aerological Apparatus,' Monthly Weather Review, vol. XLVIII, 1920, pp. 317–322.

CHAPTER XII

THE METEOROLOGICAL LABORATORY. THE STUDY OF THE ATMOSPHERIC HEAT-ENGINE AND THE CYCLE OF PHYSICAL CHANGES IN THE GENERAL CIRCULATION

c. 1757	Grass minimum thermometer	1893	Ångström's compensating pyrheliometer
1814	Leslie's differential thermometers	1893	Chvolson's actinometer
c. 1825	Herschel's black bulb	1898	Callendar's bolometric sunshine recorder
1837	Pouillet's pyrheliometer	1900	Callendar's radio-balance
1870	Marie-Davy's actinometer	1902–3	Abbot's bolometer
1874	Violle's actinometer	1902 and 1909	Silver disk pyrheliometer
1875	Crova's actinometer	1903	Wilson's radio-integrator
1880	Langley's bolometer	1903	Abbot and Fowle's absolute pyrheliometer
1885	Ångström's differential registering pyrheliometer	1908	Michelson's actinometer
1885	Crova's registering actinometer	1920	Dines's ether differential radiometer
1888	Boys' radio-micrometer	1920	Dines's aero-radiometer

THE ENERGY OF THE ATMOSPHERE, THERMAL AND ELECTRICAL

WE have reserved for a separate chapter the enumeration of the special instruments for exploring the energy-changes in the atmosphere. These changes are developed in the most conspicuous degree at surfaces of discontinuity, as for example between air and earth, or air and sea, on the one hand, or between air and water-drops or between layers of different composition and density on the other. The most notable manifestations are solar and terrestrial radiation and the thermal energy associated therewith and the energy of electrical and electro-magnetic discharges. These are naturally associated together and become merged one with another as being related to the all-pervading ether according to an older theory, or to wave-motion of varied wave-lengths according to the more modern view.

It is the more desirable in these days to recognise the association of thermal energy and electrical energy as fully as possible because the atmospherics that disturb wireless instruments tuned to definite wave-lengths have been associated with the electrical energy of thunderstorms from the earliest days of radio-telegraphy, and have more recently been associated with rainfall which is one of the chief evidences of changes of thermal energy in the atmosphere.

No doubt, in order to be strictly logical we ought to have brought the measurement of rainfall and sunshine into this chapter with solar radiation and atmospheric electricity because sooner or later we shall have to regard the complete cycle: the expenditure of the solar energy in the evaporation of water, with or without an accompanying rise of temperature of the water, the development of instability[1] by the saturation of the free air, the consequent convexion and condensation, the development of vast electrical charges by the breaking up of falling drops, the electrical discharges, and the consequent atmospherics, and the return of the condensed water to the earth again after its energetic excursion in the free air. We notice the defect in the arrangement without regret if the attention of meteorologists is thereby drawn to the fact that solar radiation and rainfall may be regarded as almost convertible terms for the expression of the energy of changes in the atmospheric structure. We thereby call attention to the circumstance that rainfall has been regarded as an indispensable meteorological measurement from time immemorial while its counterpart, solar radiation, is not measured at all except at a few special observatories.

The reason for this circumstance is sufficiently evident—the measurement of rainfall can be carried out effectively with an instrument of the simplest construction by following a few easy rules, but the exploration of the various aspects of solar or terrestrial radiation or of the manifestations of electrical or electro-magnetic energy in the atmosphere requires the resources of a laboratory and the skill of an experimenter. The apparatus needs adjustment and readjustment, not merely observation or the reading of a permanently adjusted instrument like a clock or a barometer. We have therefore brought together these and other like methods under review in a chapter with the comprehensive heading of the Meteorological Laboratory.

INSTRUMENTS FOR THE MEASUREMENT OF THERMAL RADIATION, SOLAR AND TERRESTRIAL

In the heading of this chapter we have enumerated in the briefest possible manner the preliminary steps in the progress towards the effective measurement of the changes of energy which are associated with radiation. We have not thought it necessary to recapitulate the steps of that progress in detail.

An excellent summary of the instruments for measuring radiation is given in R. S. Whipple's paper before the Optical Society[2], and a still more recent

---

[1] An ingenuous reader of the proof of this chapter has marked the word "instability" with a note of interrogation. I share his doubts. The word instability is over-worked in the literature of the mechanics of the atmosphere. Stability is used normally to denote a body or system whether at rest or in motion which is so arranged that it recovers its original condition automatically if it is displaced; but here instability means that the column of air is not in statical equilibrium. I crave permission to use a new word "sistible" to describe a condition of possible statical equilibrium, derived from *sistere* just as stable is derived from *stare*. A stork standing on one leg is obviously *sistible* and by use of its muscles is also *stable*, a drunken man is really sistible but not stable, so is a wireless-mast in the absence of its guys, so also is a layer of atmosphere in convective equilibrium. A vortex-ring on the other hand is stable but not sistible, it would lose its distinctive pressure-distribution if it lost its rotation. A cyclonic depression is regarded by some meteorologists as having a similar property. A column of atmosphere would be sistible if the density of the air diminished continually with height but not otherwise. So I would satisfy the reader's justifiable curiosity by writing *unsistibility* for *instability*. In view of the necessity of separate words for separate connotations the demand for a new word is to me irresistible.

[2] *Trans. of the Optical Society*, London, Session 1914–15, pp. 1–63.

summary is given in the *Dictionary of Applied Physics*[1]. An exhaustive survey of the position of the subject at the end of the nineteenth century is given by Jules Violle in a report to the International Meteorological Committee at St Petersburg, 1899[2]. We simply enumerate the instruments here, referring the reader to the sources mentioned for descriptions and explanations.

Solar maximum and grass minimum thermometers

For a long series of years meteorological stations have included in their equipment a solar maximum thermometer which consists of a maximum thermometer with a blackened bulb mounted in a glass vacuum-tube. The whole is exposed freely to the sky and the maximum temperature is registered day by day. It has produced some interesting facts, as for example that the solar maximum registered in the Antarctic in R. F. Scott's first expedition (1901–04), in a very cold environment, agreed very closely with that registered in Madras[3], but it has not added much to our knowledge of the energy-cycle of the atmosphere. Objection is often taken to it on the ground of want of agreement between different instruments of the same or slightly different pattern, but the basic fact about the measurements is that a single reading of the intensity of sunshine in the day does not give us a measure which can be incorporated with the physical processes of the day. Much more is wanted before the effect of solar radiation can be traced.

The same may be said of the grass minimum thermometer, another climatological instrument. It indicates the lowest temperature reached during the night; terrestrial radiation reduces the temperature of the air close to the ground below that registered within the screen, often by as much as 5 t. Such a reading is of considerable value for climatological purposes because frost close to the ground is often of real significance for the plants similarly exposed. But it is a very imperfect indication because the temperature attained depends upon the method and manner of exposure. Poynting has shown that very low temperatures can be obtained by providing suitable thermal insulation of the thermometer and protection from air-currents. S. Skinner[4] has suggested a Dewar vacuum vessel with a cup-shaped top exposed to the sky as an instrument for collecting dew with the name Drosometer; that might perhaps be effective as an integrator of terrestrial radiation but it has not been calibrated and the instrument for calibrating it is what is really required. As a contribution to the physics of the atmosphere we ought to have a measure of the loss of thermal energy due to radiation to the sky, and that the grass minimum does not give.

[1] *Dictionary of Applied Physics*, Macmillan and Co., Ltd., 1923, vol. III, s.v. E. A. Griffiths 'Radiant Heat and its Spectrum Distribution, Instruments for the Measurement of.'
[2] *Meteorological Office publication*, No. 148, London, H.M. Stationery Office, 1900, p. 42.
[3] National Antarctic Expedition 1901–04, *Meteorology*, Part I, p. 514, London, Royal Society, 1908.
[4] 'The Drosometer, or Measurer of Dew,' by Sidney Skinner, *Q. J. Roy. Meteor. Soc.*, vol. XXXVIII, 1912, p. 131.

Pyrheliometers

The instruments which are in use at the present time for the absolute measure of the thermal energy received by a blackened surface in a limited beam of direct sunlight are the compensating pyrheliometer of Ångström which has been approved as a standard instrument by various international authorities, the silver disk radiometer of Abbot and the actinometer of Michelson. The first, Ångström's instrument, depends upon the comparison of the electrical resistance of two blackened strips of thin metal, one exposed to the sun and the other receiving a compensating amount of heat from an electric current. Abbot's instrument measures the amount of heat absorbed by a water-reservoir from the solarisation of the blackened surface of a small silver disk screened from everything but direct sunlight, and Michelson's instrument is a bimetallic couple, blackened and suitably exposed to direct sunlight, the intensity of the radiation being indicated by the deformation of the couple.

Meteorologists will welcome the introduction by Messrs Richard Frères of a simple and effective self-recording pyrheliometer, designed by L. Gorczynski, which is described in the *Procès-Verbaux de la réunion à Madrid de la Section météorologique de l'Union géodésique et géophysique internationale*. The recording is by the deviation of a galvanometer-needle through the effect of radiation on a thermopile of the pattern designed by Moll.

All these instruments require calibrating or standardising in order to obtain the amount of radiation in absolute measure. It has been customary for a long series of years to express the results in gramme-calories per square centimetre per minute, but such a unit is only tolerable when radiation-measurements are regarded as belonging to a separate physical compartment and the transformations of energy in the atmosphere which are the natural result of radiation are ignored, although the comprehension of those transformations is the very purpose of the measurement of solar and terrestrial radiation so far as meteorology is concerned.

We prefer therefore, and shall endeavour to be consistent in the use of, a dynamical unit based upon the C.G.S. system for the expression of the thermal energy of solar and terrestrial radiation. The unit which we have found to be most convenient in practice is the kilowatt per square dekametre, which is indeed the same as a milliwatt per square centimetre; its relation to the customary unit is expressed by the equations:

1 gramme-calorie per square centimetre per minute = 69·7 kilowatts per square dekametre. 1 square dekametre is 1076·4 or approximately 1000 square feet.

1 kilowatt per square dekametre = ·0143 gramme-calorie per square centimetre per minute.

The actual reading of solar radiation obtained from a pyrheliometer expresses the conditions arrived at when the temperature of the receiving surface has become steady. When that state is reached the thermal energy received by the solarised surface is passed on to its environment as fast as it comes and, as in every other case of radiation, what is registered is the balance of a complicated system of exchanges. The sun's rays are the immediate cause of the rise of temperature of the receiving surface, but the receiving surface itself is also radiating outwards. Some radiation comes from the atmosphere itself in line with the sun and with that part of the sky which is included within the exposure, and the atmosphere itself acts to some extent as an absorbing screen, so much so that the measured intensity of direct solar radiation falls off quite notably as the altitude of the sun diminishes from its maximum, and a greater thickness of the atmosphere is traversed. Hence if we wish to bring the measurements of solar radiation into account in seeking an explanation of meteorological processes we want to understand clearly what we are dealing with.

Two ways of approaching the subject are open; the first is to analyse the whole complicated process into its separate parts, to estimate the original rate of radiation from the sun and the rate of simultaneous radiation from the earth and atmosphere, and work out a balance of receipts and disbursements. Since the starting-point of the balance-sheet is the rate of issue of energy from the sun beyond the confines of the atmosphere, which goes by the name of the solar constant, about 135 kilowatts per square dekametre, or 1·93 gramme-calories per square centimetre per minute, and since we are at the moment unable to make direct observations of the solar radiation beyond the confines of the atmosphere we have by some means or other to correct the reading of the pyrheliometer for the losses due to the other items of the balance-sheet. To this object a great deal of effort has been directed by the Smithsonian Institution of Washington, first under S. P. Langley and now under Abbot, Fowle and Aldrich.

Various efforts have been made by these investigators, by choosing sites at which the amount of atmospheric absorption is small, to obtain an effective measure of the residual correction. Many estimates of the value of the solar constant have been evaluated which, when corrections have been most carefully made, show residual variations of the "constant" of about 10 per cent. between 125 and 139 kilowatts per square dekametre.

Radiometers

The other way of approaching the meteorological problem is different. It takes as its fundamental observation the thermal condition of a horizontal black surface exposed to the radiation of the whole of the sky, or of a part of it. That, at any rate, helps us to define what happens at the earth's surface. We have then to make out some sort of estimate as to the effect of changes in the aspect or condition of the surface. What will be the result if instead of a black solid there is white earth, green grass, snow, water or cloud. These things have to be thought of, but the fundamental consideration is how much there is to dispose of. The instruments designed for that purpose are called radiometers and among those in use are Callendar's radiometer consisting of a platinum-resistance bridge with arms of black and bright wire. By a process of calibration it can be made to give a continuous record of the amount of radiation received on a horizontal surface from the whole sky. The record is extraordinarily intriguing, it shows remarkable fluctuations due to clouds of every kind of density. The trouble about it is that in order to protect the platinum coils from wind and weather they are covered with a hemispherical glass shade, and whatever assurance may be given that the calibration, based on comparison with an Ångström pyrheliometer, compensates for any error due to the glass, an observer with a suitable allowance of inquisitiveness wants to try it for himself.

Another instrument which fulfils the purpose of recording the radiation from the whole sky on a horizontal surface is Anders Ångström's radiometer, and in that the inconvenience of the glass cover is avoided[1].

A self-recording radiometer is described by M. Robitzsch, *Met. Z.* xxxii, 1915, p. 470.

Wilson's radio-integrator

An instrument which was introduced for the purpose of measuring the total radiation received by the ground is the radio-integrator designed by W. E. Wilson. Solar radiation can be expressed as the amount of distillation of a volatile liquid at constant temperature. The radio-integrator aims at making use of this simple relation by enclosing blackened alcohol in a retort, with two bulbs and without air, and exposing one bulb to the sky while the other is kept at an approximately uniform temperature. The instrument is not so simple in its operation as the principle which it is designed to express. One defect is that it is sensitive to the cooling effect of wind.

Pyranometers and Pyrgeometers

A type of instrument which appeals equally to the solar investigator who wishes to estimate the value of the solar constant with sufficient accuracy to register its variations from day to day and to the meteorologist who desires to bring solar and terrestrial radiation into account is the pyranometer which measures the radiation received from a portion of the sky to which it is directed. With this may be associated the pyrgeometers which are designed to measure the rate of loss of energy from a black body exposed to the sky, because the receiving surface of the pyranometer is itself a radiating body and may show a loss or gain of heat according to the condition of the sky at the time of observation. In this connexion we have only to mention the pyranometer of Abbot and Aldrich[2] which depends on the difference in the rise of temperature of two strips of manganin of different thicknesses on exposure to radiation, the difference in temperature being indicated by thermo-junctions connected with a galvanometer. The strips can be covered

---

[1] *Washington Monthly Weather Review*, vol. XLVII, 1919, p. 795, and vol. XLIX, 1921, pp. 135–138.

[2] *Smithsonian Misc. Coll.*, 1916, vol. LXVI, Nos. 7, 11.

by a glass screen which admits direct or scattered solar radiation but prevents the exchange of long-wave radiation; by removing the glass the instrument can be used to measure the outgoing radiation at night.

The pyrgeometer of Ångström[1] is a modification of the compensated pyrheliometer, in which a measure of the outgoing radiation is obtained by a comparison of the rate of cooling of polished and blackened strips of manganin, the additional cooling of the dark strip beyond that of the light strip being compensated by passing a current through the dark strip. The instrument has recently been adapted for the measurement of sky radiation[2] by using one strip covered with platinum black and the other with magnesium oxide; the absorptive power of both elements is very nearly independent of the wave-length in the case of sky radiation. Further both strips radiate almost equally for the dark waves so that these waves do not influence the temperature equilibrium between the strips, and when the instrument is covered with a glass screen in order entirely to cut out the long-wave radiation and shield the strips from air currents, the heating and cooling of the glass do not affect the readings.

The pyrgeometer of W. H. Dines, which we have called elsewhere an aero-radiometer, is an instrument based upon a special form of thermopile; by the rotation of a mirror, the radiation limited by a solid angle of sky can be compared with that from the same solid angle of a grass meadow or other terrestrial surface.

Before leaving the subject of thermal radiation we ought to mention three instruments designed by Professor Callendar which have many advantages but have not yet been brought within the reach of meteorological investigators. Two of them are pyrheliometers in which the effect of direct radiation upon a ring of thermo-electric junctions is compensated by the Peltier effect, the cooling effect of a current passed through the junctions. It has the advantage of being a null-method. The third is a pyranometer, a modification of the radiometer in which a frame of blackened platinum wire can be exposed to a certain limited area of sky with special means of compensation for the ends of the frame.

THE RELATION OF ENERGY TO WAVE-LENGTH IN SOLAR AND TERRESTRIAL RADIATION

The applications of measures of the energy of solar radiation in the problems of the atmosphere are very much complicated by the extraordinary difference in behaviour of the radiation of different wave-lengths. The most notable achievement in this section of the subject is the measure of the energy in different parts of the solar spectrum by S. P. Langley of the Smithsonian Institution. His measurements extended from the invisible ultra-violet rays on the side of short wave-length to the invisible rays beyond the extreme red on the side of longer wave-lengths. The range of wave-lengths

[1] K. Ångström, *Nova Acta Reg. Soc. Sc. Upsal.*, Ser. 4, vol. 1, No. 2, 1905.
[2] *Washington Monthly Weather Review*, 1919, p. 795.

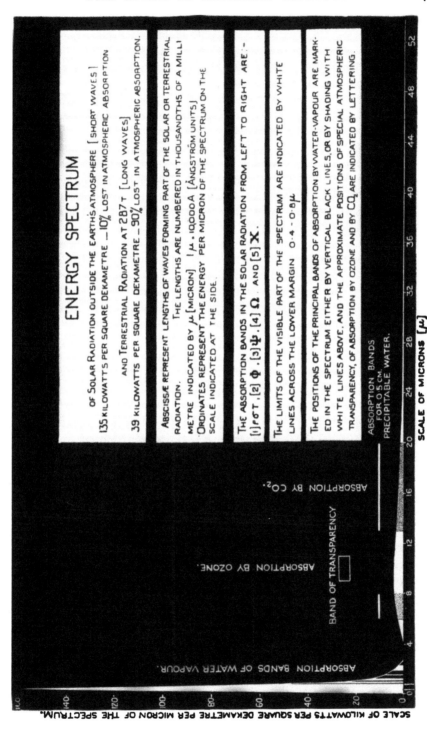

Fig. 100. Solar and terrestrial radiation.

in the thermal spectrum is from beyond ·4μ on the short side to 4μ on the long side, whereas the length of the visible spectrum is from ·4μ to ·8μ only.

This great achievement was accomplished in a spectrum from a reflexion-grating by a bolometer, or fine-wire receiver for radiation, the temperature of which was measured electrically.

One of the direct consequences of this analysis of solar radiation is to mark it off as of much shorter wave-length than the radiation which emanates from bodies at terrestrial temperatures, the laws of emission of which are now based upon Planck's quantum theory. In writing about solar and terrestrial radiation it is customary to distinguish between them as consisting of short waves and long waves respectively. The diagram (fig. 100) exhibits the meaning of this phraseology and illustrates the complication to which the meteorologist becomes liable when he attempts the application of measurements of radiation to the practical needs of the study of weather.

The wave-length is expressed in thousandths of a millimetre, called microns, a micron is commonly denoted by μ, the corresponding radiation is expressed in kilowatts per square dekametre. In this particular case two graphs are included in the same diagram, the first represents the normal intensity of radiation from the sun (corrected for atmospheric absorption) upon a square dekametre of surface normal to the sun's rays, and the second represents the radiation which is emitted from a square dekametre of a perfect or "full" radiator at a temperature of 287 tt, in a perfectly transparent atmosphere. Both are more or less idealised by disregarding, for the purposes of the diagram, the realities of the earth's atmosphere. In that sense they represent something more than an analysis of the natural conditions. The two graphs do not overlap and being on the same scale their juxtaposition exhibits some very important facts about solar and terrestrial radiation.

The use of colour-screens in the measurement of radiation

One of the most disturbing complications due to the difference of behaviour of radiation of different wave-lengths arises from the fundamental association of the properties of emission, absorption, transmission and reflexion. A perfect reflector can neither absorb nor emit any radiation at all. To all appearance snow and cloud are examples of amazingly efficient reflectors and the question at once arises whether they are correspondingly bad absorbers and radiators; but before we can think out the reality of the problem we must take account of the fact that what we can see is at best only one octave of vibrations out of four that are concerned in solar radiation and another and a different four which are concerned in terrestrial absorption and emission. The brilliantly white cloud may be absorbing dark heat and the patch of light on the waves only dealing with a part of the sun's energy.

Investigation of such questions is rendered easier now by the colour-screens made to absorb or transmit particular groups of wave-lengths, infra-red, red, orange, green, blue, violet or invisible ultra-violet.

We extract the following note on light-filters from a paper by L. Gorczynski, 'On a simple method of recording the total and partial intensities of solar radiation[1].'

I have used principally four light-filters, viz.:

(1) The so-called marmor-glass, opaque for visible rays but with a transmission up to 85 per cent. (for thickness of 7·3 mm.) between 0·9μ and 2·8μ.

(2) The red glass (Jena F. 4512, thickness 3·95 mm.) with transmission up to 84 per cent. between 0·8μ and 2·5μ, and opaque below 0·55μ, and ,for the wavelengths greater than 4·0μ.

(3) The yellow filters prepared by Dr Moll from 5·16 g. $K_2Cr_2O_7$ in 100 g. H_2O. This liquid filter, in a cell of 11 mm. thickness, transmits up to 98 per cent. between 0·7μ and 1·2μ, decreasing to 0 per cent. for 1·5μ.

(4) The blue filter of 30 g. $CuSO_4$ in 100 g. H_2O; it transmits besides short wavelengths also the radiation of long wave-lengths beginning at 1·8μ.

Further information about light-filters is given in the *Dictionary of Applied Physics*, vol. IV.

The subject needs careful consideration because the filters themselves have surfaces which are capable of a certain amount of reflexion, but when the chief object of investigation is the comparison of one source with another the effect of the reflecting surfaces may be disregarded.

Ultra-violet radiation: actinometers

The solar spectrum in the region of the ultra-violet is of little importance in meteorology because the energy which it conveys is very small in amount; but it is of great importance to humanity in other ways because of its power of penetrating the skin and affecting the underlying tissues. It is also of fundamental importance in photography because the chemical action of light is most intense in the ultra-violet region.

Actinometers have accordingly been suggested with a view of obtaining thereby a measure of the solar intensity. Sir John Herschel's actinometer was in use for some time at the Meteorological Office.

Recently Dr Leonard Hill has introduced the practice of recording the "ultra-violet" intensity at various places in England by the effect of sunlight upon a quartz tube containing a solution of acetone and methylene blue.

MEASUREMENTS OF OZONE

The action of the ultra-violet radiation from the sun is also regarded as causing the formation of ozone from oxygen in the upper regions of the atmosphere[2]. Ozone has a strong absorption band in the ultra-violet, and recently Fabry and Buisson have developed a spectroscopic method of estimating the amount of ozone present in the atmosphere by measuring the absorption coefficients of the air for wave-lengths just within the ultra-violet absorption band. The subject has been taken up in Oxford by G. M. B. Dobson, who works with a somewhat simplified apparatus, in which the light is limited to the ultra-violet by a filter of thin silver or subsequently of

[1] *J. Opt. Soc. America and Rev. Sci. Instr.*, vol. IX, 1924, pp. 455–464.
[2] See a lecture by S. Chapman to be published in *Q. J. Roy. Meteor. Soc.*, July 1926.

bromine vapour and chlorine which excludes the whole of the visible spectrum. His preliminary results[1] show an inverse relationship between the amount of ozone and the pressure at the surface, and a still closer relationship between the amount of ozone and the pressure at 10 km as obtained from soundings of the upper air.

THE ABSORPTION BY THE ATMOSPHERE AND THE SOLAR CONSTANT

It was the analysis of the intensity of solar radiation in different parts of the spectrum and in different sites that enabled Langley to find a satisfactory means of extrapolating from the variation of the radiation between Washington and Mount Whitney, to form an estimate of the absorption of the atmosphere and so correct the observed solar radiation and determine the solar constant. The approach to the present accepted value of 1·93 gramme-calories per square centimetre per minute[2] or 135 kilowatts per square dekametre has been quite gradual. Langley himself stood at a value of 3 for some time and the smaller value was only gradually arrived at by his successors with the new observatory on Mount Wilson.

INSTRUMENTS FOR THE STUDY OF VISIBILITY

Closely connected with the absorption of the solar radiation by the atmosphere is the question of the visibility, the distance at which terrestrial objects are visible. A scale of visibility has been drawn up within the last six years and after some modification has been confirmed by international agreement. The scale has been given already on p. 19. It is understood that the method of using the scale is to have a series of objects selected at the various distances and to instruct the observer to note and enter the furthest of the objects which is visible.

The observation seems to give satisfaction to those concerned though it is not intrinsically of a very satisfactory character because the visibility of a particular object depends upon many circumstances, notably the illumination of the object itself, the background and the illumination of the space between the object and the observer. A screen of muslin, for example, interposed between the observer and the object, if it were illuminated by a light from behind the observer, would render the object completely invisible.

No doubt some means of dealing with that aspect of the question will be developed when suitable statistics have been accumulated; in the meantime instruments have been devised for defining the visibility upon any occasion by counting the number of screens of unit density which must be interposed between the object and the observer to make it indistinguishable. An instrument of this kind has been described by Professor P. Gamba[3] of the Royal

[1] *Q. J. Roy. Meteor. Soc.*, vol. LI, 1925, p. 363.

[2] A more recent value for the solar constant for the interval 1912–20 is 1·946 gramme-calories per square centimetre per minute which is equivalent to 136 kilowatts per square dekametre. *Ann. of Astrophys. Obs.*, vol. IV, p. 192.

[3] 'Le osservazioni della nebbia ed il Nefelemetro Gamba,' *Bollettino bimensuale della Società meteorologica italiana*, vol. XL, 1921, pp. 29–30.

Geophysical Institute of Pavia. An instrument for a like purpose can be made by a series of holders for opalescent glass which slide on a wooden rail. Many interesting observations can be made with an instrument of that kind. The whole question of visibility and its relation to different kinds of impurity in the atmosphere has been for some years under the consideration of the Committee for the Investigation of Atmospheric Pollution, under the auspices of the Meteorological Committee. One noteworthy result obtained is that in the comparison of the beam from a parabolic reflector with the light from a standard lamp by means of an iris photometer the effect of a muslin-screen interposed in the beam is not altered by illuminating the muslin, hence it would appear that the visibility of a lamp by night is not quite the same thing as the visibility of an object viewed in the day-time by reflected light.

DUST-COUNTERS

Among instruments which have been introduced for determining the condition of the atmosphere since the observations of meteorological stations of the various orders became stereotyped by the International Organisation, we ought to refer to the instruments for obtaining information about the particles of solid or liquid, other than water, that are a notable feature of the atmosphere. This part of the subject has recently been endowed with special interest on account of its relation to the condensation of water-vapour in the atmosphere on the one hand and to the pollution of the atmosphere by solid particles on the other.

In 1880 Aitken, following on the researches of Coulier, Mascart and others, called attention to the fact that condensation did not take place in air reduced below the dew-point if all nuclei for condensation were filtered from the air by passing it through a plug of cotton-wool. He proceeded to the design of apparatus for counting the nuclei in the air by the precipitation of drops upon a glass reticule under a microscope after the condensation due to sudden rarefaction[1].

The instrument came to be known as Aitken's "dust-counter" and was used by him to determine the number of "condensation nuclei" in the atmosphere in many parts of the world. Visibility depends chiefly on them. The distance of the furthest visible object is probably controlled by the number of such particles per cubic centimetre in the atmosphere, and the humidity. For a given depression of the wet bulb thermometer, the limit of visibility multiplied by the number of particles per cubic centimetre of air was found to be roughly constant. This constant, however, increases as the air becomes drier.

For a given depression of the wet bulb, therefore, the number of particles in a column one square centimetre in cross-section and stretching from the observer to the limit of visibility is constant. J. Aitken's estimates of the

[1] *Collected Scientific Papers of John Aitken, LL.D., F.R.S.*, edited by C. G. Knott. Cambridge University Press, 1923, 'Dust Fogs and Clouds,' p. 65.

values for different degrees of humidity are shown in the accompanying table[1].

Depression of wet bulb in Fahrenheit degrees	2° to 4°	4° to 7°	7° to 10°
No. of particles of dust to produce complete haze	12×10^9	17×10^9	22×10^9

It is remarkable that an instrument so easy in its practical working and so interesting on the physical side should not have been taken into the Meteorological Observatory. It has however been used in Java by Dr C. Braak who has set out the results of his observations in a paper 'On cloud-formation. Nuclei of condensation, haziness, dimensions of cloud-particles[2],' and attention has been called to it by Wigand, who asserts that the nuclei for condensation are not dust, that the instrument gives the same number of particles in a coal cellar, purposely made dusty, and in the unpolluted air outside. Hence we are invited to conclude that the visible solid particles which we call dust, the motes in the sunbeam, are not Aitken's nuclei for condensation. It is not at all clear that Aitken ever regarded them as being so, but in the meantime an instrument has been designed by Dr J. S. Owens, and introduced into international practice by the Union Géodésique et Géophysique Internationale, which precipitates the solid particles contained within a measured volume of air, 50 cubic centimetres or more, upon the cover-glass of a microscope-slide. After being mounted the deposit can be examined under the microscope and counted. That these are actually the solid particles of the air is perhaps most safely concluded from the fact that the numbers of the counts give a representation of the dustiness of the atmosphere which is in fair agreement with that obtained from a continuous automatic filter.

Once more the opportunity arises of using two instruments, one giving the number of nuclei for condensation and the other giving the number of solid particles, as combining to give information about the atmosphere not otherwise obtainable, instead of regarding them as alternative instruments for the measurement of one quantity.

And here it may be well to reiterate a warning that has been shown to be necessary over and over again. The ordinary atmosphere is always apparently well-furnished with condensation nuclei; as evidenced by the Aitken dust-counter it requires to be carefully filtered through cotton-wool in order that it may be deprived of its nuclei of condensation. It is only when it has been so deprived that condensation upon "ions" as distinct from ordinary nuclei comes into consideration, and condensation only takes place even upon negative ions when the reduction of temperature by reduction of pressure is so great that the pressure of vapour in the atmosphere is four times that necessary for saturation at the temperature to which it is reduced. The warning, to be repeated until it is generally understood, is that negative ions are not nearly such effective nuclei for condensation as the ordinary nuclei, whatever they may be, which are found in Aitken's dust-counter by rarefaction of the atmosphere.

[1] *Meteorological Glossary*, M.O. 225, ii. H.M.S.O. London, 1918, s.v. Visibility.
[2] *K. Mag. en Met. Obs. te Batavia, Verhandelingen*, No. 10, Batavia, 1922.

ATMOSPHERICS AND THUNDERSTORM-RECORDERS

Another subject which requires the appliances of a laboratory for its study is that of the radio-telegraphic indications of lightning. Soon after the discovery of the wireless transmission of long waves of electro-magnetic disturbances lightning flashes were recognised as affecting the receivers and an instrument for the automatic recording of the occurrence of flashes was produced by Fenyi of Buda Pesth. The question soon became associated with the more general one of what is known as directional wireless which aims at identifying the direction from which any radio-telegraphic disturbance comes by the orientation of a system of coils used as receivers.

A system of this kind was worked out for the Meteorological Office by C. J. P. Cave, in conjunction with Dr Whiddington, at South Farnborough in 1916. It was extended to determine the position of the origin of disturbance by collecting the information about atmospherics or "parasites" from a number of stations in the British Isles. It has since been developed by R. A. Watson Watt. A standard form of receiving coil has been installed and now the whole subject has passed from the Meteorological Office to the Wireless Board. So far as we are aware the combination of recording and direction-finding is a subject which the meteorological laboratory has yet to solve.

LIGHTNING CONDUCTORS

The lightning rod as a piece of practical physical apparatus traces back its history to Franklin's experience of obtaining a spark from a cloud to a kite in 1752. The development of the subject by Snow Harris and others belongs to the physical rather than to a specially meteorological laboratory, though the report of the Lightning Rod Conference[1], in which meteorologists, architects and telegraph-engineers co-operated, was edited by G. J. Symons of the British Rainfall Organization. A number of very striking experiments by Sir Oliver Lodge with tea-trays representing clouds are described in his Cantor Lectures before the Royal Society of Arts, the main point of which is to show that electro-magnetic induction has an important influence in controlling the path of the discharge and not simply electrical conduction according to Ohm's law. There are still some unsolved problems connected with lightning discharges, particularly in connexion with what is known as rocket lightning represented by rocket-like off-shoots from the main flash of which Dr S. Fujiwhara has obtained some notable photographs, and also in connexion with ball-lightning which, according to E. Mathias[2], director of the Observatory of the Puy-de-Dôme, may be ascribed to the formation of very unstable compounds of nitrogen and oxygen consequent upon the concentration of a vast amount of energy in an electrical form within the extremely limited space actually occupied by lightning discharge through

[1] E. and F. Spon, London and New York, 1882.
[2] 'Traité d'Électricité Atmosphérique et Tellurique' publié sur la direction de E. Mathias, *Comité Français de Géodésie et de Géophysique*, Publications de la 6e Section, Paris, 1924.

air. The amount of energy which is concentrated in this way can be arrived at by simple calculation based upon Professor C. T. R. Wilson's estimates of the amount of energy in a lightning-flash. The explanation is the more plausible because electrical discharge is the process by which physicists and chemists have been able to reverse such natural processes as the combination of the alkali metals with oxygen or chlorine, and in fact it provides a means of overriding locally the inference to be drawn from the general tendency of the law of dissipation of energy. It is to a well-equipped meteorological laboratory that we must look for further information on these interesting subjects.

We may remind our readers that the theory of the separation of electricity in the atmosphere in quantity sufficient to form lightning-flashes is that of Dr G. C. Simpson[1], which is based upon experiments in a laboratory in connexion with the Meteorological Office in Simla. Electrical separation was shown to be a natural accompaniment of the breaking up of water-drops in the atmosphere which happens when the velocity of fall exceeds certain limits defined by Lenard. The explanation suggested was to a certain extent confirmed by observations of the charge upon drops of rain at Simla.

MEASUREMENTS OF THE EARTH'S ELECTRIC FIELD AND ITS CHANGES. IONISATION AND AIR-EARTH CURRENT

The study of the electrical energy of the atmosphere may be approached from another side which presents a much milder aspect than the lightning-flash or the thunderstorm. It has long been known than an electroscope will take an unlimited amount of charge from the atmosphere if it be connected by an insulated wire to a flame or conducting match burning in an open situation as on a roof. The process furnishes a very striking experiment with a pair of pith-balls. As the match burns the pith-balls diverge, collapse when touched with the finger, and diverge again if the match is still burning.

The burning brings the insulated pith-balls to the electrical potential of the atmosphere at the position of the match. That is different from the potential in the air close to the earth where the pith-balls are: hence the divergence is neutralised when the accumulated charge is sparked away by touching, and recovers when the insulation is re-established. The burning match is merely a device for separating particles from a conductor in a position where there is an electric field. The separation of the electricities of different sign goes on so long as there is any field at the point and when there is no field the match and its immediate environment are at the same potential.

A measurement of this kind forms the basis of the determination of the vertical potential-gradient in the atmosphere, the primary measurement of observatories for atmospheric electricity. It is an auxiliary measurement which furnishes a factor for transforming the tabulated readings of a self-

---

[1] G. C. Simpson, 'The Electricity of Thunderstorms,' *Memoirs of the Indian Meteorological Department*, vol. xx, part VIII, 1910.

THE CYCLE OF PHYSICAL CHANGES 249

recording electrometer into absolute measure—volts per metre. The plan is to take a measurement "in the open" with an instrument such as a Kelvin portable electrometer calibrated to give results in absolute measure while the self-recording instrument is running, and thus obtain the absolute potential difference between a certain height "in the open" and the ground, and use the ratio thus obtained as a factor of conversion. The factor thus determined is applicable for all the tabulated readings because the shape of the equipotential surfaces in the neighbourhood of the building which houses the self-recording instrument does not change with the intensity of the electric field in the region.

Absolute measurements have generally been made by means of a flame, fuse or radio-active collector connected to some portable type of electrometer. To avoid distortion of the equipotential surfaces, in the neighbourhood of the point where the potential is determined, by the observer and his apparatus, the collector should be fixed in the middle of a horizontally stretched wire or at the end of a horizontal conducting rod, the wire or rod being considerably longer than its height above the ground[1].

Incidentally it may be noticed that the use of a special collector, a flame, a burning match or a radio-active body, such as radium, polonium or ionium, is only necessary when a rapid determination is required, in order to avoid fluctuations in the electric field itself during the observation. The small conductivity of the air itself owing to the free ions contained in it is sufficient to bring the potential even of a sphere into agreement with that of its environment in the course of time when the conditions are steady. A metallic point like that of a lightning conductor is still more efficient and in like manner the branches and leaves of trees and herbage are persistently active in transferring electricity from the earth to the air or vice versa in order to equalise the potential of the air with that of the projecting conductor. It is well therefore for observers to be on their guard against regarding atmospheric electricity as limited to the processes taking place in the electrical instruments of the observatory or laboratory; everything that is visible from the laboratory, buildings, trees, shrubs and grass, are all as actively engaged in electrical operations as the instruments themselves; wherever we may be we are living all the time in an electrical laboratory.

We turn therefore to the apparatus for keeping a continuous record of the changes in the atmospheric electrical potential at a fixed point as a guide to the atmospheric changes which are taking place in the neighbourhood of the observatory. Lord Kelvin had brought the quadrant electrometer into satisfactory working order. He then realised that it could be used for the purpose of recording changes in the atmospheric potential by bringing it into electrical communication with an insulated metal cistern which maintained a jet of water through a pipe projecting into the open air at some distance from the wall of the laboratory. That is the water-dropper, a form of electrical collector, which Lord Kelvin used in a number of very interesting electrical experi-

[1] C. T. R. Wilson in 'Atmospheric Electricity,' *Dictionary of Applied Physics*, vol. III, p. 89, Macmillan and Co., Ltd., 1923.

ments. The authorities of Kew Observatory (Richmond) and the Royal Observatory, Greenwich, were accordingly induced to set·up self-recording electrometers with water-dropping collectors. The records have been maintained ever since; the water-dropper is still retained at Richmond, but at Greenwich it has been replaced by a radium collector.

We gather from the records referred to in the volume edited by E. Mathias that recording instruments are in operation at the following observatories: Wolfenbüttel, Ladenburg, Meiningen, Upsala, Helsingfors, Potsdam, Kew Observatory (Richmond), Paris and Val-Joyeux, Lyon, Perpignan, Munich, Bamberg, Kremsmünster, Trieste, Davos, Observatorio del Ebro (Tortosa), Moscow, Tokio, Batavia, Simla, Washington, Buenos Ayres, Helwan, Samoa; in addition observations for short periods have been made at Cap Thordsen (Spitsbergen), Danmarkshavn, Karasjok, Vassijaure, Tomsk, Petermann and Wandel Islands (Terre de Graham), Cape Adare, and Cape Evans.

Experimental work in atmospheric electricity, which had been overshadowed by the enormous advances made in galvanic electricity and electromagnetism during the third quarter of the nineteenth century, was resuscitated by Elster and Geitel, two schoolmasters of Wolfenbüttel, with the simple expedient of using the gold-leaf electroscope as a quantitative instrument by adding a means of noting the extent of the deviation of a gold-leaf. Thence has arisen a whole new field of research in atmospheric electricity which appears in physical observatories as measurement of the number and mobility of the ions in the atmosphere. "The apparatus used is generally that of Ebert who first made measurements of this kind. Some of the sources of error have been pointed out by Swann who has introduced improved apparatus[1]."

There are large ions in the atmosphere with which the name of Langevin has been associated. They "are formed by small ions becoming attached to the uncharged nuclei or dust-particles such as are made visible by Aitken's method of condensing water upon them." We are thus brought again into contact with dust-counting, but the investigation of large ions to which McClelland paid much attention is not yet out of the stage of development which may be characterised as belonging to the laboratory.

C. T. R. Wilson[2] himself has introduced apparatus which ought to stimulate activity in all meteorological laboratories. They are first an electrometer by which the earth-air current can be regularly determined, and secondly an apparatus for watching the changes in the electrical potential of the atmosphere in the immediate neighbourhood of an observatory in consequence of the discharges of lightning at a distance. The recording instrument in this case is a capillary electrometer formed of mercury electrodes with an intervening cell of sulphuric acid. The collector is a copper sphere upon a

[1] C. T. R. Wilson, *loc. cit.* p. 94.
[2] 'Investigations on lightning discharges and on the electric field of thunderstorms,' London, *Phil. Trans. R. Soc.*, A, vol. ccxxi, 1920.

4-metre pole. The effect upon the electrometer expressed as discharges in one direction or the other of ascertained quantities of electricity can be recorded on the plate of a photographic camera, a further illustration of the use of the camera in the experimental work of a meteorological laboratory.

We have already remarked that the electrical phenomena of the atmosphere are not by any means limited to the electrical instruments of an observatory; electrical changes which are constantly going on around us take their part in the transformations of energy in the atmosphere and ought not to be disregarded. Atmospheric electricity in particular affords a great field of investigation which is not at all fully explored. We have already referred to the development of an electrical field by the breaking up of water-drops and we may therefore conclude that every fire-hose and every garden-sprinkler when in operation is an electrical appliance. The development of an electrical field in blown snow is also a well-recognised phenomenon and to blown snow must also be added any other form of dust though apparently the sign of the field depends upon the nature or composition of the dust. According to W. A. D. Rudge[1] a passing locomotive produces an electrified cloud that can be easily detected at a great distance from the track.

The meteorological laboratory has therefore many calls upon its attention outside the routine of potential-gradient and ionic charges.

MAGNETIC FORCES IN THE ATMOSPHERE, ABSOLUTE AND RECORDING INSTRUMENTS

Finally the complete meteorologist cannot afford to ignore the phenomena of terrestrial magnetism. If we could regard the earth as a great magnetised sphere depending merely upon the position and permanent magnetism of rocks, we might leave the geologists and the magneticians to discuss the details. Magnetic and magnetised rocks there certainly are; but to what extent the phenomena of terrestrial magnetism are geologic, and to what extent atmospheric, is still an open question which magneticians have sought to resolve by obtaining a line integral of magnetic force round a closed curve on the earth's surface, but not yet with complete success. In the meantime the relation between magnetism and electric currents which we represent as the motion of electricity has come to be regarded as much more fundamental in the material system of the universe than it used to be.

We know that the earth's atmosphere is the seat of electrical phenomena and that electricity is there in motion. Many investigators require its upper layers to be ionised in order to account for the transmission of wireless signals, the aurora and other phenomena. Whatever may be the description of the normal phenomena in which our mind rests for the time being we must be prepared to take the atmosphere into account in dealing with the variations of the magnetic elements whether they be regularly diurnal or seasonal or irregular and transient as the magnetic storms. The meteorological laboratory

---

[1] 'On some Sources of Disturbance of the normal atmospheric Potential-Gradient,' *Proc. Roy. Soc.* A, vol. XC, 1914, p. 571.

would therefore do well to provide itself with the means of keeping the terrestrial magnetic conditions always within view and to seek every opportunity of recognising the connexion that must certainly exist between changes in the general circulation of the atmosphere and those in the earth's magnetic field.

One of the writer's ambitions, still unsatisfied, was to suspend an ordinary galvanometer mirror, nearly astaticised, in a quiet corner of an observatory and with a powerful beam of light, which now could be quite easily maintained, keep a spot of light always visible upon a scale. The ambition was never realised because the habit of a magnetic observatory was to keep its instruments rigorously free from all possibility of disturbance by local influences due to magnetic material of any kind. The idea of taking under observation a spot of light that might conceivably be disturbed by a pocket-knife or a hat-pin was repugnant to the well-trained magnetician.

But in time gone by, when the duty was to get accurate measures of electrical resistance with the aid of a mirror galvanometer, the operations had occasionally to be suspended for hours on account of the spontaneous fluctuations of the galvanometer-needle. What forced itself upon the attention when one wanted merely a constant magnetic field might very well be observed as a matter of interest. One could in fact get an index of the state of a terrestrial magnetic field, quite as easy to read and not much less accurate than an ordinary thermometer, yet I have not succeeded in obtaining it. It is however very desirable for the meteorologist to regard the fluctuations of terrestrial magnetism as a subject which he can bring under his observation if he will.

The ordinary instruments for recording the magnetic elements, declination, horizontal force and vertical force, or alternatively W-E force, S-N force and vertical force, are terribly cumbrous and require undisturbed accommodation at a uniform temperature, certainly from day to day and preferably the year through. They require a special magnetic building and many other special arrangements. A standard form of recording instrument was introduced at Kew Observatory in 1856 and has been repeated in many parts of the world. The instrument for determining vertical force is the least satisfactory of the Kew-pattern; variations have been made by W. Watson who introduced a quartz fibre suspension and a much lighter needle[1].

More portable forms of recording instruments are available but not common, those of Eschenhagen of the Potsdam magnetic observatory are the best known. Development in these matters is now largely dependent upon the activity of the Magnetic Institute of the Carnegie Institution of Washington under the guidance of Dr L. A. Bauer.

For the conversion of the reading of continuous records of the magnetic elements into absolute measure means of absolute measurement are required, and the observations must be conducted in a building free from any source of disturbance of the earth's local field and consequently, as a rule, specially

---

[1] 'A quartz-thread vertical force magnetograph,' by W. Watson, *Proc. Phys. Soc.*, vol. XIX, London, 1904, p. 102; *Phil. Mag.*, Ser. 6, vol. VII, 1904, p. 393.

erected for the purpose. Fluctuations of temperature are of less importance as they can be allowed for.

The instruments in general use for absolute measurements are the Kew-pattern magnetometer and the dip-circle; they are a commonplace of the physical laboratories of this country and need not therefore be specially described.

Electro-magnetic apparatus is gradually superseding permanent magnetic apparatus for the absolute determination of the magnetic elements. For some years now a rotating coil has been used for the determination of dip; the magnetic effect of coils of known dimensions, with known potential difference and known resistance, is a possible method of approaching the question. This was shown by W. Watson in 1901[1]. It has now been developed as a means of determining the horizontal force in an instrument constructed at the National Physical Laboratory to designs by Sir Arthur Schuster and F. E. Smith.

The original pattern is certainly elaborate and expensive, but the measurements of electromotive force and resistance are now capable of such great precision that some simplification of the apparatus under satisfactory conditions of calibration may be expected. The inherent difficulty of the situation is that for the purposes of a magnetic survey very great accuracy is required. The latest report of the Royal Observatory (1924–25) gives the declination for 1923 as 13° 35·1′, the horizontal force as ·18431 C.G.S. units, the vertical force as ·43137 C.G.S. units and the dip as 66° 51·9′. The variations from year to year are of the order of 12′ in declination, ·00016 in horizontal force, ·00039 in vertical force and 0·4′ in dip, hence the most ample precautions for accuracy are necessary. Not every magnetic observation is however necessarily directed towards the final specification of standard values and experiments in the measures of terrestrial magnetism might claim a greater freedom than is yet enjoyed.

Record of changes in the vertical force

An example of the prospect of stimulating experience with new methods of approach was given by Dr A. Crichton Mitchell at the meeting of the Union Géodésique et Géophysique at Madrid in 1924. He there showed a record of the fluctuations in the vertical force as represented by the inductive effect of the fluctuations upon a large circuit of wire laid upon the ground of the observatory at Eskdalemuir. From the point of view of the study of the general circulation of the atmosphere such observations are very acceptable because the period within which the fluctuations take place corresponds more nearly with those of the variations of meteorological elements than any which can be derived from the disturbance of heavy permanent magnets.

[1] *Phil. Trans.* 1901.

CHAPTER XIII

THE DEVELOPMENT OF ARITHMETICAL AND GRAPHICAL MANIPULATION

1512–1592	Mercator	1768–1830	Fourier
1550–1617	Napier	1777–1855	Gauss
1596–1650	Descartes	1784–1846	Bessel
1682–1716	Cotes	1792–1871	J. Herschel
1710–1761	Simpson	1801–1892	Airy
1728–1777	Lambert	1806–1871	De Morgan
1749–1827	Laplace	1811–1863	Bravais
1752–1833	Legendre	1822–1911	Galton

PROMINENT among the scientific achievements of the nineteenth century must be ranked the development of arithmetical methods of treating a series of observations such as form the staple material for the study of climate. From our point of view the most important stage in the development was marked by certain papers of Francis Galton in 1886–88[1] introducing the coefficient of correlation, or correlation ratio, as a method of expressing numerically the relation between the corresponding deviations of the quantities from their respective mean values. From that has grown a vast science of statistics and statistical methods which has already found many practical applications.

The essential part of a numerical estimate of relationship by a correlation coefficient turns upon the use of the "product-sum," that is to say the algebraic sum of the products of corresponding deviations of the values of the two quantities to be related, from their respective mean values, in comparison with the product of the so-called standard deviation of each. It is thus analogous to the relation of a product of inertia about two axes to the moments of inertia about the two axes, in the dynamics of a rigid body. The "product-sum" was introduced by A. Bravais[2] in 1846. The whole science thus set up derives from the theory of probability as applied to the choice of the best value to be adopted for the measure of a physical quantity when there are a large number of independent determinations, equally trustworthy so far as skill and care are concerned, yet differing from one another within the limits of actual measurement. Meteorology concerns itself very little with the repeated measurement of physical quantities which are supposed to be without variation; such operations belong rather to astronomy and geodesy. The earlier history of the science of correlation belongs therefore more to astronomy and geodesy than to meteorology. We find for example the idea of the arithmetic mean of a set of equally trustworthy observations as the most plausible value to be adopted for astronomical measurements as far back as Roger Cotes and Thomas Simpson, who were close followers of

---

[1] F. Galton, 'Correlations and their measurement,' *Proc. Roy. Soc.*, vol. XLV, 1888, p. 135.

[2] A. Bravais, 'Analyse mathématique sur les probabilités des erreurs de situation d'un point,' *Acad. des Sci., Mémoires présentés par divers savants*, II sér., t. IX, 1846, p. 255.

Newton. This disposes of the question as to the selection of the best value to be adopted for a series of observations of a single quantity; its extension to the method of least squares for selecting the best value to be derived by the solution of simultaneous equations was made by Legendre in 1805. That turns out to be related to the law of frequency of errors of observation when the number of observations is very large, the so-called "exponential law of errors" which was demonstrated in 1808 by Dr R. Adrain of Reading, Pa. in the *Analyst* or *Mathematical Museum*[1], by Gauss in 1809 in the *Theoria motus corporum coelestium* "thence deducing the principle of minimum squares," by Laplace in 1810, by Thomas Young in the *Philosophical Transactions* in 1819, by Hagen in 1837[2], by Herschel in 1850, and by G. B. Airy in 1861[3].

All these are mainly concerned with the theory of errors of measurement of quantities which, so far as is known to the observer, are not subject to variation in the interval within which the measurements are made. Airy, for example, on p. 114 of his book, in order to provide a test of the practical application of the exponential law of errors chooses 636 measurements in the years 1869–73 of the North Polar distance of Polaris made at the Royal Observatory. In each separate year the difference between each of these mean north polar distances and the annual mean of all was taken. But statistical science also uses numerical values of a very different character, as for example W. Palin Elderton in a chapter on 'Variates and medians' in a *Primer of Statistics*[4] chooses the measurements of the length of fifty-nine nutshells which are expected to be different, instead of fifty-nine measurements of a single shell which would show no difference if the art of measurement were perfect.

The same theory of the properties of great numbers of observations is used in meteorology and now in many other statistical inquiries in an entirely different manner. We apply the laws to observations which are designed and intended to register changes in the quantity which is measured in determining, for example, daily, monthly, seasonal or secular changes in the meteorological elements such as pressure, temperature, humidity, rainfall, sunshine or any other. We are therefore borrowing from the science of probabilities the methods which have been developed to deal with errors, and we apply them to such cases as the variation of annual temperature or rainfall in the measurement of which we assume no errors at all. We attribute the differences to natural causes which are certainly real but which are unknown to those who make the observations.

It is necessary to remember this because the theory of probability that deals with errors of observations uses a normal curve of error for which we have to substitute a curve of frequency of deviation from mean values, and the

[1] T. W. Wright, *On the adjustment of observations*, New York, Van Nostrand, 1884, p. 427.
[2] *Grundzüge der Wahrscheinlichkeitsrechnung*, Berlin, 1837.
[3] *Theory of Errors of Observations*, Macmillan and Co., London, 1861, 1875, 1879.
[4] London, A. and C. Black, 1909.

deductions from the theory of probability apply or do not apply according as the curve of deviation corresponds with the normal curve of error or not.

Many of the meteorological elements give curves of deviation which correspond sufficiently well with the normal curve of error but others, such as winds of different velocity which have a maximum on the lighter side[1], rainfall which has a J-shaped curve with a maximum at no rain, do not.

The advantage of the proceeding is that with the aid of conclusions drawn from the theory of probability, relationships can be established between quantities either of which, so far as the observer is concerned, are subject to arbitrary variations.

Thus in the development of the science of statistics as applied to meteorology in the course of the last thirty-five years we have to notice the development of a new terminology which was not necessary for the treatment of errors of observation. The arithmetic mean is not always the most plausible value of a definite physical quantity but itself creates a new idea of mean value which may be applied to any collection of observations of the same element, mean daily temperature, mean range of temperature, and so on.

When the observations extend over a series of years the mean value becomes the **normal** with the understanding that the same period of years shall be used for the observations from a whole group of stations. Thus thirty-five years is regarded as being a suitable period for normals of rainfall in view of the cycle of thirty-five years suggested by Brückner[2]. But that is only a temporary understanding depending on the fact that organised series of observations do not as a rule extend beyond that limit. The true idea of a normal for a period of years is that the mean value for every period of the same length of a very long series would give the normal and that necessarily would imply perfect recurrence at the completion of the period. We have certainly not yet attained that position for rainfall, consequently the practice is to extend the duration of the period for the computation as further observations accumulate. Normals may now sometimes be given as the mean for fifty years, or even for a hundred years.

The word **average** is often used to indicate the mean for a series of years which is not long enough to give a satisfactory approximation to an ideal normal, but in the case of new stations which have observations only for short periods the word normal is often used for the mean value for ten or twenty years.

Average is certainly a very hard-worked word, and as it is very convenient to use it for the mean values of an irregular series of years, or months, it would be better to make a practice of using other words for different kinds of grouping. Thus Dr H. R. Mill suggests "general rainfall" for expressing the mean of a number of values over a geographical area.

---

[1] *The Beaufort Scale of Wind-Force*, by G. C. Simpson, M.O. publication No. 180, London, 1906, p. 23.

[2] A table of the mean values of rainfall for each group of consecutive 35 years between 1868 and 1921 is given by J. Glasspoole in *British Rainfall* 1923, p. 241, H.M. Stationery Office.

A normal having been obtained, the next step is to obtain the deviations of the several observations, that is to say the differences between the individual observations and the normal; some will be positive and others negative. From the definition of the term it follows that the algebraic sum of the deviations is necessarily zero.

It is with these deviations that we have to do in the use of statistical methods to define the extent of relationship between the varying quantities. A name is accordingly wanted to define the extent of the deviations; the standard deviation is the term employed; it represents the square-root of the mean of the squares of the deviations and is represented by the symbol σ, where

$$\sigma = \sqrt{\frac{1}{N}\Sigma d^2},$$

N being the number of observations, d a deviation and Σ the sign of summation.

Next we ought to compare the frequency of deviations within successive equal steps of magnitude with the frequency of the normal curve of errors.

The theoretical normal curve is represented by the equation

$$y = \frac{N}{\sigma\sqrt{2\pi}}e^{-\frac{d^2}{2\sigma^2}}.$$

This equation represents the distribution of a great number of observations, classified according to their deviation from the mean value. From it, with the aid of special tables, the percentage of the whole number of observations which fall within a specified limit of deviation can be calculated. A table of percentages of that kind, for deviations which are expressed as fractions of the standard, σ, is given here.

Limits of deviation (+ or −) as fractions of the standard deviation:

·1	·2	·3	·4	·5	1	2	3	4	5

Percentage of observations which fall within the limit:

8·0	15·8	23·6	31·1	38·3	68·3	95·4	99·73	99·993	99·999

We can in this way estimate the chance that a deviation of a measure of a meteorological element will exceed any assigned limit. We note for example that 90 per cent. of the observations will be found outside the limit of a deviation of one-tenth of the standard deviation; whereas a deviation exceeding three times the standard should be found only twenty-seven times in ten thousand observations.

It is to be remembered that this conclusion is only logically justified when the number of observations of the element under consideration is indefinitely great; but the conclusion has useful significance even when the number of observations is restricted to a hundred or even fifty, provided that they are sufficiently distributed to show whether the curve of deviation corresponds with the normal curve of error.

A distribution which follows the normal curve is a symmetrical distribution about the arithmetic mean, and from that curve certain terms are derived with which the student of meteorology should become familiar. The probable error for example, or as we must say the probable deviation, is that deviation which divides the observations (positive and negative combined) into two equal groups. The word probable really means that the deviation is just as likely to be less than the "probable" as it is to be greater than that figure. In the table given above the deviation which would give a fifty per cent. grouping is ·6745σ where σ is the standard deviation, and that fraction is in consequence the "probable error." The deviations which separate the observations into quarters are called **quartiles**, and into sixths **sextiles**, and into hundredths **percentiles**, according to Galton's notation.

Not all collections of meteorological observations, by any means, group themselves in symmetrical distribution, there are many other forms which are called "skew-distributions" and which require separate consideration. The general feature of a curve representing an unsymmetrical distribution is that the mean value is not the most frequent, nor is the whole series distributed equally on either side of the mean. So well-recognised is the occurrence of distributions of that kind that a special notation has been created for them. The arithmetic mean still retains part of its utility but is shorn of its comprehensiveness; the deviation of greatest frequency is called the **mode**, and the deviation which marks the separation of the whole set of observations into two equal groups is called the **median**.

We have not occasion to pursue the further ramifications of arithmetical manipulation, and we leave it for the time being to consider some applications of graphic representation; but we shall find that that branch of the subject leads us back again into numerical computation before the end of the chapter.

GRAPHIC REPRESENTATION

While astronomy and geodesy were concerned with the mathematical treatment of errors of observation on the basis of the theory of probability as applied to large numbers, meteorology was engaged in the graphic representation of its accumulating observations, having disposed of the question of their accuracy by making such corrections and reductions as have been referred to in chapter x. The ramifications of the method of graphic representation in meteorology are almost endless, but we may make a first line of division as between the quasi-pictorial representation by a map and the merely numerical representation by a Cartesian diagram which expresses the results of a table of numerical values.

The map is a very ancient method of presenting facts to a reader, the earliest which has come down to us being Ptolemy's map of A.D. 150. The earliest pictorial meteorological map is Halley's map of the winds of the intertropical regions of the globe, which was published in 1688. We reproduce it here (fig. 102) with a modern equivalent for comparison (fig. 101).

The process of mapping labours under the difficulty of representing on a flat sheet the lines which belong to spherical surfaces. Various devices of projection are resorted to in order to minimise that difficulty.

NORMAL PRESSURE AND WINDS IN THE INTERTROPICAL REGION. JULY.

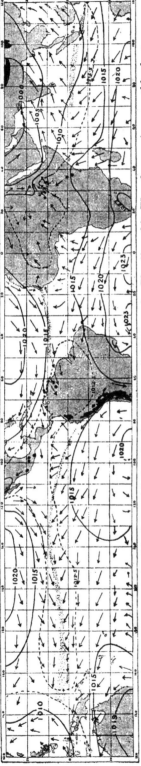

Fig. 101. The prevailing winds of the intertropical region in July, compiled in 1922 for *The Air and its Ways* from various monthly charts. The dotted areas over the sea indicate the positions of the Doldrums; the black areas positions of land over 4000 metres in height.

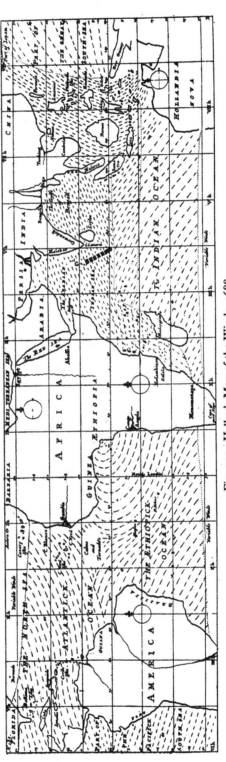

Fig. 102. Halley's Map of the Winds, 1688.

Carte des lignes Isothermes

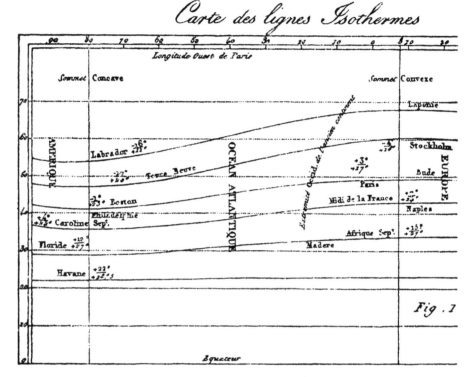

Fig. 103 (1). Chart of Isotherms of the Northern Hemisphere.

Isograms

The special maps for which meteorologists are responsible are those which display the geographical distribution of any element by a series of lines, each line being drawn through the points at which the element has the same magnitude. Such lines, according to Galton, should receive the general name of isogram, which would cover isotherms, lines of equal temperature, isobars equal pressure, isohyets equal rainfall, and so on.

As a primary map for meteorological purposes we may take that which represents the orographic features of the earth, the coast-lines, and the contour lines, or isograms of height. They are drawn to connect points at equal heights above mean sea-level selected so as to give a sufficient idea of the relief of the land or in some cases of the sea bottom. We may suppose that, in course of time, when "level-surfaces" are accepted as the proper surfaces for the display of meteorological data we shall require isograms of level surfaces in place of contours at fixed heights. For the heights within our maps the differences are not large enough to be regarded.

As an early example of representation by isograms we reproduce Humboldt's map of the distribution of temperature, and we do that with the greater interest because it includes also an example of a vertical section of

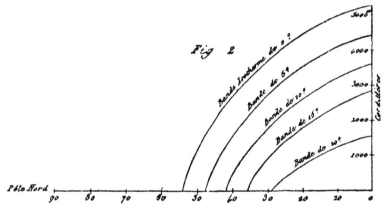

par M. A. de Humboldt

An addendum in a separate reprint of a *Mémoire de la Société d'Arcueil*, vol. III, p. 462.

Fig. 103 (2). Humboldt's sketch of the distribution of temperature in the free air based on mountain observations (*Les Bases*, vol. I, p. 187).

the atmosphere showing isotherms in the free air based upon observations on mountains near the equator and on the surface further North. Vertical sections are another kind of map, less common than the map in plan but equally necessary for the study of the atmosphere.

Projections

One of the most important points to notice with respect to mapping for the representation of meteorological data is the almost irresistible temptation to use Mercator's projection for meteorological charts.

It is perhaps the only projection which is acceptable for a single chart of the whole world because it is the only one in which the wind-directions can be said to be true. Their truth depends on the fact that, in Mercator's projection, meridians are at right angles to parallels of latitude as they are on the globe itself, and, at any given point of the map, the scale is the same in any direction. The temptation to use the projection is particularly seductive because the polar regions where the exaggeration of the map is so great as to make the representation ridiculous as well as misleading are precisely the regions for which there are hardly any data and can therefore be left blank without exciting adverse criticism.

But for that very reason the projection must be pronounced unsuitable for general meteorological purposes. The polar regions are the regions for which data are very urgently required and the adjustment of new data to those already obtained should not be hampered by an unsuitable projection. In the Meteorological Office I used to have a very striking demonstration of the importance of this consideration. An artist who had been employed in making relief-maps conceived the idea of exhibiting the geographical results of Scott's first expedition, 1900–04, upon a map of the world and took Mercator's projection as his base. In consequence the land about Ross Island assumes gigantic proportions, and only the failure to reach the pole on that occasion made it possible to get the results on the map at all.

Many other projections for the whole world have been attempted by: Lambert, Aitoff, Mollweide, Babinet, Goode, Lagrange, Guyou, August, Sanson, Flamsteed, Bonne, Werner, Collignon, but none affords a really effective representation. In preparing a representation of the distribution of the meteorological elements for the Réseau Mondial, we were led ourselves to a division of the whole area of the globe into five sections, two polar circles from 90° to 66½°, one equatorial belt for the part between the tropics, Cancer in the North and Capricorn in the South, and two parts on Albers' equal-area conical projection for the regions from the tropic to the polar circle in each hemisphere. These maps are represented in the first volume of the Réseau Mondial that was issued, namely that for 1911[1]. Subsequently Professor V. Bjerknes approached the same problem independently and arrived at a very similar conclusion[2]. He proposed three conformal projections, two polar planes extending to latitude 67½°, an adjusted conical projection for the region between latitudes 67½° and 22½°, and for the equatorial regions between 22½° N and 22½° S projection on a cylinder adjusted to be conformal.

[1] *British Meteorological and Magnetic Year Book*, 1911, Part v, Réseau Mondial, 1911, M.O. publication No. 207 g (Charts), London, 1917.
[2] 'Sur les projections et les échelles à choisir pour les cartes géophysiques,' by V. Bjerknes, *Geografiska Annaler*, vol. ii, 1920, p. 1.

In neither map does the distortion at any point exceed 5 per cent. A comparison of measurements made on the maps of the Réseau Mondial with the tables given in Bjerknes' paper shows that the differences are very small.

In this book we have thought it desirable as a rule to represent the world as made up of Northern and Southern hemispheres, each represented as circles with the meridians divided into equal intervals of ten degrees of latitude. There is, of course, considerable distortion in that representation, the equatorial circle acquires a radius equal to one-quarter of the earth's circumference instead of half its diameter, but the representation gives a proper idea of the relation of the data to the pole, and the circumpolar arrangement of meteorological elements is a fundamental characteristic of the meteorology of the globe.

For the intertropical regions however we use a separate representation of the intertropical belt similar to that proposed by Bjerknes. It has the advantage of presenting the features related to the equator in the most effective way.

For the daily charts issued by Weather Offices some modification of the conical projection is most suitable. In choosing a new chart for the *Bulletin International du Bureau Central Météorologique*, M. Angot[1] adopted a projection similar to that proposed by Albers in 1805.

Dans ce mode de projection, la portion de surface terrestre considérée est représentée sur un cône droit ayant pour axe la ligne des pôles; la distance des parallèles est déterminée par les conditions suivantes:

(1) Les surfaces sur le cône sont égales aux surfaces correspondantes de la sphère.

(2) Les longueurs de deux parallèles de latitudes λ_1 et λ_2 sont les mêmes sur le cône et sur la sphère.

The limits of latitude which he wished to represent were 30° and 70° and he therefore chose λ_1 and λ_2 to be 40° and 60°. The errors of longitudinal distances and angles of azimuth were then found to be less than the uncertainties in the tracing of the isobars, isotherms and wind-arrows.

International agreement with regard to details is in the highest degree desirable. In its absence the compilation of weather-charts for an extended region from the separate maps of the separate regions is an exasperating occupation.

Representation by graphs—the Cartesian method

For the diagrammatic representation of quantities we must go back presumably to Descartes who introduced the method of referring the variation of physical quantities to two axes at right angles, the method of Cartesian co-ordinates. The most representative and typical diagram in Cartesian coordinates is a curve referred to axes at right angles when the one co-ordinate represents a series of measurements of one physical quantity and the other the corresponding measurements of another quantity to which the first is known to be related. Within the last half-century a curve of this kind has

[1] 'La Nouvelle Carte du Bulletin International du Bureau Central Météorologique,' par M. Afred Angot, *Ann. bur. cent. météor.*, 1896, Part 1, Paris, 1898, p. B151.

STRATOSPHERE AND TROPOSPHERE

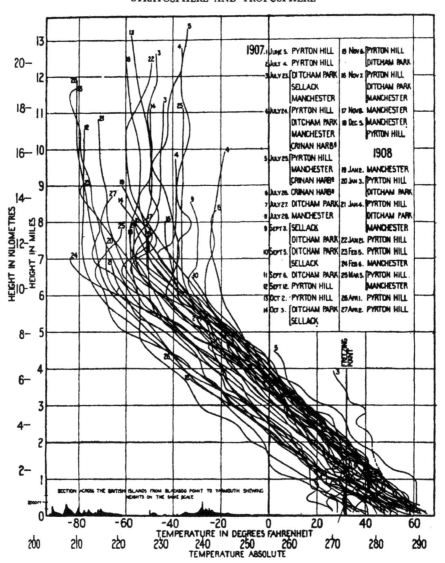

Fig. 104. Polygraph of changes of temperature with height above sea-level obtained from balloon ascents 1907–8.

The separate curves represent the relation between temperature, in degrees Fahrenheit or on the tercentesimal scale, and height in miles or kilometres in the atmosphere. The numbers marking the separate curves indicate the date of ascent at the various stations as shown in the tabular columns. The general aspect of the curves shows the great complexity of the temperature variations within the first two miles from the surface, and a very nearly uniform rate of fall of temperature above the two-mile limit until the stratosphere is reached, at from six to eight miles. The difference of height at which the stratosphere is reached, and the difference of its temperature for different days or for different localities, is also shown on the diagram by the courses of the lines.

come to be known as a graph and its use has become a very common exercise for schools. The graph then represents a curve of relation which may in some cases be expressed algebraically as a Cartesian equation. In the expression of the winds on the Beaufort scale we have already given an example of a curve which could be expressed empirically by an equation in that way, but the relationship exhibited by the observations is evidently applicable whether a Cartesian equation can be found for it or not. The graph is indeed so well known as an artifice for presenting a relationship to the eye, that in this place we need not do more than refer for an example to the dual graph of solar and terrestrial radiation, p. 241.

Polygraphs

Our next example (fig. 104) shows a number of graphs on the same ground or "grid," each one representing the relation of temperature to height in a sounding of the air by a registering balloon in 1907–8. There are forty-five curves in all. In the original diagram the graphs of soundings from the several stations are represented in different colours; the run of the separate curves can therefore be distinguished. With reproduction in black and white this is not possible but the diagram conveys some general ideas about the variation of temperature with height which would require many words to describe. We note for example the isothermal condition of the stratosphere in spite of large variations of temperature from day to day, the general smoothness in the run of each curve in the upper part of the troposphere; and the irregularities displayed in the first three kilometres. The combination of graphs in this manner has accordingly a usefulness of its own and we propose to give it a place in the meteorologist's tool-chest of graphical manipulation with the name of polygraph.

Logarithmic scales

The Cartesian method of graphs is now so fully recognised in all branches of science that squared paper for the purpose of constructing graphs is to be obtained in all shops that sell educational stores, as millimetre-paper or inch-paper, and sometimes two-millimetre paper or half-inch paper. Two-millimetre paper has some advantages for general work in meteorology because the interval of two millimetres makes plotting by eye to tenths of the divisions much easier than with the spacing of single millimetres.

Further avenues in the graphical study of the relationships of quantities by the Cartesian method were opened by the introduction of paper ruled according to logarithmic scale either singly with a linear scale for the other co-ordinate, or doubly with both scales ruled logarithmically[1]. The logarithmic scale is ruled generally to correspond with a slide-rule and has the same properties as the rule. The chief advantage is that for an equation such as $y = ax^n + b$ the graph with x on logarithmic scale and y on linear scale is a straight line, and with the double logarithmic paper such a curve as $y^m = ax^n$ is represented by a straight line. We have made use of this property in fig. 2.

[1] 'Scale Lines on the Logarithmic Chart,' by C. V. Boys, *Nature*, vol. LII, 1895, p. 272.

As an example of the double logarithmic paper we reproduce a specimen of paper prepared for the reduction of observations of the upper air with ballons-sondes. One of the advantages of this paper is that the curve when plotted is identical, except as regards its position on the sheet, whatever units be employed, provided of course that the measurement is properly adjusted to the zero.

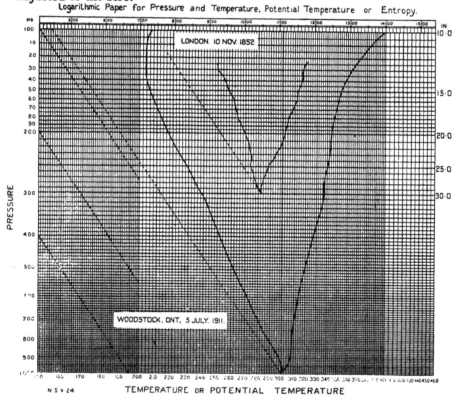

Fig. 105. Examples of graphs of soundings of the atmosphere referred to pressure-temperature and pressure-potential-temperature respectively (1) for an ascent in a manned balloon by J. Welsh at Vauxhall in 1852, and (2) for a sounding with a registering balloon in Canada in 1911. The two graphs for each occasion meet on the line of standard pressure. The observations of pressure in (1) are in inches and in (2) millibars; both are plotted direct on to the paper without conversion as curves in full line. The faint straight pecked lines are adiabatic lines for dry air.

As an example of single logarithmic paper we show the paper prepared for exhibiting the "curve of vertical section of the atmosphere," or tephigram, obtained by ballon-sonde referred to temperature on a horizontal linear scale and to potential temperature on a logarithmic vertical scale. The advantages of this form of representation will be dealt with more fully in the consideration of the thermodynamics of the atmosphere in Vol. III. On both these forms, figs. 105 and 106, we give the representation of the results obtained on the balloon-ascent by Welsh on November 10, 1852, and by ballon-sonde at Woodstock, Ont., on July 5, 1911.

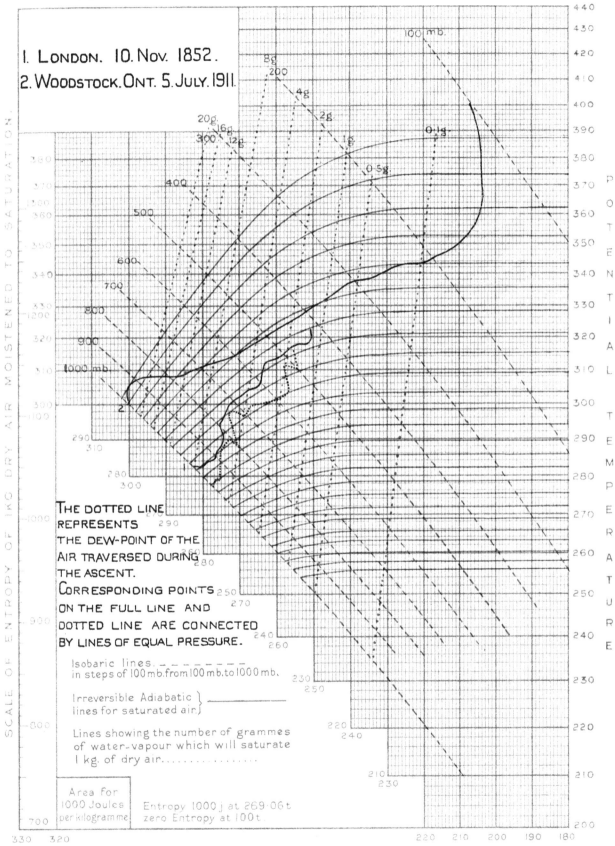

N S FORM B. FOR TEPHIGRAM (1926)

LᎾ INDICATOR DIAGRAM: I SQ. CM.= 372 JOULES PER KG.

1. LONDON. 10. NOV. 1852.
2. WOODSTOCK. ONT. 5. JULY. 1911.

THE DOTTED LINE
REPRESENTS
THE DEW-POINT OF THE
AIR TRAVERSED DURING
THE ASCENT.
CORRESPONDING POINTS
ON THE FULL LINE AND
DOTTED LINE ARE CONNECTED
BY LINES OF EQUAL PRESSURE.

Isobaric lines. _ _ _ _ _ _ _
in steps of 100 mb. from 100 mb. to 1000 mb.

Irreversible Adiabatic }
lines for saturated air. }

Lines showing the number of grammes
of water-vapour which will saturate
I kg. of dry air.................

Area for
1000 Joules
per kilogramme

Entropy 1000 j at 269·06t
zero Entropy at 100t

SCALE OF ENTROPY OF 1 KG DRY AIR MOISTENED TO SATURATION.

POTENTIAL TEMPERATURE

TEMPERATURE

Fig. 106.

ISOPLETH DIAGRAMS

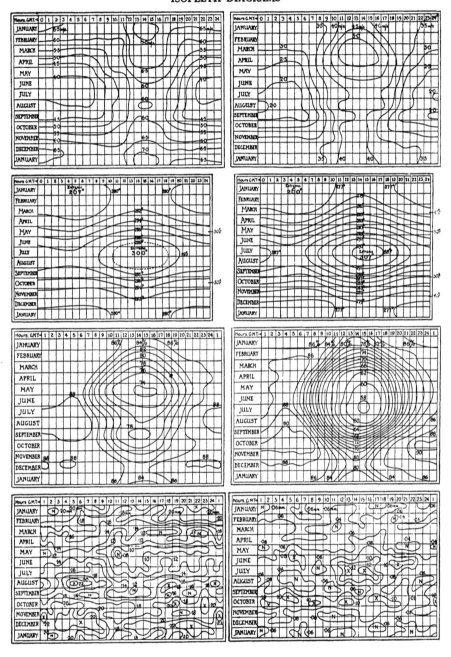

Fig. 107. Chrono-isopleths of the meteorological elements at Kew Observatory (Richmond) right-hand column and at Valencia Observatory (Cahirciveen) Ireland, left-hand column.

The elements represented are: first row wind-velocity in metres per second; second, temperature on the tercentesimal scale; third, relative humidity as percentage of saturation, and fourth, rainfall in millimetres. In the fourth row × indicates a period of maximum rainfall and N a period of minimum. (From *The Weather Map*, M.O. 225 i.)

Isopleths

Another application of the Cartesian method is to bring under inspection the variations of some physical quantity that is represented, for example, by hourly tabulations.

We give examples of the most common and most effective application of this method to show the mean values of observations at the several hours of the day in the several months of the year. The means are set out on squared paper, the values at the several hours of the day in rows and those of the same hour in the several months in columns. Lines are then drawn to divide all the points with values above a certain limit from those with values below that limit. In this way a series of isopleths is drawn which indicate the time of day in the several months of the year at which certain values have been obtained. These particular curves are called chrono-isopleths. We give here chrono-isopleth diagrams for Kew Observatory in Eastern England and Valencia Observatory in Western Ireland. (Fig. 107.)

HYPSO-CHRONO-ISOPLETHS OF TEMPERATURE OF THE ATMOSPHERE

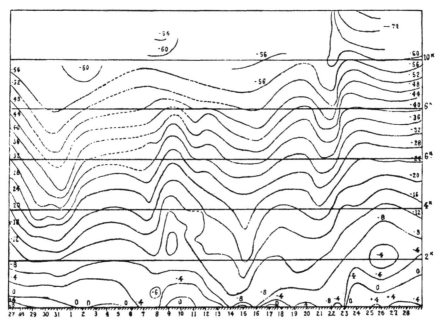

Fig. 108. Isothermes au-dessus de la région de Paris dans l'atmosphère libre, du 27 janvier au 1er mars, 1901.

Other curves, also called isopleths, are obtained in like manner when one of the co-ordinates is time and the other is height. Such a diagram gives hypso-chrono-isopleths. We give the first example of such a diagram for the upper air which was made by L. Teisserenc de Bort in 1902[1]. (Fig. 108.)

[1] 'Étude des variations journalières des éléments météorologiques dans l'atmosphère,' *Comptes Rendus de l'Académie des Sciences*, Paris, vol. CXXXIV, 1902, pp. 253–6.

Climatic diagrams

As another example of the use of a Cartesian method of exhibiting the relation between different quantities we give a graph showing the relation between temperature and humidity which, so far as we are aware, was devised by Dr John Ball of the Egyptian Physical Department of Cairo[1]. Somewhat similar curves have been employed by Dr Griffith Taylor.

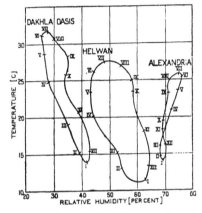

The purpose of the climatic diagram is to show the change in the relationship between two selected climatic elements in the course of the year by plotting the monthly values. In the examples selected (fig. 109) the elements are mean temperature, and mean relative humidity. The values for the several months are marked by points with a roman numeral and the points are connected by a continuous line.

Fig. 109. Climatological diagrams of normal temperature and relative humidity in the several months of the year at three Egyptian stations (J. Ball).

The nomogram

Graphs provide also a convenient substitute for some of the more elaborate meteorological tables. An example is provided by the nomogram, which by the use of a straight edge and two graduated curves gives the solution of a complicated equation by reading the intersection of the straight edge with a third graduated curve. An account of the method is given under the heading 'Nomography' in the *Dictionary of Applied Physics*, vol. III. We give one example by Dr J. Ball[2] which furnishes a substitute for Glaisher's tables for the reduction of the readings of the dry and wet bulb thermometers. The straight edge is set to connect the reading of the dry bulb as set out in the curve on the left-hand side and that of the wet bulb on the middle scale, with the third graduated line on the right on which the dew-point and vapour-pressure are read off.

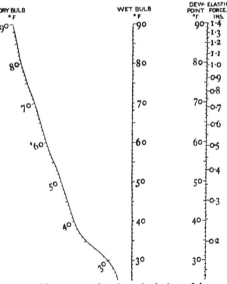

Fig. 110. Nomogram for the calculation of dew-point and vapour-pressure (J. Ball).

[1] 'Climatological Diagrams,' by John Ball, *Cairo Scientific Journal*, vol. IV, 1910, p. 280.
[2] 'A rapid method of finding the elastic force of aqueous vapour and the relative humidity from dry-bulb and wet-bulb thermometer readings,' by Dr John Ball, *Q. J. Roy. Meteor. Soc.*, vol. XXXII, 1906, p. 47.

THE TREATMENT OF VECTOR-QUANTITIES

In the consideration of the treatment of meteorological observations by numerical or graphical methods we have tacitly supposed them to be scalar quantities, each observation defined by a single number without any reference to direction. Vector-quantities like wind, or pressure-gradient, introduce difficulties of their own. If we consider first the graphic representation we are met at the outset with the difficulty of an adequate representation of the results of observations of a pilot-balloon. It is a common practice to represent the horizontal trajectory of the balloon on paper ruled for angles and radii from a centre, relying upon figures marked at points of the curve to indicate height at definite times and inferentially the velocity of the balloon in transit. Such a diagram is inconvenient for general work as the paper used for the representation is only a small fraction of a large sheet and sometimes, in the most interesting examples indeed, unmanageable in size. We have occasionally set out the results of pilot-balloons on plates of glass, using for successive levels separate plates which can be superposed. We have found the plan useful for the purposes of exhibition, but it does not lend itself to reproduction. A method in use at the Aerological Observatory at Lindenberg is to draw from the point of observation the vector for the air at the start and the finish and to connect the extremities of these two vectors with a line of points each representing the extremity of a vector from the point of observation, which is not actually drawn.

Another method used by C. J. P. Cave in his *Structure of the atmosphere in clear weather* is to give two graphs, height-velocity and height-direction respectively. The diagrams are not easy to interpret when the wind is very light, because in that case the direction-graph takes on an appearance of wild excursions, when, as a matter of fact, there is practically no wind at all.

For our own purposes we have found the best way out of the difficulty is to resolve each wind into its components, W to E and S to N, and to set out a separate graph for the West-East component and the South-North component. We have used that plan in Part IV, and more recently for displaying the results of observation of the travel of a ballon-sonde with those obtained from its meteorograph for pressure, temperature and humidity.

For the numerical manipulation of the observation of wind the process of resolution has been regarded as the best mode of procedure for a long time. On that principle is based a formula which was given by J. H. Lambert in the eighteenth century for combining the observations of wind-frequency from eight points of the compass to give the direction of a resultant wind as follows:

$$\tan \alpha = \frac{E - W + (NE + SE - NW - SW) \cos 45°}{N - S + (NE + NW - SE - SW) \cos 45°}.$$

The extension of the statistical methods of correlation to vector-quantities has been worked out by J. I. Craig, of the Egyptian Weather Service.

Wind-roses and vector-diagrams

For summarising upon maps the results of observations of winds especially at sea *wind-roses* are used in considerable variety. We need mention only two, first the wind-rose of the monthly charts of the Deutsche Seewarte and of Washington, and second that of the more recent issues of marine charts of the Meteorological Office in London.

In the former the percentage frequencies of winds, grouped according to eight or sixteen points of the compass, are indicated by the lengths of arrows drawn towards a centre; and the *mean force* of the wind on the Beaufort scale, for the selected points, is indicated by the number of feathers at the outer end of the arrows. In the latter more detail is attempted on the ground that it is desirable to distinguish between light winds, moderate winds, and gales. Accordingly the arrow is a composite one, the innermost portion a single line showing the percentage frequency for light winds, a middle portion is a double line for moderate winds, and the outer portion is filled up black for gales. Examples of these are given in chapter VI of Vol. II.

With wind-roses of either kind it is not easy to compare the lengths of the several portions. For that purpose it is better to abandon the representation of direction and set lines as parallel ordinates from a common base. Some examples are given in *The Weather of the British Coasts*.

To connect together the results of wind-observations for the months of a year, Hildebrandsson uses a step by step chain of links of equal length, which show the resultant or prevalent direction for the successive months marked on consecutive links. Such diagrams give a good idea of the steadiness or alternations of the wind in the course of the year. The same device may be used with the elements of a wind-rose to exhibit the steadiness or variability of wind in any locality.

Diurnal or seasonal variation of wind can be represented by a vector-diagram of the kind already indicated on p. 270. An example is given in Vol. II, chapter VI.

CONTINUOUS RECORDS

The Cartesian method finds its most natural expression in the autographic records of the meteorological elements which are traced by self-recording instruments as up and down displacements of a pen or recording spot of light on a revolving drum. If the motion is not directly up and down but along the arc of a curve round the pivot of the pen-arm, as it is with many portable barographs, thermographs and hygrographs, a very awkward complication is produced which ought whenever possible to be avoided. In the absence of self-recording instruments a Cartesian diagram analogous to a summary of the recording curve is obtained by plotting such quantities as the daily, monthly or annual means of meteorological elements over a long period of years.

We take special note of this method of plotting in two important relations. It has been used times without number as a means of tracing a relation between the concurrent variations of the separate elements, and perhaps with equal frequency it has been employed to suggest the recurrence of salient features of the various elements of weather and indeed to indicate the period of recurrence. We propose therefore to treat these types under separate headings as periodicity and curve-parallels.

PERIODICITY

A single curve representing the successive values of a meteorological element at successive times can be dealt with from the aspect of possible periodicity of occurrence of the values of the element; but the detection of periodicities is full of pitfalls for the unwary. The appearance of the curve of values itself is deceptive and the estimation of periods by cursory inspection, using the interval between prominences as a guide is apt to give quite illusory results. The difficulty arises from the interference of oscillations of different periods. For example if a quantity be subject to two oscillations concurrently, one of m years and the other of n years, the values will only recur in a number of years which is the least common multiple of both m and n, and if m and n have no common divisor, the period is mn years.

In practice there is another difficulty arising from our ignorance of the life-history of the oscillation. It is usual tacitly to assume that a periodic oscillation in any meteorological element is due to some influence external to the oscillating system, which is of the same period as the oscillation itself—in other words that the oscillation is a forced oscillation[1].

But we are not entitled to assume that the oscillating system has no free period of its own. An oscillation in temperature, for example, may be regarded as an index of a periodic change in the general circulation of the atmosphere, and on the analogy of a typical vibrating system we must allow that if the general circulation is disturbed by some serious convulsion, an unusual extent of ice in the polar regions or a great volcanic eruption, the circulation will have natural periods of vibration of its own, and even if the disturbing cause is only temporary the oscillation, once set up, will take a long time to die away; and if the disturbing cause is itself periodic there will be all the complication that is incidental to the coexistence of forced and free vibrations. It seems impossible for us to determine the natural free periods of vibration of the circulation except by the long process of prolonged observations and the disentangling of the free and forced vibrations. To that, at present, we have hardly made an approach. In the present stage of the subject we have to determine the periods of the components of the vibrations as recorded in the observations, but while that stage is in progress we ought not rashly to assume that the phase of a vibration of particular period necessarily remains the same throughout the whole period of observation.

In the analysis of the periodicity of sunspots, for example, the most conspicuous period for one group of observations is eleven years, yet if the analysis is extended over a hundred and fifty years the importance of the eleven year period becomes much reduced[2]. That would certainly be the result if

[1] *The Theory of Sound*, Rayleigh, Macmillan and Co., London, 1894, chaps. IV and V.
[2] 'On Sunspot Periodicities,' Schuster, *Proc. Roy. Soc.*, A, vol. LXXVII, 1906, pp. 141–45.

the period were a natural period of the sun and was excited from time to time by a cause that was not strictly periodic in the same time.

Fourier's theorem

The fundamental theorem bearing upon the resolution of the natural variations of meteorological elements into periodic components is Fourier's theorem that any curve whatever representing variation with time can be resolved into a series of pure sine or cosine curves, that is to say of curves which are represented by the equation

$$y = A_n \cos\left(\frac{2\pi n}{\tau} t - \alpha_n\right).$$

Thus the complete variation at any time is given by

$$y = A_0 + A_1 \cos\left(\frac{2\pi}{\tau} t - \alpha_1\right) + A_2 \cos\left(\frac{4\pi}{\tau} t - \alpha_2\right) + A_3 \cos\left(\frac{6\pi}{\tau} t - \alpha_3\right) + \dots.$$

A_0 is the mean value; τ the period; $A_1 \cos\left(\frac{2\pi}{\tau} t - \alpha_1\right)$ the first component, A_1 its amplitude, α_1 the phase of its maximum, and so on for the other components.

Harmonic analysis

Fourier's theorem gives a theoretical method of determining $A_1, A_2, A_3, \dots$ and $\alpha_1, \alpha_2, \alpha_3, \dots$. How many terms will be necessary to represent any particular variation cannot be determined without trial. For meteorological purposes the analysis is sufficient if the combination of two, three or four terms gives a satisfying reproduction of the original curve. The process is applied generally to diurnal or seasonal variations and more particularly to mean values of those variations. In such cases the element may be supposed to be periodic within the day or the year, and the problem to be solved is to determine the constants A and α of the primary oscillation, and of its second, third, and fourth harmonic components.

Here we may note that the form of component

$$A \cos\left(\frac{2\pi}{\tau} t - \alpha\right)$$

is not convenient for arithmetical computation; we can express it as

$$A \cos\frac{2\pi}{\tau} t \cos\alpha + A \sin\frac{2\pi}{\tau} t \sin\alpha \quad \text{or} \quad A \sin\alpha \sin\frac{2\pi}{\tau} t + A \cos\alpha \cos\frac{2\pi}{\tau} t$$

and assume the form

$$S \sin\frac{2\pi}{\tau} t + C \cos\frac{2\pi}{\tau} t,$$

where $S = A \sin\alpha$ and $C = A \cos\alpha$ and $\tan\alpha = S/C$.

This form is as general as the original one, but instead of a single component of specified amplitude and phase it represents the variation as consisting of two components of the same period one of which has its maximum when $t = 0$ and the other is zero at that epoch.

If we assume that the variation which we are studying can be completely represented by a combination of cosine curves, a few observations will suffice for the determination of the constants.

Take for example the seasonal variation of temperature at Jerusalem as indicated by alternate monthly values: Jan. 280, March 284·5, May 292·5, July 297, Sept. 295·5, Nov. 287. Assume that the variation is represented as

$$y = A_0 + S_1 \sin \frac{2\pi t}{\tau} + C_1 \cos \frac{2\pi t}{\tau} + S_2 \sin \frac{4\pi t}{\tau} + C_2 \cos \frac{4\pi t}{\tau};$$

we desire to find S_1, C_1, S_2, C_2, t being measured from the middle of January. τ is the full period, namely one year.

The interval of two months may be taken as one-sixth of the whole year, and as the sequence of events runs through 2π or $360°$ in the course of the year, the times of the successive values counting from the middle of January will be:

for the first component $t = 0$, $\frac{1}{6}\tau$, $\frac{2}{6}\tau$, $\frac{3}{6}\tau$, $\frac{4}{6}\tau$, $\frac{5}{6}\tau$

the angles will be 0, $\frac{\pi}{3}$, $\frac{2\pi}{3}$, π, $\frac{4\pi}{3}$, $\frac{5\pi}{3}$

or in degrees $0°$, $60°$, $120°$, $180°$, $240°$, $300°$

and for the second component $0°$, $120°$, $240°$, $360°$, $480°$, $600°$

We have to use the sines and cosines of these angles. For the sake of brevity let us call sin $60°$, s, and cos $60°$, c, instead of the numerical values $\sqrt{3}/2$ and $1/2$. Then if the two cosine curves represent the variation adequately, remembering that the values of the sines and cosines will always be s or c, but with different signs for different quadrants, we get the following six equations for determining S_1, C_1, S_2, C_2:

Jan. $y_0 = 280$ $= A_0 + 0 + C_1 + 0 + C_2$ (0)
March $y_1 = 284·5 = A_0 + sS_1 + cC_1 + sS_2 - cC_2$ (1)
May $y_2 = 292·5 = A_0 + sS_1 - cC_1 - sS_2 - cC_2$ (2)
July $y_3 = 297$ $= A_0 + 0 - C_1 + 0 + C_2$ (3)
Sept. $y_4 = 295·5 = A_0 - sS_1 - cC_1 + sS_2 - cC_2$ (4)
Nov. $y_5 = 287$ $= A_0 - sS_1 + cC_1 - sS_2 - cC_2$ (5)

We next multiply equation (0) by sin $0°$, (1) by sin $60°$, s, (2) by sin $120°$, s, (3) by sin $180°$, 0, (4) by sin $240°$, $-s$, (5) by sin $300°$, $-s$, and add. Thus we obtain $4s^2 S_1 = \Sigma y_n \sin n \times 60°$, or $S_1 = \frac{1}{3}\Sigma y_n \sin n \times 60° = -1·59$.

Next multiply the equations in like manner by the cosines instead of the sines and add; we get $C_1 (2 + 4c^2) = \Sigma y_n \cos n \times 60°$, or $C_1 = \frac{1}{3}\Sigma y_n \cos n \times 60° = -8·42$.

Thirdly multiply the equations in like manner by the series of sines of $0°$, $120°$, $240°$, $360°$, $480°$, $600°$ and add; we get $4s^2 S_2 = \Sigma y_n \sin n \times 120°$, or $S_2 = ·144$.

Lastly multiply in like manner by the series of cosines of $0°$, $120°$, $240°$, $360°$, $480°$, $600°$ and add; we get $C_2 (2 + 4c^2) = \Sigma y_n \cos n \times 120°$, or $C_2 = -·92$.

We now remember that we can put the relation in the form

$$y = A_0 + A_1 \cos \left(\frac{2\pi}{\tau} t - a_1\right) + A_2 \cos \left(\frac{4\pi}{\tau} t - a_2\right),$$

where $A_1 = \sqrt{C_1^2 + S_1^2}$, $A_2 = \sqrt{C_2^2 + S_2^2}$,
     $\tan a_1 = S_1/C_1$, $\tan a_2 = S_2/C_2$.

Substituting the values which we have obtained for S_1, C_1, S_2, C_2 we get

$$8·57 \cos \left(\frac{2\pi}{\tau} t - 191°\right) \text{ for the first component}$$

and

$$·93 \cos \left(\frac{4\pi}{\tau} t - 351°\right) \text{ for the second.}$$

For the first component the maximum will be when $2\pi t/\tau = 191°$ or eleven days after the middle of July. For the second component the maximum will be when $4\pi t/\tau = 351°$; $t/\tau = 351°/720°$ or 178 days from the middle of January.

We have used only six values for the determination of two components. The values for the alternate six months could be used in like manner to give independent values which ought to agree with those we have obtained. But with twelve values we can find constants for four components by a process which is similar in principle to that which we have used but is too voluminous for presentation here. The details are explained in the various books on the treatment of observations.

The analysis of a curve for unknown periodicities. Method of residuation

The usual method of harmonic analysis which has been sketched in what precedes is only applicable when we are justified in assuming a definite interval such as the day or the year as the period within which the values will recur. That is not the case when we are dealing for example with a long series of annual values as of temperature or rainfall. In that case we may obtain some simplification by the "method of residuation" which consists in annihilating variations of selected period while leaving unchanged, or only slightly modified, the other periodic changes[1].

We may take for example a component variation $a_n \sin \frac{2\pi}{n} t$, where n is the period of the component in years and t is the time in years measured from a node of that particular component. If we add the ordinates with an interval of m years we get for the component of period n in the new curve

$$a_n \sin \frac{2\pi}{n} t + a_n \sin \frac{2\pi}{n} (t + m)$$

or

$$2a_n \cos \frac{\pi m}{n} \sin \frac{2\pi}{n} \left(t + \frac{m}{2}\right).$$

The amplitude is zero if $n = 2m$, and thus in a curve of complex periodicity a component n is eliminated by adding the ordinates with an interval of $n/2$ years. All the other components have their amplitudes altered by the factor $2 \cos \pi m/n$ and the phase of each is put forward by $m/2$ years. This process serves very well to eliminate disturbing components leaving the residual variation to be more easily identified provided that the interval m can be selected at discretion. That would obviously be the case with a continuous curve, but when only annual values are available and its operation is limited to periods of years its utility is similarly limited.

In our own experience the most striking example of the resolution of an apparently irregular curve into harmonic components is that of the curve for the yield of wheat in Eastern England from 1885–1906. The original curves and the harmonic components of the long period term of eleven years are shown in the figure which illustrates very well the point from which we

[1] 'An investigation of the Seiches of Loch Earn by the Scottish Lake Survey. Part I. Limnographic Instruments and Methods of Observation,' by Professor G. Chrystal, *Trans. Roy. Soc. Edin.*, vol. XLV, part II, no. 14, 1906, pp. 382–87.

started, namely the unwisdom of relying upon the prominences of a curve of long series as a means of detecting the periods which make up the resultant variation. The final test of periodicity is the actual recurrence of successive values after a complete period, but with complex periodicities such as occur with nearly all natural phenomena that form of examination is not to be expected to be free from difficulty. It is remarkable that in the case referred

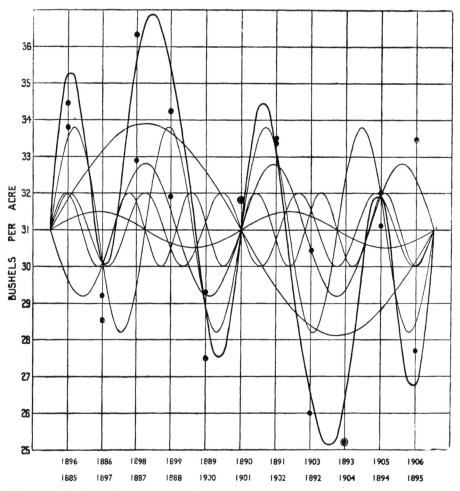

Fig. 111. Six harmonic components of an eleven years' period and their resultant curve, marked by a thickened line, with figures for the yield of wheat in Eastern England 1885–1906. The upper line of dates refers to the upper row of dots and the lower line to the lower row.

to in fig. 111 a test of that form held in an extraordinary manner, the values repeating themselves after eleven years with an astonishing degree of accuracy until 1906, considering the crudity and the vicissitudes of the data. Moreover the way in which deviations on either side of one point in the curves compensated each other was equally astonishing. These remarkable results are most easily explained by periodic variations concurring in a node at the point of reversal and having a master-period of eleven years. The relation has not continued undisturbed. Constructed in 1904 the figure for 1905 was a successful forecast, that for 1906 was not; and there are reasons why it should not be. The most interesting from the point of view of periodicity is that the natural period of the subject under investigation is eleven years, but the period of the "cause" is nearly but not exactly eleven years, in which case there are "beats" between the cause and the effect, one of the maxima of which was in process of demonstration during the years in question.

The periodogram

On the analogy of the analysis of light in the spectrum, arranged according to the frequency of oscillation of the waves of which it is composed, Sir Arthur Schuster has developed a method of dealing with the periodic oscillations, in the curves of observations of a quantity, to which he gave the name of periodogram. This method is in a sense the opposite of the method of residuation in that it proceeds by adding ordinates in such a manner as to retain one component and annihilate all the others. It depends on the fact that if in the equation of p. 275 m is equal to n, the resultant curve is $2a_n \sin \dfrac{2\pi}{n} t$, in other words the amplitude of the resultant is doubled. If we take half the sum the amplitude will be the amplitude of the vibration in the original curve. But the vibration of every other period which is not a harmonic component of n will be reduced and if the process be repeated by the addition of ordinates for a great number of successive periods n, the vibrations of periods not harmonic sub-multiples will be annihilated.

Hence by dividing the original curve into equal intervals and adding the ordinates of all the intervals and taking the mean, we get a curve which represents the amplitude of the component oscillation of that period in the original. Final values for this oscillation being obtained its first harmonic is determined and that gives us the amplitude of the vibration of that period in the original curve. This process is repeated for the whole series of possible components—with a series of yearly values components of two years, three years, four years, etc., and for monthly values components of two months, three months, four months, and so on.

The process is laborious. Sir Arthur Schuster[1] applied it to the periodicity of sunspots and found components with periodic times of 4·81, 8·38, and 11·125 years, the latter 11-year period being noticeable in the years 1825–1900 but not in the years 1750–1825. More doubtful periods of 5·625, 3·78, 2·69 and 4·38 years were also indicated. Sir William Beveridge[2] applied the method to the periodicity of adjusted wheat prices and found components 3·415, 4·415, 5·100, 5·671, 5·960, 8·050, 12·840, 19·900, 35·500, 54·000 and

[1] 'On Sunspot Periodicities,' by Arthur Schuster, *Proc. Roy. Soc.* A, vol. LXXVII, 1906, p. 141.
[2] 'Weather and Harvest Cycles,' *Economic Journal*, vol. XXXI, 1921, pp. 429–52; 'Wheat prices and rainfall in Western Europe,' *Journ. Roy. Stat. Soc.*, vol. LXXXV, 1922, pp. 412–78.

68·000 years, and it has been applied by Captain D. Brunt[1] to the periodicity of Greenwich temperatures which show components of one year with the second and third harmonics, 5 years, 594·5 days for part of the interval only, and a doubtful period of 26·21 months.

CURVE-PARALLELS

As an illustration of the method of curve-parallels we give a reproduction (fig. 112) of part of a diagram from W. J. Humphreys' *Physics of the Air* which draws "parallels" between pyrheliometric values of solar radiation, 1883–1913 (*P*), sunspot numbers 1750–1913 (*S*), a hybrid curve (*P* + *S*),

CURVE-PARALLELS

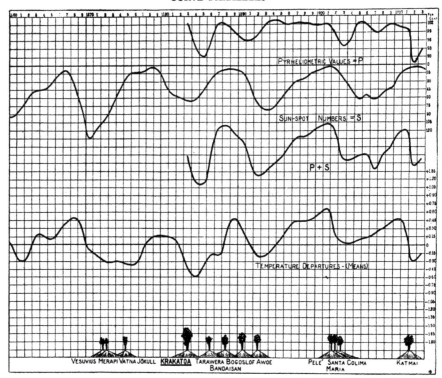

Fig. 112. Relation of pyrheliometric values and mean temperature-departures to sunspot numbers and violent volcanic eruptions (W. J. Humphreys).

and temperature-departures as evaluated by Köppen (1750–1913). The most striking similarity is curiously enough between the hybrid curve and temperature. Below the curves are symbols representing notable eruptions of volcanoes the dust from which may have affected the temperature-curve shown next above them.

[1] 'A periodogram analysis of the Greenwich temperature records,' *Q. J. Roy. Meteor. Soc.*, vol. XLV, 1919, pp. 323–38.

Curve-parallels for pressure are given by W. J. S. Lockyer extending over forty-five years (fig. 113) from which a see-saw of pressure in complementary parts of the world was inferred[1].

One of the most assiduous exponents of this method of representation is H. H. Hildebrandsson in five papers on 'Quelques Recherches sur les Centres d'Action de l'Atmosphère.' We reproduce in fig. 114 two pairs of parallels, the first for the winter rainfall of the Faröe Islands and the summer rainfall of Newfoundland and Berlin, the second for the winter rainfall of British Columbia and the subsequent autumn rainfall of the mid-Atlantic[2].

CURVE-PARALLELS

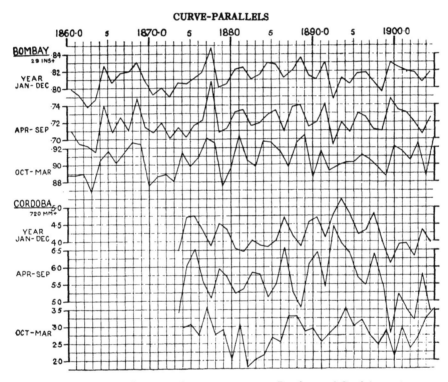

Fig. 113. The variations of pressure from year to year at Bombay and Cordoba, 1860–1905. (W. J. S. Lockyer.)

It is necessary to remark that the comparison of curve-parallels is sometimes suggestive of relationships that cannot be rigorously defended. When there is a succession of notable ups and downs in two curves one under the other the succession of prominences often bears a superficial aspect of concurrence which on closer examination is somewhat impaired by differences of a year or more in the maxima while others give no sign of lag.

[1] 'Barometric Variations of Long Duration over Large Areas,' *Proc. Roy. Soc.* A, vol. LXXVIII, 1906, p. 43.
[2] *K. Svenska Vetensk. Akad. Handlingar,* vol. XXIX, no. 3, vol. XXXII, no. 4, vol. XLV, no. 2, vol. XLV, no. 11, vol. LI, no. 8, 1897, 1899, 1909, 1910, 1914.

We revert to the methods which statisticians have developed to guard against a false impression of that kind. They deal with the numbers themselves, thus denying any opportunity for the deception, willing or unwilling, of the eye. The most instructive process is the construction of a Cartesian dot-diagram of which the co-ordinates are the corresponding departures of the several observations from their mean value. Thus in the case of the pressure at St John's and Berlin we should form a set of departures or deviations of the several observations from their mean values and set them out on a diagram. The dot-diagram may be drawn for simultaneous values of the departures of

CURVE-PARALLELS

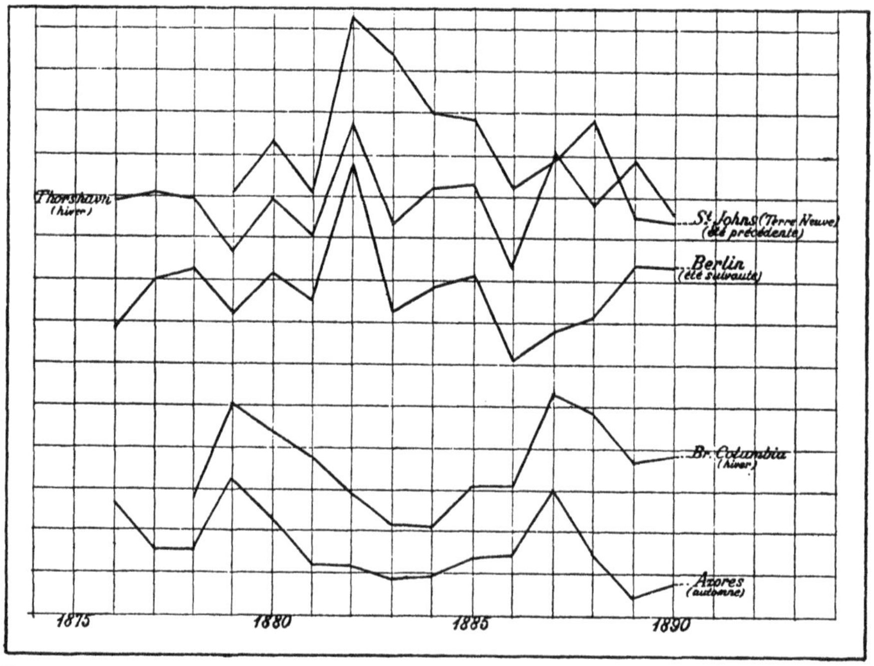

Fig. 114. Curves of variation of rainfall from year to year: (1) for winter at Thorshavn (Faröe Islands), for the preceding summer at St John's (Newfoundland), and for the succeeding summer at Berlin, and (2) for the winter in British Columbia and the succeeding autumn at the Azores (Hildebrandsson).

the two elements or the value of the departure of one of the elements may be plotted against that of the second element after a definite period of time; in either case the time does not appear in the diagram the formation of which is the geometrical equivalent of the algebraical process of elimination of a common co-ordinate from two equations, although in the geometry we have no knowledge of the algebraic equation which would represent the curves.

The mere arrangement of the dots in the diagram will show whether there is any *prima facie* reason to consider the two sets of events to be connected. If all the points lie on a straight line through the origin it is evident that there is direct proportionality between the two. If the straight line passes from the left lower quadrant to the right upper one the corresponding relation is evidently direct proportion, but if the line connects the other pair of quadrants the relation is inverse in the sense that an increase in the one element corresponds with a proportionate decrease in the other.

In so far as the points do not lie definitely on a straight line but show some "scatter" over the area the relationship is correspondingly ill-defined and therefore uncertain.

The dot-diagram leads us back again to the consideration of arithmetical manipulation by statistical methods based upon the theory of large numbers of observations with which the chapter began.

REGRESSION EQUATION AND CORRELATION

In so far as the scatter of the dots in a dot-diagram is restricted and suggests little deviation from proportionality between the two quantities we are induced to look upon a linear equation as being the best expression of the relation between them. And the next step is to determine the straight line which expresses the relationship in the best manner. For that purpose we assume an equation to a straight line passing through the origin of co-ordinates $y = bx$ and we seek the best value of b among all the various possibilities. A line can be drawn by eye, but if we desire to adhere to arithmetic in order to avoid bias, on the part of the investigator, or guessing, which is by some regarded as even more heinous than bias, we may note that if the equation $y = bx$ were a true representation of the conditions, $y - bx$ would be zero, and we may therefore seek arithmetically to choose b so that when all the values are taken into account the difference of $y - bx$ from zero shall be numerically as small as possible. Since we are not here concerned with the sign of the difference we make the differences least if we choose b so that the sum of the squares of the differences is as small as possible. Thus $\Sigma (y - bx)^2$ is to be a minimum. By the ordinary rule of differentiating, for the minimum value of any quantity X depending upon b, $\dfrac{dX}{db}$ must be taken as zero.

Differentiating therefore, we get for the best value of b

$$\Sigma (y - bx) x = 0$$

or

$$\Sigma xy = b\Sigma x^2$$

or

$$b = \frac{\Sigma xy}{\Sigma x^2}.$$

If we had approached the question from the other aspect of the relation of $x = b'y$, we should clearly have arrived at the result

$$b' = \frac{\Sigma xy}{\Sigma y^2},$$

and we can represent the best relationship either by

$$y = \frac{\Sigma xy}{\Sigma x^2} x \qquad \text{or by} \qquad x = \frac{\Sigma xy}{\Sigma y^2} y.$$

DOT-DIAGRAMS AND REGRESSION-LINES

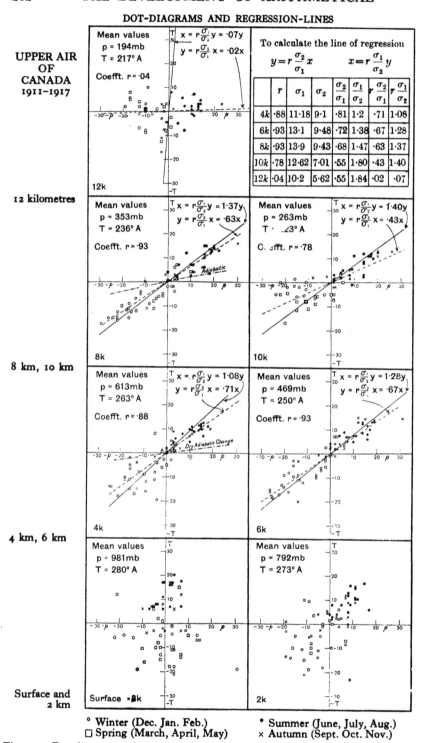

Fig. 115. Dot-diagrams and regression-lines of pressure and temperature at the surface and at heights of 2, 4, 6, 8, 10 and 12 kilometres. The mean values of pressure and temperature at each level are taken as axes.

We can see the relation between these two more clearly if we remember that $\sqrt{\frac{1}{n}\Sigma x^2}$ is the standard deviation of x, which we will denote by σ_x and $\sqrt{\frac{1}{n}\Sigma y^2}$ is the standard deviation of y, which we will denote by σ_y.

We then get as the best equation

either
$$\frac{y}{x} = \frac{\Sigma xy}{n\sigma_x\sigma_y} \cdot \frac{\sigma_y}{\sigma_x}$$

or
$$\frac{x}{y} = \frac{\Sigma xy}{n\sigma_x\sigma_y} \cdot \frac{\sigma_x}{\sigma_y}.$$

Either of these equations is called a regression equation and the common factor $\frac{\Sigma xy}{n\sigma_x\sigma_y}$ is called the correlation-coefficient. Calling this ratio r the regression equations are

$$y = \frac{\sigma_y}{\sigma_x} rx \quad \text{and} \quad x = \frac{\sigma_x}{\sigma_y} ry,$$

or
$$\frac{y}{\sigma_y} = r\frac{x}{\sigma_x} \quad \text{and} \quad \frac{x}{\sigma_x} = r\frac{y}{\sigma_y}.$$

These two equations represent different straight lines on the diagram, which become identical if $r = 1$, and diverge at wider angles the smaller the value of r.

An instructive series of dot-diagrams with the graphs of the regression equations is shown in fig. 115 which represents the relationship between pressure and temperature-changes in the upper air of Canada as determined by the results of observations with registering balloons. The lines representing the regression equations are very wide open at the ground-level and at the level of two kilometres, so wide indeed that we have not ventured to insert them in the diagrams; but from two kilometres they gradually close up like a pair of scissors as greater elevations come under calculation, to open out again at still higher levels. At the level of twelve kilometres they are so wide apart and so near the axes of the diagram that the reader may have some difficulty in recognising them.

We can illustrate the numerical method of expressing the relationship which is suggested to the eye in curve-parallels, by drawing the dot-diagram and obtaining the correlation-coefficient and the regression equations.

We deal in this manner with the curve-parallels of water-level in Victoria Nyanza and sunspots (fig. 116) which give a very high coefficient of correlation. In this case the arrangement of the figures by which the computation of Σxy and σ_x and σ_y is determined is set out in detail.

Further details of the methods of economising labour in such processes may be sought in the various works on the subject:

Computer's Handbook, Section v, M.O. publication 223. London, 1915.
D. Brunt. The combination of observations. Cambridge University Press, 2nd impression, 1923.
G. Udny Yule. An introduction to the theory of statistics. London, 1911.
W. Palin Elderton. Frequency Curves and Correlation. London, 1917.

CURVE-PARALLELS AND DOT-DIAGRAMS

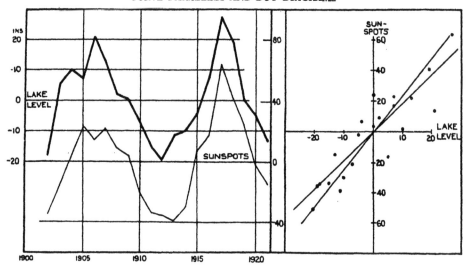

Fig. 116. Curve-parallels for the variations in the mean number of sunspots and the levels of Victoria Nyanza year by year from 1902 to 1921 together with the dot-diagram representing the same variations as departures from the respective mean values and the regression lines derived from the observations.

Calculation of the correlation-ratio between the mean annual level of Lake Victoria Nyanza and the number of sunspots.

Let X denote the level of the lake above or below some fixed standard, and x the deviation from the mean value for 1902–21.

Let Y denote the number of sunspots and y the deviation from the mean value for 1902–21.

Year	X	Y	x	y	x^2	y^2	$+$ xy	$-$
1902	− 10	5	− 18	− 35	324	1225	630	—
1903	13	24	5	− 16	25	256	—	80
1904	18	42	10	2	100	4	20	—
1905	15	63	7	23	49	529	161	—
1906	29	54	21	14	441	196	294	—
1907	21	62	13	22	169	484	286	—
1908	10	49	2	9	4	81	18	—
1909	8	44	0	4	0	16	—	—
1910	1	19	− 7	− 21	49	441	147	—
1911	− 7	6	− 15	− 34	225	1156	510	—
1912	− 11	4	− 19	− 36	361	1296	684	—
1913	− 3	1	− 11	− 39	121	1521	429	—
1914	− 2	10	− 10	− 30	100	900	300	—
1915	4	47	− 4	7	16	49	—	28
1916	15	57	7	17	49	289	119	—
1917	35	104	27	64	729	4096	1728	—
1918	27	81	19	41	361	1681	777	—
1919	8	64	0	24	0	576	—	—
1920	3	38	− 5	− 2	25	4	10	—
1921	− 5	25	− 13	− 15	169	225	195	—
Sum	169	799			3317	15025	6308	108
							6200	
Mean	8	40			166	751	310	

Correlation coefficient $= \Sigma xy / \sqrt{\Sigma x^2\, \Sigma y^2} = 6200 / \sqrt{3317 \times 15025}.$

The figures of the preceding page give the following results:

Standard deviation of lake level X $\sigma_x = \sqrt{\dfrac{\Sigma x^2}{n}} = \sqrt{166} = 12 \cdot 9.$

Standard deviation of sunspots Y $\sigma_y = \sqrt{\dfrac{\Sigma y^2}{n}} = \sqrt{751} = 27 \cdot 4.$

Correlation coefficient $r = \dfrac{\Sigma xy}{n} \cdot \dfrac{1}{\sigma_x \cdot \sigma_y} = 310/(12 \cdot 9 \times 27 \cdot 4) = 0 \cdot 88.$

Regression equations $x = r \dfrac{\sigma_x}{\sigma_y} y = \dfrac{\cdot 88 \times 12 \cdot 9}{27 \cdot 4} \, y = \cdot 414 y = \dfrac{6200}{15025} y = \dfrac{\Sigma xy}{\Sigma y^2} y,$

$y = r \dfrac{\sigma_y}{\sigma_x} x = \dfrac{\cdot 88 \times 27 \cdot 4}{12 \cdot 9} \, x = 1 \cdot 87 x = \dfrac{6200}{3317} x = \dfrac{\Sigma xy}{\Sigma x^2} x.$

Partial correlation coefficients

In recent years multitudes of correlation coefficients have been calculated. E. H. Chapman set out a vast number in 1919 as a supplement to Section v of the *Computer's Handbook*. The list includes a number of partial correlation coefficients which were introduced in a discussion of the relation between weather and crops by R. H. Hooker with the assistance of G. Udny Yule[1]. The purpose of partial coefficients is to separate the issues in case the relationship between two quantities is contingent upon, or influenced by, variations in a third quantity. In the case referred to the subject under consideration was the yield of crops as affected by variations in rainfall, temperature and sunshine. If one took the crude relation, for example, between sunshine and the wheat-crop, the immediate and direct effect of solarisation might be enhanced or prejudiced by the increase of temperature or the absence of rain. A method has therefore been elaborated of using the correlation between the influences themselves to correct the correlation-coefficients for the indirect influence. We cannot enter into the details of the calculation but we quote the result in the notation which has been accepted.

r_{12}, r_{23}, ... are the crude coefficients between the quantities 1 and 2, 2 and 3, etc.

$r_{12.3}$ is the partial coefficient of relation between 1 and 2, that is to say the coefficient corrected for the influence of 3 which it is wished to eliminate:

$$r_{12.3} = \frac{r_{12} - r_{13} r_{23}}{\sqrt{(1 - r^2_{13})(1 - r^2_{23})}},$$

and so for any number of variates

$$r_{12.3...n} = \frac{r_{12.4...n} - r_{13.4...n} r_{23.4...n}}{\sqrt{(1 - r^2_{13.4...n})(1 - r^2_{23.4...n})}}.$$

As an arithmetical check upon the possible illusions of the curve-parallel the correlation-coefficient is a useful addition to the discipline of meteorological science; but it is quite easy to push it beyond the limits within which it is worth while. On the basis of the theory of probability a large number of developments have been made. First we may mention that if the probable error does not exceed one-quarter of the correlation-coefficient the probability of that ratio being accidental is so small that the risk may be neglected, in other

[1] 'Correlation of the weather and crops,' by R. H. Hooker, *Journ. Roy. Stat. Soc.*, vol. LXX, part I, March 1907.

words a real relationship is proved. On the other hand some statisticians are very insistent that unless the number of observations is very large the possibility of the ascertained coefficient being an accident of sampling cannot be excluded. It has been said that the correlation-coefficient of ·5 between sunspots and Gambia-rainfall must not be regarded as conclusive because with the number of years of observation that are available ·5 is not beyond the limit of the correlation which might result from random sampling[1].

Another conclusion from the theory of probability is that a coefficient of correlation r means that of the two quantities the variation in one controls a fraction of the variation of the other which is expressed by r^2 and *vice versa*. It follows therefore that any correlation less than ·3 means that less than one-tenth of the variations in the two quantities are mutually dependent; the remaining nine-tenths at least are attributable to other causes. The possible inaccuracy of the observations, or of their treatment, makes the detailed pursuit of relationships of small magnitude in most cases of very problematical advantage. The survey of the curve-parallels would be sufficiently informing.

Here we put in a plea in favour of the curve-parallel in addition to the correlation-coefficient or the dot-diagram. We have explained that the formation of a dot-diagram amounts to the algebraical process of eliminating a variable between two equations. By doing so the numerical and pictorial relationship of the two variables to the common third is lost. Such relationship is often suggestive of lines of fruitful inquiry which are lost in the dot-diagram. In the process of correlation everything is sacrificed to a single numerical value which no doubt expresses the influence of the whole but gives no information about its parts. When the data have passed through the correlation machine only the machined product is available. Hence we would regard the correlation-coefficient as a useful addition to the curve-parallel but not a substitute for it.

We opened this chapter with a list of pioneers in the science of statistics, we will close it with a warning from W. H. Dines[2] in the form of a list of standard errors between which the computer has to discriminate.

" The proofs mostly lie beyond the range of an elementary book.

Standard error of a mean of n observations $\dfrac{\sigma}{\sqrt{n}}$.

Standard error of a standard deviation $\dfrac{\sigma}{\sqrt{2n}}$.

Standard error of a correlation coefficient, total or partial $\dfrac{1 - r^2}{\sqrt{n}}$.

Standard error of a regression coefficient $b_{12} = \dfrac{\sigma_{1 \cdot 2}}{\sigma_2 \sqrt{n}}$, $b_{12 \cdot 3} = \dfrac{\sigma_{1 \cdot 23}}{\sigma_{2 \cdot 3} \sqrt{n}}$.

Standard error of the amplitude in any term in a Fourier's series obtained from

n well-distributed observations $\dfrac{\sigma \sqrt{2}}{\sqrt{n}}$.

" The normal distribution is assumed and in such a case the probable error is two-thirds of the standard error."

[1] An exposition by Sir Gilbert Walker of the need for discipline in this connexion will be found in *Discovery*, vol. VI, p. 99, March 1925.

[2] *Computer's Handbook*, v (2), M.O. publication 223, London, 1915.

CHAPTER XIV

THE ANALYSIS OF AIR-MOVEMENT INTO THE GENERAL CIRCULATION AND THE CYCLONE

*Dates of the first daily weather-maps published by the meteorological services
of the several countries of the world*

1863	France	1883	Japan
1871	United States	1886	Spain
1872	Great Britain	1890	Holland
1873	Russia	1895	Canada
1874	Denmark	1896	Greece
1874	Sweden	1899	Mexico
1876	Belgium	(1922)	,,
1876	Germany	1902	Argentine
1877[1]	Algeria	1902	Egypt
1877	Australia	1905[2]	Hungary
1877	Austria	1906	China (Zi-ka-wei)
1878	India	1908	Roumania
1880	Italy	1912	Norway
1880	Switzerland	1922	Poland
1882	Portugal		

[1] Possibly 1875. The issue from 8 April, 1882, is marked 8th year.
[2] Possibly 1888. The issue for 1905 is numbered XVIII.

Daily charts of the oceans

1861. Synoptic weather charts of the Indian Ocean for the months of Jan., Feb. and March. Mauritius

Sept. 1873–Nov. 1876. Hoffmeyer's charts of the North Atlantic

1877–87. Charts of the Northern Hemisphere published by the U.S. Signal Service

1880–1910. Daily synoptic charts of the N. Atlantic published by the Danish Institute and the Deutsche Seewarte jointly

Aug. 1882–Aug. 1883. Daily charts of the N. Atlantic published by the Meteorological Office, London

1900–14. Dekaden Bericht of the Deutsche Seewarte—daily charts of the North Atlantic and adjacent continents

1912–14. Daily charts of the North Atlantic issued with the Weekly Weather Report of the Meteorological Office

1914. Daily charts of the Northern Hemisphere. U.S. Weather Bureau

1923. Daily charts of the Pacific Ocean published by the Imperial Marine Observatory, Kobe

Daily charts of the South Atlantic Ocean for the period of twelve consecutive days, 1909 Feb. 20 to March 3, are given on the back of the *Monatskarte für den Nordatlantischen Ocean* for February 1911.

THE subject of this chapter has been treated by Hildebrandsson and Teisserenc de Bort[1] in a manner which is exemplary in every sense of the word. The first volume of their treatise is devoted to it and is illustrated by lithographic reproductions of maps from the original documents which make the work of peculiar value for the student of meteorology.

Meteorological publications are of a special character: they are issued as official documents and are distributed much more by presentation and exchange than by sale; they are international and the international exchange extends only to the few libraries in any single country that can reciprocate the exchange. The library of the British Museum, for example, is not

[1] *Les bases de la météorologie dynamique*, tome 1, Gauthier-Villars et Fils, Paris, 1907.

entitled to receive the official publications of the meteorological services of foreign countries, neither are the libraries of the provincial towns nor those of the universities. Such publications may be found in a few specialised libraries but even there the multiplicity, and variety in form, are so great that the recipients do not always keep their rare documents accessible to the student. Consequently we may pay a tribute of admiration to the public spirit of the two distinguished meteorologists which led them to put together in book-form specimens of many of the original documents of which the copies are now out of reach. It is a contribution to science of a kind which is rare, and quite intelligibly so, because no one can suppose that it is re-munerative in any other way than from the satisfaction of having made a permanent contribution to a science of great interest but of no less great difficulty. Here we must restrict ourselves to a brief abstract of the story which is set out in the first volume of *Les Bases*, and advise the reader to refer to the volume itself for further particulars and details.

EARLY CHARTS OF THE DISTRIBUTION OF THE METEOROLOGICAL ELEMENTS

The story begins with Halley's map of the winds of the globe (fig. 102) taken from his celebrated memoir in the *Philosophical Transactions* of 1686, entitled 'An historical account of trade-winds and monsoons observable in the seas between and near the tropics with an attempt to assign the phisical cause of the said winds.' The ordinary facts about trade-winds and monsoons are now a commonplace of physical geography, but Halley adds to the conventional description two notes that are worth repeating; first that off the coast of Africa in the Gulf of Guinea, close to the region of the South East trade-winds, the predominant direction of the wind is from West, and secondly, that in the China Sea the alternation of wind in the monsoon is between South in the summer season and North in the winter, instead of between South West or West and North East or East as in the Indian Ocean. Con-necting these with the etesian winds of the Mediterranean and Egypt which blow during the summer months from the North, we get three sides of a summer circulation which proclaims the monsoon to be part of the chief cyclonic circulation of the whole globe.

Halley's proposed physical cause is the ascent of warm air in the hotter regions of the globe, or more generally the distribution of solar radiation which causes a belt of maximum temperature near the equator and consequent ascent of warm air with a flow thitherward along the surface from either side, and calm or doldrums in the belt itself. An extension of the idea of air rising in the hotter regions to be replaced by air flowing along the surface expresses Halley's explanation of the West winds of the Guinea coast, and the alternation of the monsoons in correlation with the alternation of the relative temperatures of land and sea in the East Indian region.

Halley's leading idea of the rising of warm air also passed into the litera-ture of science; thermal convexion is its technical name. It has been regarded as a sufficient basis of many expositions of meteorological phenomena, including indeed those of the tropical revolving storm. The study of the upper air has taught us to ask how far in ordinary circumstances such convexion can be expected to extend, and further to ask for the study of details in respect of that displacement of the risen air of the trade-winds and monsoons; and altogether the convexion of warm air has proved to be a subject requiring all the investigation that is to be expressed in Vol. III of this book instead of being a simple process that even a school-boy could easily understand and apply. As between the tropical "high-pressure" and the equator a sufficient proximate cause is the relatively lower pressure of the equatorial belt, but to claim that the low pressure of the equator compared with the high pressure of the tropic is the simple expression of the difference of temperature between the two would contravene all the laws of meteorological procedure.

The direction of the trade-wind was the next subject for inquiry. The replacement of Halley's convected air by flow inwards from both North and South would, it was supposed, give rise to a North wind in the Northern hemisphere and a South wind in the Southern, but the winds were North East and South East, not North and South. Halley merely suggested that the wind followed the sun from East to West. An explanation was offered by George Hadley in 1735, also in a paper in the *Transactions of the Royal Society*. The physical cause in this case was given as the rotation of the earth from West to East with the consequent more rapid motion of the equatorial belt as compared with the tropics. The explanation was subsequently re-invented by John Dalton, who writes as follows[1]:

Now in order to perceive the reason of these facts, it must be remembered, that the heat is at all times greatest in the torrid zone, and decreases in proceeding Northward, or Southward; also, that the poles may be considered as the centres of cold at all times: hence it follows, that, abstracting from accidental circumstances, there must be a constant ascent of air over the torrid zone, as has been observed, which afterwards falls Northward and Southward, whilst the colder air below is determined by a continual impulse towards the equator. And, in general, wherever the heat is greatest, there the air will ascend, and a supply of colder air will be received from the neighbouring parts. These, then, are the effects of the inequality of heat.

The effects of the earth's rotation are as follow: the air over any part of the earth's surface, when apparently at rest or calm, will have the same rotatory velocity as that part, or its velocity will be as the co-sine of the latitude; but if a quantity of air in the Northern hemisphere receive an impulse in the direction of the meridian, either Northward or Southward, its rotatory velocity will be greater in the former case, and less in the latter, than that of the air into which it moves; consequently, if it move Northward, it will have a greater velocity Eastward than the air, or surface of the earth over which it moves, and will therefore become a SW wind, or a wind between the South and West. And, *vice versa*, if it move Southward, it becomes a NE wind. Likewise in the Southern hemisphere, it will appear the winds, upon similar suppositions, will be NW and SE respectively[2].

---

[1] *Meteorological Observations and Essays*, by John Dalton, D.C.L., F.R.S., 2nd edition, Manchester, 1834, p. 85.

[2] "M. De Luc is the only person, as far as I know, who has suggested the idea of the earth's rotation altering the direction of the wind, which idea we have here pursued more at large. Vide *Lettres Physiques, etc.* Tom. V, Part 2, Let. cxlv."

In the Preface to the same volume Dalton remarks:

The second essay, containing the theory of the trade-winds, was, as I conceived when it was printed off, original; but I find since, that they are explained on the very same principles, and in the same manner, in the *Philosophical Transactions* for 1735, by George Hadley, Esq., F.R.S....

I cannot help observing here, that the following fact appears to be one of the most remarkable that the history of the progress of natural philosophy could furnish. Dr Halley published in the *Philosophical Transactions*, a theory of the trade-winds, which was quite inadequate, and immechanical, as will be shown, and yet the same has been almost universally adopted; at least I could name several modern productions of great repute in which it is found, and do not know of one that contains any other. The same gentleman published, through the same channel, his thoughts on the cause of the *aurora borealis* [associated with the earth's magnetic field], as mentioned above, which must then have appeared the most rational of any that could be suggested, and yet I do not find that anybody has afterwards noticed it, except "Amanuensis." On the other hand, G. Hadley, Esq. published in a subsequent volume of the said *Transactions*, a rational and satisfactory explanation of the trade-winds; but where else shall we find it?

We have already noticed a glimpse of a circulation in the upper air, opposite to that at the surface, in the power of attracting clouds attributed by Theophrastus to Kaikias, the North Easter. The next step in our progress to a knowledge of the general circulation is the repetition of this curious behaviour of the North Easter in the case of the North East trade-wind, which by the observations of von Humboldt in 1798 and subsequently by Leopold von Buch in 1815 and others was found to be reversed at the Peak of Tenerife. Similar phenomena were observed by Fouqué on the mountains of the Açores and by Goodrich and others on the peak of Mauna Loa in the Sandwich Islands.

This led to the further suggestion, we know not exactly on what grounds, that the South Westerly wind over the top of the North Easter came to the earth's surface where there were known regions of prevailing Westerlies and South Westerlies, and that the main currents over the earth's surface were from South West and North East. In this manner H. W. Dove, at one time a meteorological king, "the founder of the entire superstructure of accurate climatological knowledge," was led to look upon the general circulation of the atmosphere as consisting of an equatorial current and a polar current, and to regard the vicissitudes of weather in the temperate zone as incidents in the conflict between these two currents. The theory is given in the very celebrated meteorological book *Das Gesetz der Stürme* or *Law of Storms*.

A special ground for taking this view of the importance of the conflicting currents is the law of gyration which shows how during the passage of a storm the wind changes from the equatorial to the polar side. We are thus confirmed in the suggestion that the local circulations, which may be tropical hurricanes or cyclonic depressions with all the complications of their form, are to be regarded rather as incidents in the general circulation than as separate and independent phenomena.

In spite of the fact that the general circulation must always be regarded as having its part in every meteorological problem we shall take the liberty of passing over the representations of the general circulation which were elaborated by Maury[1], Ferrel[2] and James Thomson[3]; first because they divide the earth into zones of latitude in which parallel transverse winds are shown all round the several zones, regardless of the difficulty or impossibility of adjusting such an arrangement to any admissible scheme of pressure; and secondly because they show, similarly distributed, an exchange between upper and lower air. In presenting a notion of the general circulation of the atmosphere we desire to rest upon observations, and we regard observations of pressure as among the most important.

Dove rendered a signal service to the observational representation of the general circulation by producing monthly maps of isotherms of the globe which are to this day the foundation of the maps of temperature in many atlases. His monthly isotherms[4] which reviewed observations of temperature from 1729 onwards were preceded, as we have already noticed, by Humboldt's chart (fig. 103), also by a chart of mean isotherms for the year by L. F. Kämtz in his *Lehrbuch der Meteorologie* (Halle, 1832), which included isogeotherms by Kupffer[5]. Kämtz also noted the existence of two poles of cold in the Northern hemisphere, and in this view he was in agreement with Sir David Brewster and Sir John Herschel. Dove however controverted the view. We also find two charts by Mahlmann, 1836 and 1840, two others by Berghaus[6], 1838 and 1839, and preliminary charts by Dove in 1848.

Here we add the titles of other illustrations of the normal distribution of the meteorological elements which have been reproduced by Hildebrandsson and Teisserenc de Bort. They represent successive steps in the advancement of our knowledge of the normal general circulation of the atmosphere. The general result of these contributions forms in great part the subject of Vol. II in which a representation of the present state of our knowledge will be attempted.

Temperature:

Mean annual temperature by Buchan 1868; monthly and annual charts 1871 and 1889 (*Challenger* Report).

Teisserenc de Bort 1881 (4 maps).

Annual and monthly charts by J. Hann 1895 and 1901.

North polar charts, Mohn 1905.

Monthly isanomalies, Teisserenc de Bort and Wild 1879.

Annual amplitudes, Supan and Wild (Hann's edition of Berghaus's Atlas).

Isotherms and Isanomalies for Russia and Europe by Wild 1881, for Spain by Teisserenc de Bort 1880, for Europe by Ekholm 1899.

Chart illustrating the march of the isotherm of 0° C and 12° C in spring and autumn, Hildebrandsson and Högbom 1880 and 1883.

Annual isotherms and isanomalies for the sea-surface by W. Köppen.

[1] *The physical geography of the sea*, by M. F. Maury, London, Sampson Low, Son & Co. 1855.

[2] 'Essay on the winds and currents of the ocean,' by W. Ferrel, *Nashville Journal of Medicine and Surgery*, vol. XI, nos. 4 and 5, 1856. The paper is reprinted in *Professional papers of the Signal Service*, no. 12, Washington, 1882.

[3] 'The grand currents of atmospheric circulation,' British Association meeting, 1857.

[4] H. W. Dove, *Verbreitung der Wärme*, 1852.

[5] 'Über die mittlere Temperatur der Luft und des Bodens auf einige Punkten des östlichen Russlands,' *Pogg. Ann.* 1829.

[6] *Physikalischer Atlas*, Gotha, 1838; 2te Auflage, 1849.

Wind:

Wind charts for the Atlantic by L. Brault 1874–1880.

Windverhältnisse des Indischen Ozeans der Monate Januar-Februar und Juli-August, with similar charts for the Atlantic and Pacific Oceans. Köppen, Segel-handbücher.

Pressure:

Isobaric lines for France by M. E. Renou 1864.

Isobaric lines for the world. Buchan 1868.

Monthly charts of isobars and prevailing winds, Buchan 1871.

Monthly charts, Teisserenc de Bort 1881.

Isobars and winds, Mohn 1903.

Isobars of the North Polar regions, Mohn 1905.

Isobars of Spain, Teisserenc de Bort 1879.

Isobars of the Mediterranean, J. Hann 1887.

Isobars of Sweden, Hamberg 1898.

Isobars of the Atlantic Ocean, G. Rung 1894.

Mean range of the barometer, winter and summer, Köppen.

Isobars and isanomalies, January and July, Teisserenc de Bort 1879.

Cloudiness:

Teisserenc de Bort 1884–87. Isonephs, January and July.

Cloudiness on the Atlantic Ocean between 20° W and 30° W

    (*a*) from the official publications of the Seewarte and the Meteorological Office,

    (*b*) from the Netherlands observations.

Rainfall:

Rain over the land-areas of the globe, Supan 1898.

Seasonal variation of rain in various countries.

To these we must add:

Buchan's charts of pressure over the globe from the *Challenger* volume.

Herbertson's rainfall over the land, 1901.

Hann's charts in recent editions of Berghaus's Physikalischer Atlas.

Bartholomew's Physical Atlas, Meteorology, 1899.

The sea has been the subject of many special charts compiled by the British Meteorological Office, the Deutsche Seewarte, the Meteorological Institute of the Netherlands, the United States Signal Service, Weather Bureau and Hydrographic Office. The publications of the Meteorological Office, which followed on certain "Pilot Charts" of the Board of Trade, cover all oceans in respect of currents and sea-temperature in the four months, February, May, August, November. Separate volumes represent the results of observations of the meteorological elements over the equatorial Atlantic square 3 (1874) and nine ten-degree squares (1876), North Atlantic Ocean, monthly from 1901–1923, South Atlantic Ocean and Coastal Regions of South America 1902, Red Sea 1895, Coastal Regions of South Africa 1882, Cape to New Zealand (3rd edition) 1917, Indian Ocean and East Indian Seas, monthly from 1906–1923, Mediterranean 1917, and Davis Strait and Baffin Bay 1917.

It is upon marine charts such as these that we depend for our knowledge of the main features of the normal distribution of the meteorological elements over the globe. Yearly charts of ice in the North polar regions have been published for many years (1900–date) by the Danish Meteorological Office[1]. More recently the Imperial Marine Observatory at Kobe, Japan, has published daily charts of the North Pacific.

TROPICAL REVOLVING STORMS

We pass to the more specialised consideration of those disturbances of the general circulation which are represented by the revolving storms of the tropical regions and the milder but more extensive disturbances which through the use of daily charts have come to be recognised as at least so far similar in form as to share the name cyclone, or its milder variant cyclonic depression. We may note here in advance that the name of cyclone was coined by H. Piddington in the *Sailor's Horn-book* as derived from the Greek κύκλος, the coil of a snake, to indicate the combined circular and centripetal movement which was, at one time at least, thought to be characteristic of all central systems of low pressure, and is still regarded as appropriate to tropical revolving storms. It is perhaps desirable at the outset of our consideration of the subject to bear in mind the manner in which a motion of rotation appears as a means of conserving kinetic energy which would otherwise be rapidly lost in friction or transformed into the potential energy of pressure difference. Conservation of that character appears in every form of eddy or vortex. The vortex-ring, the dust-eddy and the tornado all preserve the identity of the kinetic energy of the motion of one part of a fluid relative to the adjacent parts or a solid obstacle, which would be lost almost immediately if the form of rotation were not assumed. In this connexion we may refer to papers by S. Fujiwhara[2] on vortical systems, not so much for the purpose of dealing with their immediate meteorological significance as to ask for a recognition of rotation as an artifice, so to speak, of the most general kind for conserving kinetic energy which, without such assistance, would become degraded through friction into a slight rise of temperature of the fluid.

We may remind the reader that in his satire of the meteorological views of the philosophers, Aristophanes accused them of dethroning Zeus and setting up vortex in his place. Our reading of classical and mediaeval literature is not sufficiently extensive to enable us to say how far the idea of revolving fluid in the atmosphere was accepted before the development of physical science in the seventeenth century.

If Halley gives us the first idea of the winds as a general circulation of the air over the surface of the globe, his contemporary, William Dampier, master seaman, gives us the first description of a typhoon, the name given to the tropical hurricanes of the China Sea, and therefore an example of the more conspicuous local cyclonic circulation which forms, with the general

---

[1] 'Isforholdene i de arktiske Have,' *Publikationer fra Det Danske Meteorologiske Institut, Aarbøger*, København.

[2] 'The natural tendency towards symmetry of motion and its application as a principle of meteorology,' *Q. J. Roy. Meteor. Soc.*, vol. XLVII, 1921, p. 287; 'On the growth and decay of vortical systems' and 'On the mechanism of extratropical cyclones,' *Q. J. Roy. Meteor. Soc.*, vol. XLIX, 1923, pp. 75–117.

circulation, a sort of ground-work of the general dynamical or thermo-dynamical problem of meteorology. We quote from Dampier's *A new voyage round the world* the description which has become classic[1].

1687, 4th July. Some of our Men went over to a pretty large Town on the Continent of *China*, where we might have furnished our Selves with Provision, which was a thing we were always in want of, and was our chief business here; but we were afraid to lye in this Place any longer, for we had some signs of an approaching Storm: this being the time of the Year in which Storms are expected on this Coast; and here was no safe Riding. It was now the time of the Year for the S.W. Monsoon, but the Wind had been whiffling about from one part of the Compass to another for two or three Days, and sometimes it would be quite calm. This caused us to put to Sea, that we might have Sea-room at least; for such flattering weather is commonly the fore-runner of a Tempest.

Accordingly we weighed Anchor, and set out: yet we had very little Wind all the next night. But the Day ensuing, which was the 4th day of *July*, about 4 a clock in the afternoon, the Wind came to the N.E. and freshned upon us, and the Sky look'd very black in that quarter, and the black Clouds began to rise apace and mov'd towards us; having hung all the morning in the Horizon. This made us take in our Top-sails, and the Wind still increasing, about 9 a clock we rift our Main-sail and Fore-sail; at 10 we furl'd our Fore-sail, keeping under a Main-sail and Mizen. At 11 a clock we furl'd our Main-sail, and ballasted our Mizen; at which time it began to rain, and by 12 a clock at night it blew exceeding hard, and the Rain poured down as through a Sieve. It thundered and lightned prodigiously, and the Sea seemed all of a Fire about us; for every Sea that broke sparkled like Lightning. The violent Wind raised the Sea presently to a great heighth, and it ran very short and began to break in on our Deck. One Sea struck away the Rails of our Head, and our Sheet-Anchor, which was stowed with one Flook or bending of the Iron, over the Ship's Gunal, and lasht very well down to the side, was violently washt off, and had like to have struck a hole in our Bow, as it lay beating against it. Then we were forced to put right before the Wind to stow our Anchor again; which we did with much ado: but afterwards we durst not adventure to bring our Ship to the Wind again, for fear of foundring, for the turning the Ship either to or fro from the Wind is dangerous in such violent Storms. The fierceness of the Weather continued till 4 a clock that morning; in which time we did cut away two Canoas that were towing astern.

After four a clock the Thunder and the Rain abated, and then we saw a *Corpus Sant* at our Main-top-mast head, on the very top of the truck of the Spindle. This sight rejoyc'd our Men exceedingly; for the heighth of the Storm is commonly over when the *Corpus Sant* is seen aloft; but when they are seen lying on the Deck, it is generally accounted a bad sign.

A *Corpus Sant* is a certain small glittering light; when it appears as this did, on the very top of the Main-mast or at a Yard-arm, it is like a Star; but when it appears on the Deck, it resembles a great Glow-worm. The *Spaniards* have another Name for it (though I take even this to be a *Spanish* or *Portuguese* Name, and a corruption only of *Corpus Sanctum*) and I have been told that when they see them, they presently go to Prayers, and bless themselves for the happy sight. I have heard some ignorant Sea-men discoursing how they have seen them creep, or as they say, travel about in the Scuppers, telling many dismal Stories that hapned at such times: but I did never see any one stir out of the place where it was first fixt, except upon Deck, where every Sea washeth it about. Neither did I ever see any but when we have had hard Rain as well as Wind; and therefore do believe it is some Jelly: but enough of this.

We continued scudding right before Wind and Sea from 2 till 7 a clock in the Morning, and then the Wind being much abated, we set our Mizen again, and brought

[1] *A new voyage round the world*, by William Dampier, 5th edition corrected, London, 1703, chap. XV, p. 413. The storm apparently occurred just after leaving the Isle of St John on the coast of the province of Canton.

our Ship to the Wind, and lay under a Mizen till 11. Then it fell flat calm, and it continued so for about 2 Hours: but the Sky looked very black and rueful, especially in the S.W. and the Sea tossed us about like an Egg-shell, for want of Wind. About one a clock in the Afternoon the Wind sprung up at S.W. out of the quarter from whence we did expect it: therefore, we presently brail'd up our Mizen, and wore our Ship: but we had no sooner put our Ship before the Wind, but it blew a Storm again, and rain'd very hard; though not so violently as the Night before: but the Wind was altogether as boysterous, and so continued till 10 or 11 a Clock at Night. All which time we scudded, or run before the Wind very swift, tho' only with our bare Poles, that is, without any Sail abroad. Afterwards the Wind died away by degrees, and before Day we had but little Wind, and fine clear Weather.

I was never in such a violent Storm in all my Life; so said all the Company. This was near the change of the Moon: it was 2 or 3 days before the change. The 6th day in the Morning, having fine handsom Weather, we got up our Yards again, and began to dry our Selves and our Cloaths, for we were all well sopt. This Storm had deadned the Hearts of our Men so much, that instead of going to buy more Provision at the same place from whence we came before the Storm, or of seeking any more for the Island *Prata*, they thought of going somewhere to shelter before the Full Moon, for fear of another such Storm at that time: For commonly, if there is any very bad Weather in the Month, it is about 2 or 3 days before or after the Full, or Change of the Moon.

The same subject is specifically treated in another work by Captain Dampier.

Tuffoons are a particular kind of violent Storms, blowing on the coast of *Tonquin*, and the neighbouring Coasts in the months of *July*, *August*, and *September*. They commonly happen near the full or change of the Moon, and are usually preceeded by very fair weather, small winds and a clear Sky. Those small winds veer from the common Trade of that time of the year, which is here at S.W. and shuffles about to the N. and N.E. Before the Storm comes there appears a boding Cloud in the N.E. which is very black near the Horizon, but towards the upper edge it looks of a dark copper colour, and higher still it is brighter, and afterwards it fades to a whitish glaring colour, at the very edge of the Cloud. This Cloud appears very amazing and ghastly, and is sometimes seen 12 hours before the Storm comes. When that Cloud begins to move apace, you may expect the Wind presently. It comes on fierce, and blows very violent at N.E. 12 hours more or less. It is also commonly accompanied with terrible claps of Thunder, large and frequent flashes of Lightning, and excessive hard rain. When the Wind begins to abate it dyes away suddenly, and falling flat calm, it continues so an hour, more or less: then the wind comes about to the S.W. and it blows and rains as fierce from thence, as it did before at N.E. and as long[1].

As early as the year 1650 hurricanes had been treated as whirlwinds by Varenius in his book *Geographia Naturalis*; and in 1698 Captain Langford contributed to the Royal Society of London a memoir on West Indian hurricanes which included descriptions of five storms within the ten years 1657–67[2]. His observations led him to anticipate some days in advance a hurricane which devastated Nevis on August 19, 1667, and to give instructions for the manoeuvring of a British squadron which were followed with success. Hurricanes, regarded as violent whirlwinds, were also discussed within the

[1] *Voyages and Descriptions*, vol. II, in Three Parts, by Capt. William Dampier, 2nd edition, London, printed by James Knapton, at the Crown in St Paul's Churchyard, 1700; Part I, The Supplement of the Voyage round the World, p. 36.

[2] 'Observations of his own experience upon hurricanes and their prognosticks,' *Phil. Trans.* xx, p. 407.

succeeding century by Sir Gilbert Blane, Colonel J. Capper, Horsburgh, Romme, Farrar, Mitchell and Bardenfleth.

In 1828 Dove[1] expounded the phenomena of tropical revolving storms as travelling systems and recognised the clockwise rotation of the air in hurricanes of the Southern hemisphere in contradistinction to the counter-clockwise rotation of the Northern hemisphere.

His law was invoked for the investigation of some remarkable storms which visited Europe in the middle of the nineteenth century. A typical "Stausturm," a severe storm visited Europe on January 23–29, 1850, and was accompanied by a terrible fall of snow in which on January 29 hundreds of persons perished in Sweden alone. The storm was discussed by Dove in Germany and by Siljeström in Sweden. The "Royal Charter" storm of October 25–26, 1859, was discussed by Admiral FitzRoy, at that time Superintendent of the Meteorological Department of the Board of Trade. A series of maps were constructed, on a plan introduced by Dove of graphs of pressure and temperature along parallels and meridians; and a violent storm of November 5, 1864, was studied by Commandant B. Lilliehöök of the Swedish Navy and resolved into a "tempête tournante" or revolving storm between an equatorial current from the SW over the Baltic and a polar current from North or North East over Sweden.

The authors of *Les Bases* conclude their notice:

Nous avons brièvement esquissé les célèbres théories des tempêtes de Dove. Grâce à l'autorité éminente et bien méritée de son auteur, ces théories, en somme mal fondées, ont été pourtant longtemps presque unanimement acceptées des météorologistes de toute l'Europe. Par ses recherches longues et pénibles et par ses découvertes brillantes dans le domaine vaste de la climatologie, Dove sera toujours considéré à juste titre comme un des savants les plus distingués du siècle actuel. Mais il a commis la faute d'employer pour étudier les phénomènes sans cesse changeants des perturbations atmosphériques, les mêmes méthodes d'investigation dont il s'était servi avec tant de succès dans ses études climatologiques. Il est évident que ni les moyennes de cinq jours, dont il s'est servi ordinairement pour étudier ce qu'il appelait "les variations non périodiques," ni les moyennes diurnes n'ont suffi pour de telles recherches. C'est seulement par la méthode *synoptique* qu'on a réussi à préciser l'étude de ces phénomènes compliqués et à trouver les lois générales qui les régissent.

While Dove was engaged with the revolving storms of Europe William C. Redfield was enabled to give a more fundamental account of the phenomena of rotation and translation based upon observations of West Indian hurricanes which he collected during ten years[2]. From these observations he constructed a series of synoptic charts. He defined a cyclone as consisting of a large mass of air with a rapid motion of rotation counter-clockwise, a calm centre and a motion of translation. By subsequent researches he identified the region between the equator and the tropic as the place of origin of the storms and traced their paths Westward until the Northern limit of the trades about 30° N had been reached, and thence, with a recurvature to the North and East, into the region of prevailing Westerly winds in which they travelled

---

[1] 'Ueber barometrische Minima,' *Pogg. Ann.* vol. XIII, 1828, p. 596.
[2] *American Journal of Science and Art*, vol. XX, 1831.

sometimes to Europe[1]. Redfield recognised a "vorticose convergence" of the air towards the centre and discussed the height of the revolving disk of air which formed the storm. He remarked that there might nearly always be observed a layer of clouds, of which the movement was from South or from West in the region of the United States, above the nimbus clouds and scuds which carried the rain. He estimated the height of this layer and consequently the superior limit of the cyclone at an English mile, 1600 metres. Redfield's method of investigation, which included the reduction of barometer-readings to sea-level and the formation of charts on that basis, is represented in the reproduction of his charts of the Cuba hurricane[2].

Following Redfield, Colonel Reid took up the investigation. His work is set out in *The progress of the development of the law of storms and of the variable winds*[3]. It included a study of the hurricanes of the Indian Ocean, South of the equator in the region of Mauritius and Réunion, which were found to have not only the reversed direction of rotation but also a reversed line of curvature of path. Reid is said to have been the first to give the rules for the manoeuvring of ships in order to avoid the worst dangers of cyclones and therein to have introduced the distinction between the navigable semicircle and the dangerous semicircle[4]. For our part we have never found sailors willing to regard the word dangerous as an acceptable nautical term: things may be difficult but danger belongs to another category, and its use is perhaps the more inappropriate in view of one of Colonel Reid's rules for sailors, namely on the approach of a hurricane to leave port and put to sea as being in any case less "dangerous" than the protection of a harbour.

One of the results of Reid's work was the establishment for the first time of a system of warning-signals which was set up in 1847 at Carlisle Bay, Barbados, with the aid of a self-recording barometer at Bridgetown.

Reid handed on the torch of inquiry to Captain H. Piddington in 1838 by asking the directors of the East India Company to organise the study of the cyclones of the Bay of Bengal and the Arabian Sea. There followed in 1839 a memoir on a "Hurricane in the Bay of Bengal, June 1839[5]," and forty other memoirs within the next fifteen years and finally *The Sailor's Horn-book for the Laws of Storms in all parts of the world*[6]. It is in this book that the word cyclone first appears. In it are also laws for finding the centre and for the manoeuvring of ships with reference thereto. Piddington recognised the incurvature of the winds of a cyclone which was subsequently stated more definitely by Charles Meldrum who, as director of the Observatory at Mauritius for forty years and an interested student of the cyclones of the South Indian Ocean, became one of the chief authorities upon tropical revolving storms.

[1] *On three several hurricanes of the Atlantic, etc.*, New Haven, 1846.
[2] *Les bases de la météorologie dynamique*, by H. H. Hildebrandsson et L. Teisserenc de Bort, Gauthier-Villars et Cie, Paris, 1907, vol. I, p. 32.
[3] Weale, London, 1849.
[4] Instructions of this kind appear in Dove's *Law of Storms* and in the successive editions of the *Barometer Manual for the Use of Seamen* which began with FitzRoy, and has passed through nine editions issued by the Meteorological Office.
[5] *Journal of the Asiatic Society of Bengal*, vol. VIII.
[6] Williams and Norgate, London and Edinburgh, 2nd edition, 1855.

The origin of storms of this character is still far from being understood. It has been discussed by many meteorologists, among others Thom of the British Army Medical Service who was stationed at Mauritius, and Keller a French engineer. Following Dove they attributed the formation of the gyratory motion to the juxtaposition of currents travelling in opposite directions. The interaction of these currents is too difficult a subject for an introductory section: we can only say that viewed quite dispassionately there seems to be a world of difference in the way the opposing currents are arranged. If the equatorial current finds itself North of the polar current (in the Northern hemisphere) nothing happens, but if it is South of the polar current everything happens. This theory of the origin of opposing currents was controverted by the supporters of a centripetal theory of cyclonic depressions which we shall mention in connexion with the cyclonic depressions of temperate latitudes. In the meantime we may note that nothing has yet been said about later information concerning the typhoons of the China seas and tropical and other revolving storms in various parts of the world.

There can be no doubt that the phenomena as described by Dampier are characteristic of tropical revolving storms in any part of the world. For our immediate purpose of tracing the historical sequence of their study we do not require a description in detail of storms in the several localities. We confine ourselves accordingly to a bibliography, mainly of the older works on the subject, which is appended to this chapter.

There is a considerable literature on the subject of Tornadoes, which are revolving storms of extreme violence but very small diameter, less than one-hundredth of that of the tropical revolving storms over the sea. They are specially frequent in the Mississippi basin but in less violent form they are experienced in many parts of the world. The tornadoes of the Mississippi are discussed by de Courcy Ward, *Q. J. Roy. Met. Soc.*, vol. XLIII, 1917, p. 317. For corresponding phenomena in other countries we may mention: Reye, *Die Wirbelstürme, Tornados und Wettersäulen in der Erd-Atmosphäre*, Hannover, 1872; Faye, *Sur les tempêtes*, Paris, 1887; C. L. Weyher, *Sur les Tourbillons, trombes, tempêtes et sphères tournantes*, Paris, 1889; and more recently A. Wegener, *Wind und Wasserhosen in Europa*, Braunschweig, 1917.

The centripetal theory

The convergence of air towards the barometric minimum, or centre, of a tropical revolving storm was not so obvious as to escape without challenge from those who thought the motion circular rather than spiral, among whom, though erroneously, Piddington was reckoned, because his horn dial showed tangential motion. It was an important feature from the practical point of view because it affected the sailor's determination of the position of the centre from his knowledge of the direction of the wind, but even within our own official experience we have received protests against Meldrum's representation of incurved spiral paths as being erroneous. Incurvature has from the first however been regarded as an obvious characteristic of the cyclonic depressions of Europe. These were first charted by H. W. Brandes about whom Hildebrandsson and Teisserenc de Bort write[1]:

Il y a loin de la conception d'une idée à sa réalisation. C'est là une vérité qui est démontrée à chacune des grandes étapes du genre humain sur la route de la vérité. Elle est prouvée aussi par l'histoire de la question qui nous occupe.

Avant que la relation entre la pression barométrique et la direction du vent eût été formulée par la loi fondamentale de Buys Ballot et que les premières cartes synoptiques fussent dessinées à l'Observatoire de Paris, il y avait eu de longues et pénibles recherches faites dans la même direction par des savants isolés. Les résultats trouvés par eux s'accordent d'une manière surprenante avec ceux qui ont été acceptés 20 ou 30 ans plus tard comme des découvertes nouvelles. Mais sous les attaques énergiques de Dove,—dont l'autorité fut du reste bien méritée par ses recherches brillantes et fondamentales dans le domaine de la climatologie,—ces pionniers de la météorologie dynamique n'ont pas pu gagner du terrain dans le monde scientifique de l'Europe.

Il y a surtout trois savants à considérer ici: Brandes en Allemagne, Espy et Loomis en Amérique.

The researches here referred to led to the centripetal idea of a cyclone in which the motion towards the centre was regarded as the primary movement of the air caused by spontaneous convexion in the central region, whereas the tangential component of the motion was regarded as a secondary effect due to the rotation of the earth which caused the centripetal air to deviate to the right and miss its objective.

Brandes has the credit of two important meteorological publications *Beiträge zur Witterungskunde*, Leipzig, 1820, and *Dissertatio physica de repentinis variationibus in pressione atmosphaerae observatis*, Leipzig, 1826. In the first he discusses the weather over Europe for each day of the year 1783, making use of collections of observations for that year in *Mannheimer Ephemeriden*, and those published by Cotte. From the observations he constructed charts with lines of equal deviation of pressure from normal and arrows to show the direction of the wind. The charts have not been preserved but the material used is available and a specimen chart as constructed therefrom by Hildebrandsson (fig. 117) is taken from *Les Bases*. The

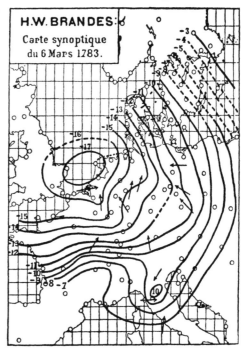

Fig. 117. Synoptic chart of Western Europe for March 6, 1783, constructed by Hildebrandsson from the data used by Brandes. Lines of equal deviation of pressure from normal are drawn, with arrows to indicate the direction of the wind.

[1] *Les bases de la météorologie dynamique*, tome 1, p. 45.

point to which special attention is drawn is the motion of the air towards the centre of low pressure. In the later publication he treated in the same manner two depressions, one which passed up the English Channel towards Norway on December 24 to 26, 1821, and the other a depression of February 2 to 3, 1823, with barometric charts of departures from normal, but before calculating the departures he reduced the pressures to sea-level. Thereby he confirmed his conclusion that "A barometric depression, produced by some unknown cause, advances from West to East above the earth's surface, and the air blows generally from all sides and constantly towards the centre of the depression to establish equilibrium." "Indicavi—Procellam ortam esse ex aere in vacuum irrumpente, aut ad loca aere rarefacto repleta ruenti" (I have demonstrated: That a cyclonic storm arises from air breaking through into a vacuum, or rushing towards regions occupied by rarefied air).

To his conclusions he adds a proposal for the organisation of a meteorological service for the study of storms.

Similar ideas were further elaborated by James Pollard Espy who, after publishing various researches in the *Journal of the Franklin Institute* and in the publications of the Philosophical Society of Philadelphia, came to Europe in 1840 to explain his ideas in London and Paris. He presented a memoir to the British Association in September of that year and in the following year presented to the French Academy of Sciences a note on thunderstorms, *Brief outline of the theory of storms*, which was reported upon by Arago, Pouillet and Babinet[1].

In the same year he published *Philosophy of Storms*, Boston, 1841, and two years afterwards he was appointed Chief of the Meteorological Bureau of the War Department; until then he had been President of the Meteorological Commission of the State of Pennsylvania.

He organised a service of daily synchronous observations which became subsequently the meteorological bureau of the Signal Service. He compiled synoptic charts for 1100 days showing the direction of the wind, the local amounts of rain or snow, lines of minimum barometric pressure and lines of maximum. With these he studied barometric depressions and drew up an exposition of their behaviour; he studied also the phenomena of tornadoes and summarised the results of his investigation of central cyclonic systems as follows:

1. The movement of the air is towards the centre.
2. A barometric depression in the centre.
3. A central ascensional current of air.
4. The formation of cloud at a certain height and its radial dispersal accompanied by rain and hail after the cloud has reached some prodigious height.
5. The travel of the whole meteor with the upper currents of the atmosphere.

---

[1] *Comptes Rendus*, 1841, p. 454.

The theory which Espy gave to account for these characteristics is summarised in *Les Bases*, and we give a translation of the summary because it expresses clearly and with brevity the view of the nature of cyclones which was accepted by meteorologists, with or without some mental disquietude, until the close of the nineteenth century.

He said that it was Dalton's researches on hygrometry which had made a remarkable impression upon him. Reading his magnificent discoveries he was convinced "that in these was to be found the lever by the help of which meteorologists would be able to move the world!" According to him water-vapour is the principal cause of atmospheric motion. He thought that if a very extensive layer of warm, moist air covers the surface of some region of the earth or sea, and by some cause, for example a lower density locally, an ascending current is set up in this mass of moist air, the upward force, instead of being diminished as a result of the elevation of the raised column would grow with the height of the column, just as a current of hydrogen ascending through the ordinary air would be pushed to the top of the atmosphere with a force and velocity the greater the greater its height. One can thus liken this column of warm air to a chimney or the ventilating shaft of a stove the up-draught of which is greater the greater the height of the chimney containing the warm air. What cause is it then that makes the ascending current of warm moist air always lighter in each of its parts than the air which is at the same height as the several portions of the ascending column? By calculations which were very exact for that period, Espy proved that the cause was to be found in the constantly higher temperature which the ascending column preserves, a temperature which arises from the heat given out by the partial precipitation of the vapour mixed with the air which makes the ascending column a true column of warm air, that is to say of a lighter gas. If in this way the cooling by expansion is counterbalanced by the heat derived from the condensation of the water, this air will remain constantly warmer than the surrounding air.

It will therefore always be lighter, and the more the column is raised the greater will be the upward force. Among the consequences that Espy draws from this theory one must notice the following, that *a current of descending air can never produce cold*, because the current will be warmed by compression as it descends, and for this reason neither a shower of rain nor even a cloud can ever be formed in the midst of a descending current.

Finally he remarks that if in tornadoes and water-spouts air is drawn in by the lower part of the column and not by the upper layers, it is because the difference between the pressure of the warm column and that of the surrounding air is so much more pronounced at the bottom of the column, which has a smaller density but equal elasticity, so that in the case of equilibrium, at the lowest point this difference will be exactly the total difference of weight of the whole warm column and the column of air of the same height which surrounds it.

The subject was further pursued by Elias Loomis, professor at Yale College, New Haven. His researches began with three memoirs on certain storms of 1836 and 1842. He traced the progress of the storms on maps with six-hour intervals showing isobars, winds, lines of "equal thermal oscillation," and regions of clear sky, cloud, rain and snow. The main principle that he derives from his investigation is that above the United States there is an immense current of air from West to East, that storms are formed in this current and carried with it Eastward. The theory of the formation of the depression is hardly different from Espy's. It commences with thermal convexion, reinforced by the latent heat of condensation: the reduction of pressure is accompanied by dilatation, and the surface-winds represent the inflow of air to re-establish equilibrium. The general circulation carries the depression from West to East, but the supremacy of the North Westerly wind of the West side overcomes the South Easter of the East side and the depression becomes enfeebled and ultimately dies. Loomis concludes his memoir by a proposal for the construction of semi-diurnal charts for the United States for a year.

THE WEATHER-MAP

This brings us to the middle of the nineteenth century when the invention of the electric telegraph had made the construction of a daily chart of the existing state of the weather over large areas possible. The idea had long been under consideration.

L'idée même du service météorologique germa à la fin du siècle dernier dans le cerveau de Lamarck et celui de Lavoisier. Ce dernier, vers 1780, frappé par les premières observations de Borda à ce sujet, s'entendit avec lui pour avoir des conférences auxquelles prirent part de Laplace, d'Assy, Vandermonde, de Montigny, etc.

Il s'agissait d'établir des instruments, et surtout des baromètres comparables, sur un grand nombre de points de la France et de l'Europe et même de l'univers. Nombre de ces instruments furent distribués par Lavoisier.

Dans une note, Lavoisier dit "que la prédiction des changements qui doivent arriver au temps est un art qui a ses principes et ses règles, qui exige une grande expérience et l'attention d'un physicien très exercé: que les données nécessaires pour cet art sont: L'observation habituelle et journalière des variations de la hauteur du mercure dans le baromètre, la force et la direction du vent à différentes élévations, l'état hygrométrique de l'air. Avec toutes ces données il est presque toujours possible de prévoir un ou deux jours à l'avance, avec une très grande probabilité, le temps qu'il doit faire; on pense même qu'il ne serait pas impossible de publier tous les matins un journal de prédiction qui serait d'une grande utilité pour la société." Quelques années après, Romme proposa d'employer le télégraphe aérien de Chappe pour renseigner les physiciens sur l'arrivée des tempêtes et pour en communiquer l'avis aux ports et aux agriculteurs. Comme on le voit, Lavoisier et Romme ont défini d'une manière très précise la prévision du temps telle que nous pratiquons aujourd'hui.

Un grand nombre d'observations météorologiques furent recueillies d'après le plan de Lavoisier dans de grands centres; vers la même époque, la Société Météorologique du Palatinat se préoccupa aussi de réunir des observations météorologiques régulières[1].

In 1842 Carl Kreil proposed the use of the electric telegraph for transmitting observations and as a basis of forecasting at Prague; a more elaborate proposal was made to the British Association in 1848 by John Ball, and the method was first employed in the United States at the suggestion of Henry, Secretary of the Smithsonian Institution, in 1849, to transmit to the Signal Service an indication of the weather at each telegraph office at the opening of business for the day. [See note, p. 324.]

In the same year James Glaisher organised the collection of reports by telegraph for the *Daily News*. Publication of reports of wind and weather at the stations began in that Journal on June 14. At the Exhibition of 1851 a map was prepared daily from August 8 to October 11 and lithographed copies were sold to the public at a penny each.

On November 14, 1854, a hurricane swept over the allied fleets in the

---

[1] *Les bases de la météorologie dynamique*, tome 1, p. 63.

Black Sea, resulting in the loss of the *Henri IV*; Le Verrier, Director of the Observatory of Paris, was appealed to by the Minister of War to study the conditions in which such disasters could be produced. Le Verrier appealed to the astronomers and meteorologists of all countries, asking them to send him information about the atmospheric conditions of November 12–16, 1854, and on February 16, 1855, he submitted to the Emperor a proposal for the establishment of a meteorological réseau, with the object of warning shipping of the approach of storms. The proposal was approved on the following day. Two days later Le Verrier presented to the Academy a chart of the weather-conditions over France at 10 a.m. on the same day. The meteorological stations of France were quickly organised, the réseau was complete in 1856, in 1857 Brussels, Geneva, Madrid, Rome, Tunis, etc. were brought in, and some time afterwards Vienna, Lisbon and St Petersburg. A daily bulletin was organised which was improved as from January 1, 1858, and from the autumn of 1863 it gave a chart of isobars over Europe.

On September 3, 1860, Admiral FitzRoy began the collection in London of daily reports by telegraph and shortly afterwards made public in the news-papers the conclusions which he drew from the maps as to the prospects of weather, and organised a system of storm-warnings.

The first organisation of a service of forecasts and storm-warnings in Europe appears however to have been in Holland by Buys Ballot, Professor of Physics in the University of Utrecht. Notifications to ports commenced in 1860 under a Government order dated May 21. They were based upon telegrams from six points and the warnings were guided by pressure-gradients. From January 8, 1864, the signals devised by FitzRoy were employed, but after the suppression of FitzRoy's scheme on December 6, 1866, Buys Ballot introduced a new instrument, called the aeroclinoscope, which indicated the position of the centre of a depression and the gradient.

This was in accordance with the ideas which are commemorated by the naming of Buys Ballot's law. Its formulation took some time. We quote a translation from the *Jaarboek* of the Meteorological Institute of the Nether-lands for 1857.

La règle trouvée, que des grandes différences barométriques dans les limites de notre pays sont suivies par des vents plus forts, est que le vent est en général perpen-diculaire ou à peu près perpendiculaire à la direction indiquée par la plus grande pente barométrique, de manière qu'un décroissement du Nord au Sud de la pression barométrique est suivi d'un vent d'Est, et un décroissement du Sud au Nord d'un vent d'Ouest.

We take the opportunity of this reference to Buys Ballot's services to meteorology to note some points in connexion with the development of the weather-map as a mode of representing the conditions of weather over a large area. Buys Ballot compiled daily maps of a large part of Europe from 1852 onwards and published them in the *Jaarboek*. We give a reproduction (fig. 118) of four of these charts. The dates that we have chosen are October 31, and November 11–13, 1852, the choice being guided by the fact that the balloon ascent at Vauxhall by J. Welsh, the results of which are represented in figs. 105 and 106, is dated November 10, 1852; and the juxtaposition of the two historic records can hardly fail to be of interest.

BUYS BALLOT'S WEATHER CHARTS, 1852

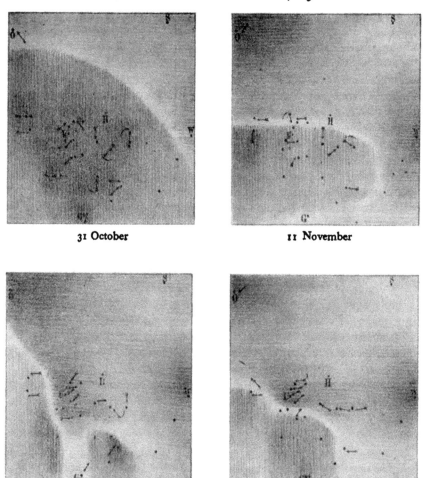

31 October   11 November

12 November   13 November

Fig. 118. Four examples of charts of the distribution of wind and temperature over North West Europe. Six stations are marked by initial letters: O Orkneys, S Stockholm, U Utrecht, H Hamburg, W Warsaw, G Geneva; twenty-one other stations by dots.
Wind-direction is indicated by arrows with an arc showing change during the day; wind-force by the length of the arrows. Shading shows departure of temperature from the normal, vertical excess, horizontal defect. The shading is closer the greater the departure.

The form of Buys Ballot's map is not at all suggestive of the modern weather-map, which may be said to date from the maps of the *Bulletin International* issued from the Paris Observatory in 1863. At that time the method of lithographic transfer had not been invented. A good deal of ingenuity was accordingly expended in devising maps which could be printed by ordinary letterpress. We give a specimen of the map devised by Francis Galton (fig. 119) and with it a letter of July, 1861, which he issued to meteorologists inviting them to send him by telegraph daily reports of their observations in consideration of his undertaking to compile, print and issue the maps at his own expense. The result of this invitation was expressed by Galton in a work entitled *Meteorographica or methods of mapping the weather*, "illustrated by upwards of 600 printed and lithographed diagrams referring to the weather of a large part of Europe during the month of December, 1861." Two years later another effort was made by a party whom Galton once referred to as "the other side." It took the form of a proposal for a Limited Company with a capital of £4000, the object of which was to compile and issue a daily weather-map. We give a reproduction of the prospectus and specimen map (fig. 120). It is not dated; but as the act which authorised the formation of companies with limited liability only became law in August, 1862, the date was probably 1863. James Glaisher, Secretary of the Meteorological Society which was founded in 1850, was apparently the leader of this movement. We have no information about it beyond the prospectus which we have reproduced at length in order to represent the remarkable enthusiasm for weather mapping in the sixties of last century.

We add one other reproduction of a historic document, a map (fig. 121) for March 16, 1859, constructed on quite modern lines and published in 1868 by Alexander Buchan to show the travel of a cyclonic depression across the Atlantic. We are thus brought into the period of modern meteorology with the weather-map for its main feature and forecasting its avowed object.

The reference to Galton's enterprise in mapping the weather and distributing copies of the maps for the use of students of meteorology recalls the efforts which he made subsequently in conjunction with his colleagues of the Meteorological Committee of the Royal Society to analyse and study the details of the phenomena of weather with the object of explaining the cyclone and other features of the weather-map on a physical basis. The Committee commenced its operations in 1867. The plan was to develop as a meteorological method the use of the autographic records of pressure, temperature, humidity, wind, and rainfall, with auxiliary eye-observations at seven observatories in the British Isles, Kew, Falmouth, Stonyhurst, Aberdeen, Glasgow, Armagh and Valencia; for this purpose arrangements were made for publishing facsimile reproductions of the records. The details of the arrangements were mostly of Galton's devising and the results are to be found in twelve years' issues of the *Quarterly Weather Report* of the Meteorological Office. The reports contained in addition to the reproductions of the records notes on the behaviour of each cyclone and anticyclone indicated on the daily maps of the period. The publication constituted the most stupendous effort at the interpretation of the weather-map ever attempted by a single institute, and one of the ironies of the life-history of meteorology is that there

CIRCULAR LETTER TO METEOROLOGICAL OBSERVERS

(Translations in French and in German are sent to the Continent.)

SYNCHRONOUS WEATHER CHARTS.

THE accompanying Map is printed with types I have designed and had cast for the purpose of representing synchronous weather data under a geographical arrangement and in a partly pictorial form, desiring to afford that intelligible picture of the meteorology of a large region which mere printed lists are wholly inadequate to supply.

In the scale I adopt, the centre of each symbol can be adjusted to within two English statute miles of any station, and, as the symbols occupy a space of one-third of a degree both in latitude and longitude, very numerous stations may be employed, the only limit to their number being, that every two of them should lie apart at least 20 English geographical miles in latitude and 13⅔ in longitude. An enormous number of observations, extending over large areas, might thus be printed with ease and collated with accuracy, mapping out broad eddying currents of air, heat, and moisture which determine our climate, but of whose directions, shapes, and mutual relations we are at present in lamentable ignorance.

As a basis to future efforts, I here invite Meteorologists who have been in the habit of contributing observations to any Society, and are therefore familiar with methods of observing, to co-operate with me during the whole of next December, in order to obtain a series of aërial charts of Northern Europe between the latitudes 42° 25' on the South (including all France and Perugia) and 61° on the North (including Shetland, Bergen, and Christiana), and extending from the westernmost limit of the British islands to the meridian 20° 30' East from Greenwich (including Konigsberg and Pesth, and even reaching Warsaw). I propose to print a few charts, containing one of the most prominent series of weather changes that may occur, on the scale and plan of that which is here annexed, but covering a sheet more than six times its size, and to write an analysis of the rest, aided by lithographs. A copy of these will be presented and forwarded, by book post, gratuitously to every Contributor who will send me, *postage free,* a series of *reduced* observations and other information, according to the subjoined conditions.

The result of a wide system of co-operation such as I propose, will be the accomplishment of a valuable piece of scientific work, that will also help to afford an answer to the question whether synchronous charts may hereafter be printed regularly, with success.

The trouble of preparing a list of observations, such as I ask for, will be an exceedingly small addition to the every-day work of an habitual Observer. I am obliged to insist upon the condition that the observations should be *reduced* ready for printing without further trouble, because a labour which is not worth consideration when divided among some hundred Observers and spread over a monthly time, becomes more serious than I care to undertake single-handed and at once. (I mean that the barometer should be corrected for temperature; reduced to the mean sea level; and its reading, if in millimètres* or other foreign scale, should be converted into *English inches* and decimals; also,

that the reading of the thermometers, if in Reaumur's or Centigrade scale,* should be converted into that of *Fahrenheit.*) Also that the observations should be entered on one of the enclosed blank forms, as want of uniformity causes an enormous waste of labour. Moreover, the postage of letters is so onerous that I cannot accept any save those that are prepaid. Neither can I undertake to correspond with individual Observers, although I shall be most happy to give any information that may be required to the representatives of Meteorological Societies. I feel sure that every Meteorologist will, on reflection, see the reasonableness of my reservations, and will excuse them. I need not add, that in any case the self-imposed cost and labour to myself will be considerable.

CONDITIONS OF CO-OPERATION.

1. Every intending Contributor to send me in a *prepaid letter* as soon as convenient, and not later than the 1st of December next, the name of himself and of his station, its latitude, longitude *from Greenwich,* and its height above the sea level, *in English feet.* Any particulars (written in English, French, or German) about the aspect of the station would be acceptable. It is particularly requested that all this may be written very legibly in a large hand. It will be sure to prevent mistakes if the names are written twice over, once in *printed* characters.

2. Every Contributor to despatch to me on the 1st of January 1862, in a *prepaid letter,* his series of observations entered in the blank form, sent herewith.

It is incomparably more important that the observations should be trustworthy than that they should be numerous or continuous. Attention is particularly requested to the amount of cloud, as its symbol is a prominent and interesting feature in the Map, though Observers are frequently somewhat careless in recording it.

Observations even of cloud and wind alone are very acceptable, especially when made at lighthouses, it being always understood that they are accurately noted.

The observations will be printed precisely as they are furnished, according to the subjoined system; that is to say, no attempt will be made to correct apparent errors of record or reduction. If, however, the results from any station should appear on comparison with those of the stations adjacent to it to be frequently faulty, they will be altogether discarded.

I cannot promise to present complete sets of charts to the contributors of very imperfect series of observations.

In addition to the copies presented gratuitously to Contributors, others will be issued for sale, to lighten in some small degree the heavy expenses of printing.

FRANCIS GALTON,
42, Rutland Gate,
July 1861.   London.

* See Tables on other side of this page.

EXPLANATION OF THE SYMBOLS.

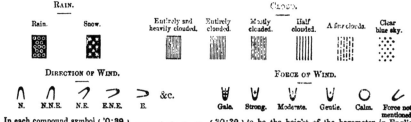

RAIN.

Rain.   Snow.

CLOUD.

Entirely and heavily clouded.   Entirely clouded.   Mostly clouded.   Half clouded.   A few clouds.   Clear blue sky.

DIRECTION OF WIND.

N.   N.N.E.   N.E.   E.N.E.   E.   &c.

FORCE OF WIND.

Gale.   Strong.   Moderate.   Gentle.   Calm.   Force not mentioned.

In each compound symbol the upper figures, as { '0·39 } { '9·72 } respectively signify { 30·39 } { 29·72 } to be the height of the barometer in English inches.

The lower figures 34—1 mean 34° is the height of Fahrenheit thermometer, and 1° is its excess over thermometer with moistened bulb.

Fig. 119 (i). Galton's appeal for daily reports by telegraph.

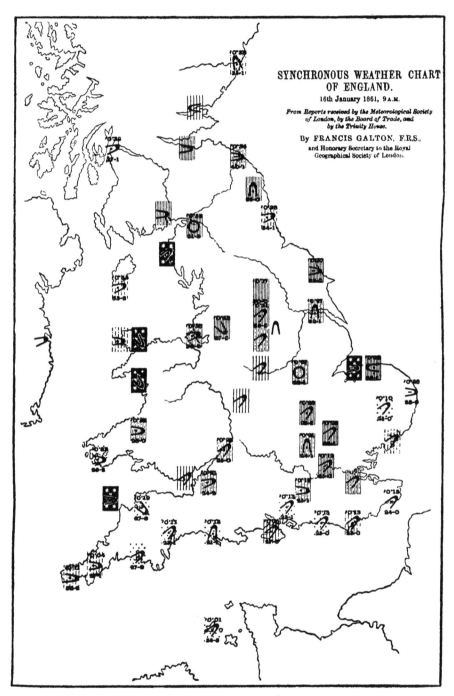

Fig. 119 (ii). Galton's weather-map.
The original is printed in two colours,

20–2

THE DAILY WEATHER MAP COMPANY

(LIMITED).

CAPITAL £4000, IN 400 SHARES OF £10 EACH.

THIS Company is formed for the purpose of raising capital to carry on the publication of the *Daily Weather Map*. All the preliminary expenses are already provided for. The patent is sealed; the instruments have been prepared, corrected by the Greenwich standard, and distributed to the various stations; the Map is engraved, the Symbols completed, and every arrangement made for commencing the publication. Only so much additional capital is required as would suffice to defray current expenses until the returns from Sale and Advertisements can be realised.

These returns are likely to be speedy and ample. The novelty of its design will secure for the *Weather Map* immediate celebrity. It will, beside, contain intelligence specially interesting to very large and numerous classes of the community—among others, 1, the Sailor; 2, the Merchant and Shipowner; 3, the Agriculturist; 4, all who travel by land or by water; 5, the Scientific public, who present a large, influential, and rapidly increasing body; 6, every one who observes, studies, or talks about the Weather—the subject which proverbially forms the first subject of conversation between Englishmen.

Being sold at so moderate a price, and addressing so large a section of the community, the sale of the *Weather Map* will in all probability be very extensive; while it will present, in proportion to the number of its readers, a first-rate medium for Advertisements. The revenue from both sources cannot fail to be very considerable. The expense, on the other hand, are comparatively moderate. Including the cost of Telegraphic Despatches, of printing (exclusive of double numbers), of Literary and Scientific Contribution, and of Office Expenses, the outlay will amount to between £8 and £9 per diem, or about £53 per week. If printed on paper of the best quality, the net proceeds of sale, after deducting the cost of paper, machining, and allowance to the trade, will leave a profit of fully £3 10s. per thousand. A circulation of 3000 copies would therefore pay all expenses, leaving the receipts from advertisements clear gain. If 5000 circulation were attained—and the number is not large for a paper sold at so moderate a price—the annual profits from sale alone would exceed £2000 per annum, or more than 50 per cent on the investment. But even a much more moderate circulation would constitute an extensive display of high-class advertisements, furnishing an ample source of profit to the Shareholder.

In order to limit individual responsibility, the proprietorship in the publication has been constituted into partnership under the Limited Liability Act. No shareholder can consequently incur any risk beyond the amount of the shares which he may think proper to take. Half the nominal amount of such shares will be required upon allotment, the residue being called up if wanted, and as wanted, to meet the current expenses of the publication.

Further particulars may be obtained from, and applications for Shares addressed to, "The Manager of THE WEATHER MAP COMPANY (Limited)," 110, Strand, W C

SHORTLY WILL BE PUBLISHED,
No. 1 OF THE

DAILY WEATHER MAP,

AND JOURNAL OF NEWS, LITERATURE, AND SCIENCE.

Published Daily before Noon. Price 2d.

THIS publication will comprise a Map of the British Isles, drawn upon a sufficient scale, showing, at a large number of stations scattered over the United Kingdom, the state of weather, wind, &c., at each locality by a simple and semi-pictorial combination of symbols. The observations for this purpose will be taken everywhere at 9 a.m., transmitted to London by telegraph, and transferred to the Map for general circulation by 11 o'clock. The Map will thus show at a glance the condition of the weather, the force and direction of the wind, &c., all over the country at an advanced hour of the morning of publication. It will by this means not only serve to indicate the present state of the weather, but will afford the best data wherefrom to anticipate coming changes, to trace the variations of the wind, and foretell the approach of rain, storms, or sunshine. The information so conveyed will be interesting to the general public—since everyone takes of and studies the weather—and peculiarly valuable to the Merchant, the Sailor, the Traveller, and the Agriculturist.

The instruments by which the daily observations are taken have been manufactured expressly for the *Weather Map* by Messrs. Negretti and Zambra, under the supervision of J. Glaisher, Esq., F.R.S., Secretary of the Meteorological Society, and by him certificated as corresponding with the Greenwich Observatory standard.

The value and interest of Meteorology, not only as a branch of scientific research, but in its practical utility to the Mariner, the Merchant, the Agriculturist, and the Traveller, have recently attained a wider recognition as the extension of the telegraphic system has made it possible to collect observations regarding the state of the wind, weather, &c., &c., over a larger area, and to transmit the results to any quarter with greater rapidity and certainty.

Much in this way has been accomplished under the supervision of the Board of Trade, and by the aid of an annual Parliamentary vote, with highly useful results. In the *Daily Weather Map* the process will be carried out far more fully and extensively, the observations collected at a large number of stations in different parts of the British Isles being telegraphed to London, and reproduced for publication within two hours, not only in tables which are intelligible only to a few readers, but also delineated upon a Map or Chart, so as to appeal to the eye, and present a sort of profile outline of the wind and weather as existing at 9 o'clock every morning in every part of the United Kingdom to which the means of telegraphic communication are now provided.

Besides the Map, the publication will contain two pages of letterpress, presenting a carefully prepared Epitome of the News of the Day, foreign and domestic, with Original Articles upon topics relating to Commerce, Agriculture, Literature, and Popular Science. In this department the columns of the *Weather Map* will furnish the means of intercommunication between all classes of the scientific public at home and abroad, and provide a medium through which all discoveries can obtain a wide and immediate publicity. It will thus supply what Science has long wanted, and the vast increase in the number of its professors and students so amply deserves, a DAILY ORGAN.

Double Numbers will be published from time to time, whenever there is an opportunity for describing some new discovery or interesting phenomenon, especially such as may be connected with Meteorological Science and its practical applications. These descriptions will be accompanied, when necessary, by Illustrative Designs.

Being addressed to a large and influential class of readers, and circulating extensively through places of commercial resort, the *Daily Weather Map*, &c., will present an eligible medium for advertisements.

Subscription:—4s. per Month; 12s. per Quarter; £2 12s. per Annum

Address, Editor of the *Weather Map*, No 110, Strand, W C

Fig. 120 (i). Prospectus of the Daily Weather Map Co.

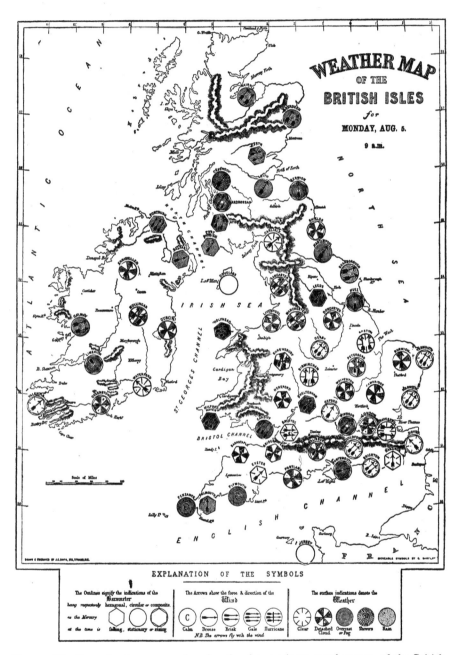

Fig. 120 (ii). Reproduction on a reduced scale of a specimen weather-map of the British Isles issued with the prospectus of the Daily Weather Map Co. about 1863. The chart is an example of the use of the movable types employed until the introduction of lithographic transfer-printing.

GENERAL CIRCULATION AND THE CYCLONE

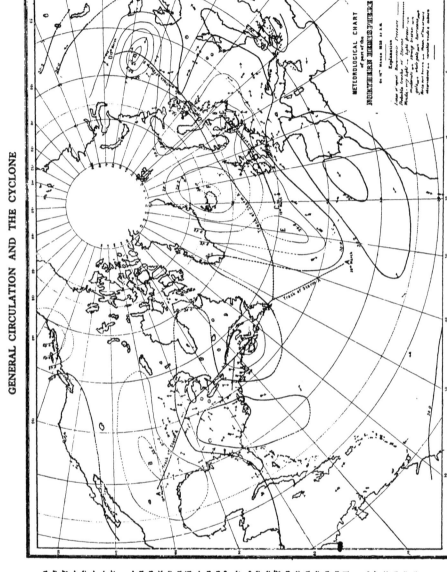

Fig. 121. Reproduction on a reduced scale of one of the charts prepared by Alexander Buchan to show the travel of depressions over the Atlantic. (From the *Journal of the Scottish Meteorological Society*, October. 1868.)

"Dès l'année 1865 M. Buchan publia dans les *Transactions de la Société royale d'Édimbourg* un Mémoire remarquable intitulé 'Examination of the storms of wind which occurred in Europe during October, November and December 1863.'....

"Trois années plus tard M. Buchan publia un nouveau Mémoire remarquable sur les tempêtes qui avaient sévi aux États Unis sur l'Atlantique Nord et en Europe du 13 au 22 mars, 1859. Il avait réuni un grand nombre d'observations faites en différents pays et sur la mer; avec ces données il a construit plusieurs cartes."

"Thus the wind blows neither in a circle round the centre, nor directly towards the centre but in a direction between the two. In effect, the direction of the wind in any place makes an angle of 60° to 80° with the line which would be drawn from the place to the centre of the depression. That is nothing else than Buys Ballot's law which can be formulated thus:

"Turn the back to the wind, stretch out the left arm slightly forward in a direction almost perpendicular to the direction of the wind, the hand will be more or less in the direction of the centre."

is hardly any reference to the publication in meteorological literature. Very few among working meteorologists have ever seen a copy. We can easily imagine that when it becomes necessary to put to a final test a really effective physical or dynamical theory of the sequence of surface-weather, Galton's reproductions, representing continuous changes as distinguished from the discontinuities of isochronous charts, may find their use; and meanwhile the neglect of them is a sure indication that, hitherto, the general theory which we hope for as the counterpart of the theories of the solar system has not yet been formulated.

After sixty years of experience of weather-maps and only very moderate success in forecasting in comparison with that achieved by astronomers in predicting the position of the heavenly bodies, it is hardly possible for us to realise the enthusiasm which the possibility of compiling daily charts excited in the sixties of the last century. Comparing Buys Ballot's chart with a modern weather-map we realise something of the effectiveness of the modern system in reducing the complication of the meteorological situation; it is not even yet in a form for effective mathematical computation, but Buchan's discussion showed maps based upon a system of isobars which has been found acceptable by the whole world.

FitzRoy's attempt at forecasting and warning for storms met with severe criticism in scientific circles in England; and after his death, when the British meteorological service was placed under the control of the Royal Society, the issues were suppressed in order that the information contained in the maps might be confronted with that obtained at observatories with self-recording instruments; in that manner, as we have already indicated, the physical processes of the atmospheric changes might be understood and brought to account. The storm-warnings were restored in response to popular appeal, but forecasts were discontinued for twelve years. In 1879 it was thought that the principles were understood, daily charts were produced and issued and a scheme of forecasting approved, the method of which was set out by R. H. Scott in *Weather charts and storm-warnings*[1], by Clement Ley in *Aids to the study and forecast of weather*[2], and by R. Abercromby in *Principles of forecasting by means of weather-charts*[3].

At the same time corresponding developments took place in other parts of Europe and in the United States; and the method of the weather-chart has now spread over the whole globe. The sea alone remains as yet hardly touched by the system, only the Atlantic has been regularly charted, and that long after the event. More recently Japan has issued daily maps of the North Pacific, but the advent of wireless telegraphy makes it possible to hope that in time even the sea may come into direct association with the daily chart.

The idea of the cyclone as formulated by Espy and Loomis received its

[1] *Weather charts and storm-warnings*, by R. H. Scott, M.A., F.R.S., Director of the Meteorological Office, 2nd edition, 1879, London, C. Kegan Paul and Co.
[2] M.O. publication No. 40, London, 1880.
[3] M.O. publication No. 60, London, 1885.

finishing touch from Galton when he brought into the scheme the areas of high pressure and gave them the name of anticyclone.

Most meteorologists are agreed that a circumscribed area of barometric depression is usually a locus of light ascending currents, and therefore of an indraught of surface-winds which create a retrograde whirl (in our hemisphere)....Conversely we ought to admit that a similar area of barometric elevation is usually a locus of dense descending currents, and therefore of a dispersion of a cold dry atmosphere, plunging from the higher regions upon the surface of the earth, which, flowing away radially on all sides, becomes at length imbued with a lateral motion due to the above-mentioned cause, though acting in a different manner and in opposite directions.

Proc. Roy. Soc., vol. XII, 1863, p. 385.

So far as can be judged by meteorological publications, these ideas formed the physical and dynamical background of meteorologists who were using maps for the purpose of forecasting. As a matter of fact the maps were used almost entirely empirically or geographically, and any conclusions derived from them were independent of physical theory. There was unfortunately no reaction between the physical processes and the practice of forecasting. The theory was, in fact, unsatisfactory in certain particulars which will be set out in Vol. II following a review of our present knowledge of the structure of the atmosphere.

We are indeed always apt to forget that what is represented on a map is two-dimensional and each successive layer may and indeed must have a plan which differs from that of the layer above and the one beneath. The scraps of information which we get about the upper layers are too meagre for us to discriminate between the various theories. That has been the characteristic defect of all the discussions about storms.

Within the last twenty-five years the phenomena of the cyclone have been subjected to a much more searching analysis beginning with the *Life-history of surface air-currents*[1] and continued with remarkable success by J. Bjerknes and others of the Norwegian school of meteorologists. The recent work of that school has indeed been so true to nature as to raise again some of the enthusiasm with which the first plotting of daily observations on maps was received. It remains to be seen whether the more detailed study of the structure of the atmosphere will redeem the promise of sixty years ago. [The conclusion which is foreshadowed at the completion of the four volumes is that in the study of weather-maps attention has been paid too exclusively to cyclones and anticyclones and too little to straight isobars. Also the entropy of air or potential temperature, as depending on its pressure as well as temperature, is a better independent variable for maps than reduced pressure. Accurate physical work is not really possible when temperature and wind at the surface are associated with isobars which may be found crossing Greenland, the Rocky Mountains, the Alps or even the Himalaya at an imaginary sea-level.]

[1] M.O. publication No. 174, London, 1906.

BIBLIOGRAPHY OF TROPICAL REVOLVING STORMS
General works

1828 *H. W. Dove.* Ueber barometrische Minima. Pogg. Ann. vol. XIII, p. 596.

1846 *Lt.-Col. W. Reid.* An attempt to develop the law of storms and the variable winds. Weale, London, pp. 572.

1847 *F. A. E. Keller.* Des ouragans, tornados, typhons et tempêtes. Imprimerie Royale, Paris, pp. 26.

1876 *H. Piddington.* The sailor's horn-book for the law of storms. Norgate, London, pp. 408.

1883 *H. Mohn.* Grundzüge der Meteorologie. Dietrich Reimer, Berlin, pp. 359.

1889 *W. Ferrel.* A popular treatise on the winds. New York, pp. 313.

1902 *S. M. Ballou.* The Eye of the storm. American Meteorological Journal, vol. IX, pp. 67–84 and 121–127.

1922 *Meteorological Office.* Hurricanes and tropical revolving storms. Mrs E. V. Newnham. Geophysical Memoir No. 19, London, 1922, pp. 213–333.

1922 *S. S. Visher.* Tropical Cyclones in Australia and the South Pacific and Indian Oceans. Washington Monthly Weather Review, vol. L, pp. 288–295.

—— Notes on Typhoons, with charts of normal and aberrant tracks. Washington Monthly Weather Review, vol. L, pp. 583–589.

General information may be found in the following works:

1894 *W. M. Davis.* Elementary Meteorology. Ginn, Boston, pp. 355.

1907 *H. H. Hildebrandsson and L. Teisserenc de Bort.* Les bases de la météorologie dynamique. Gauthier-Villars, Paris, 2 vols., pp. 228, 206.

1908–11 *J. v. Hann.* Handbuch der Klimatologie. Engelhorn, Stuttgart, 3 vols. pp. 394, 713, 426.

1919 A barometer manual for the use of seamen. Meteorological Office, London, pp. 106.

1920 *W. I. Milham.* Meteorology. Macmillan, New York.

North America and West Indies

1831 and 1846 *W. C. Redfield.* Remarks on the prevailing storms of the Atlantic coast and of the North American States. American Journal of the Sciences and Arts, New York, vol. XX, pp. 1–36.

—— On three several hurricanes of the Atlantic and their relations to the northers of Mexico and Central America. New Haven, pp. 118.

1862 *A. Poëy.* Table chronologique de quatre cents cyclones. Paul Dupont, Paris, pp. 49.

1898 *Rev. B. Viñes.* Cyclonic circulation and the translatory movements of West Indian hurricanes. United States Weather Bureau, Washington, pp. 34.

1900 *E. B. Garriott.* West Indian hurricanes. Bulletin H, United States Weather Bureau, pp. 69.

1913 *O. L. Fassig.* Hurricanes of the West Indies. Bulletin X, United States Weather Bureau, pp. 28 +plates 42.

1915 *E. H. Bowie and R. H. Weightman.* Types of storms of the United States. Monthly Weather Review, Supplement I, pp. 37 +plates 114.

1916 *E. Lopez.* Influencia de los ciclones tropicales sorbe el estado del tiempo en el valle de Mexico. Boletín mensual del Observatorio Meteorológico y Seismológico Central de Mexico, pp. 203–206.

1916 *Maxwell Hall.* Notes of Hurricanes, Earthquakes and other physical occurrences in Jamaica up to the commencement of the Weather Service, 1880, with brief notes in continuation to the end of 1915. Jamaica, Government Printing Office.

1917 *Maxwell Hall.* West Indies hurricanes as observed in Jamaica. Monthly Weather Review, Washington, vol. XLV, pp. 578–588.

1920 *I. M. Cline.* Relation of changes in storm-tides on the coast of the Gulf of Mexico to the centre and movement of hurricanes. Monthly Weather Review, vol. XLVIII, pp. 127–146.

1924 *C. L. Mitchell.* West Indian hurricanes and other tropical cyclones of the North Atlantic Ocean. Washington Monthly Weather Review, Supp. No. 24.

Additional information for these regions together with the tracks of hurricanes is given in the monthly charts published by the U.S. Weather Bureau, U.S. Hydrographic Office, the Meteorological Office, London, and the Deutsche Seewarte.

Monthly Meteorological Charts of the North Atlantic, Meteorological Office, London.
Monthly Pilot Charts of the North Atlantic, Hydrographic Office, Washington.
Monthly Weather Review, U.S. Weather Bureau, Washington.
Monatskarten für den Nordatlantischen Ocean, Deutsche Seewarte, Hamburg.
Habana. Boletín del Observatorio Nacional.

Indian Ocean

1845 *A. Thom.* An enquiry into the nature and course of storms in the Indian Ocean south of the equator. Smith, Elder and Co., London, pp. 351.

1869 and 1872 *C. Meldrum.* On the rotation of wind between oppositely directed currents of air in the Southern Indian Ocean. Proceedings of the Meteorological Society, London, vol. IV, pp. 322–324.

—— On a periodicity in the frequency of cyclones in the Indian Ocean south of the equator. Report of the British Association, pp. 56–57, or Nature, vol. VI, pp. 357–358.

1876 *H. Bridet.* Étude sur les ouragans de l'hémisphère austral. Challamel aîné, Paris, pp. 211.

1886 *Vice-amiral G. Cloué.* L'ouragan de juin, 1885, dans la Golfe d'Aden. Ann. Hydrog. Paris, vol. VIII, pp. 41–119.

1891 *R. H. Scott.* Cyclone tracks in the South Indian Ocean from information compiled by Dr Meldrum. Meteorological Office publication No. 90, London, plates 18.

1891 *W. L. Dallas.* Cyclone memoirs, Part IV. Meteorological Department of the Government of India, Calcutta, pp. 301–424 + plates 16.

1892 *W. Köppen.* Die Bahnen der Orkane im südlichen Indischen Ozean. Ann. der Hydrog. und Marit. Met., Hamburg, vol. XX, pp. 275–279.

1900–1901 *Sir J. Eliot.* Handbook of cyclonic storms in the Bay of Bengal. Meteorological Department of the Government of India, Calcutta, pp. 212 + plates 76.

1904 *T. F. Claxton.* The climate of the Pamplemousses in the Island of Mauritius. Report of the 8th International Geographical Congress, Washington, pp. 352–379.

1906 *Sir J. Eliot.* Climatological atlas of India. Meteorological Department of the Government of India, Calcutta.

1910 *A. Walter.* The sugar industry of Mauritius. Humphreys, London, pp. 228.

1924 *S. K. Banerji.* On the cyclones of the Indian Seas and their tracks. Proc. 10th Ind. Sci. Congr., Sec. Phys. Math., pp. 49–76.

1925 *C. W. B. Normand.* Storm-tracks in the Bay of Bengal. A series of monthly charts for the period 1891–1923. Indian Meteorological Department.

Periodical publications

Indian Monthly Weather Review. Meteorological Department of the Government of India, Calcutta.
Results of Magnetical and Meteorological Observations at Mauritius. Government Printing Establishment, Mauritius. (Annual.)
Annual Report of the Port Alfred Observatory, Mauritius.

Pacific Ocean

1892 *A. Schuck*. Segelhandbuch für den Indischen Ozean. Deutsche Seewarte, Hamburg.

1893 *E. Knipping*. Die Tropischen Orkane der Südsee zwischen Australien und den Paumotuinseln. Aus dem Archiv der Deutschen Seewarte, Hamburg.

1894 *Rev. F. S. Chevalier*. The typhoons of the year, 1893. Meteorological Society, Shanghai, pp. 97.

1904 *W. Doberck*. The law of storms in the Eastern Seas. Noronha, Hongkong, pp. 38.

1904 *Rev. J. Algué*. Cyclones of the Far East. Manila, pp. 283.

1911 *C. H. Knowles*. Hurricanes in Fiji. Department of Agriculture, Fiji, Bulletin No. 2, Suva, 1911.

1914 *Government-General of Formosa*. The climate, typhoons and earthquakes of the island of Formosa. Taihoku.

1920 *Rev. J. Coronas*. The climate and weather of the Philippines, 1903–1918. Manila, pp. 195.

1920 *Rev. L. Froc*. Atlas of the tracks of 620 typhoons, 1893–1918. Zi-ka-Wei Observatory, Shanghai.

1920 *Griffith Taylor*. Australian Meteorology. University Press, Oxford, pp. 305.

1923 *S. S. Visher and D. Hodge*. Australian hurricanes and related storms. Official Year Book of the Commonwealth of Australia No. 16, pp. 80–84.

Periodical publications

Bulletins of the Philippine Weather Bureau, Government of the Philippines, Manila. (Monthly.)

Admiralty Sailing Directions: Pacific Islands.

Annual Reports of Agriculture. Legislative Council, Fiji.

Australian Monthly Weather Report. Commonwealth Bureau of Meteorology, Melbourne.

Monthly Meteorological Charts of the East Indian Seas. Meteorological Office, London.

West Africa

1875 *A. Borius*. Recherches sur le climat de Sénégal. Gauthier-Villars, Paris.

1890 *A. v. Danckelmann*. Beiträge zur Kenntniss des Klimas des deutschen Togolandes. Mitteilungen aus den deutschen Schutzgebieten, Berlin, vol. III, pp. 1–44.

1908–1919 *H. Hubert*. Sur le mécanisme des pluies et des orages au Soudan. Comptes Rendus, Paris, vol. CLII, 1911, pp. 1881–1884.

—— Mission scientifique au Dahomey. Emile Larose, Paris, 1908, pp. 568.

—— Mission scientifique au Soudan. Emile Larose, Paris, 1916, pp. 319.

—— Sur la prévision des grains orageux en Afrique occidentale. Comptes Rendus, Paris, vol. CLXVIII, 1919, pp. 567–570.

1913 *Capt. v. Schwartz*. La formation des orages dans les régions montagneuses de l'Afrique occidentale française. La Géographie, vol. XXVII, pp. 208–210.

West Coast of Tropical North America

1922 *S. S. Visher*. Tropical cyclones in the North-East Pacific between Hawaii and Mexico. Washington Mon. Wea. Rev., vol. L, pp. 295–297.

1923 *W. E. Hurd*. Pilot Chart of the North Pacific Ocean (May). U.S. Hydrographic Office.

CHAPTER XV

METEOROLOGICAL THEORY IN HISTORY

Newton	1642–1727	Helmholtz	1821–1894
Euler	1707–1783	James Thomson	1822–1892
Lagrange	1736–1813	Kelvin	1824–1907
Laplace	1749–1827	C. A. Bjerknes	1825–1903
Coriolis	1792–1843	Guldberg	1836–1902
J. D. Forbes	1809–1868	Rayleigh	1842–1919
Colding	1815–1888	Margules	1856–1920
Ferrel	1817–1891	Hertz	1857–1894

THROUGHOUT the whole of the history which we have briefly reviewed in this volume meteorological theory has been invariably hampered by want of facts. That is quite evident in the case of Aristotle's work which, as we have seen, was accepted as the standard text-book of meteorology for nearly two thousand years. It is also true for the period following the invention of the barometer during which thermodynamics and the motion of fluids were subjects of special study; and it is true to-day in spite of the fact that the volumes containing the facts about the atmosphere are so numerous as to be quite overwhelming. There is still room for reflexion upon a saying attributed to Raphael Aben Ezra in Charles Kingsley's *Hypatia*, "No wonder that his theory fits the universe when he has first clipped the universe to fit his theory."

By meteorological theory we understand the treatment of the phenomena of weather after the manner of a fully organised science. Let us take as the model of such a science that of geometry as set out in the Greek form familiar to us as Euclid's elements. It has its definitions of the quantities which are to be treated as measurable, its axioms, principles or laws, which are based upon experience or, in other words, are derived by the process known as induction from observation and experiment. They are to be assumed in the course of the argument and are the major premiss of the syllogism. They can, of course, be denied in a disputation; but the denial is a denial of alleged facts not the exposure of a mistake in logic. There are besides postulates which prescribe the experimental or manipulative processes which will be allowed in the discussion, and finally there are the theorems, statements which can be shown to be implicitly contained within the axioms and postulates, and which can be confronted with the actual facts of nature in the process of observation.

Next to geometry the science of the solar system is the most perfectly organised. Its definitions are those of mass, velocity and acceleration. The axioms are the laws of motion "axiomata sive leges motus," and the law of gravitation; its postulates are the method of fluxions, and its theorems are the predictions of the positions of the heavenly bodies with a precision that is the admiration and envy of all the other sciences.

And the general science of dynamics of particles and rigid bodies, as well as of fluid bodies, is also very far advanced. It can use the same laws and axioms as the science of the solar system, and the method of fluxions with its latest developments can equally be invoked; but the theorems are less exhaustive of the observed facts than those of astronomical science.

Regarded from the point of view of theory, meteorological science is usually taken to be a special case of the science of dynamics, its definitions are those of dynamics together with the various meteorological elements; its axioms are again the laws of motion and the behaviour of atmospheric air under certain specified conditions as determined in the laboratory, and its postulates are likewise the method of fluxions with its extensions; its theorems are in the end the computations of the state of the atmosphere of to-day from that of yesterday. L. F. Richardson's work on *Weather Prediction by Numerical Process* is the most explicit example of envisaging the science of meteorology in that way. General theories have been propounded by Maury, Ferrel and James Thomson, but they are intended to be general also in their application and are not applicable in the detail which Mr Richardson attempts and which the science of the solar system actually achieves.

There are a vast number of contributions to meteorological theory which are fragments based on limited assumptions and limited conditions selected by the author as being sufficiently near to nature to make it worth while to elaborate the reasoning; but there is no theory of the general circulation of the atmosphere and its changes which can be treated in the same way as a question of eclipses, for example, in the science of the solar system. The contributions are indeed, as has been said already on more than one occasion, marginal notes to a text that does not exist.

One of the great services which Hann has rendered to the science of meteorology is the publication of two celebrated *Lehrbücher*, one on Climatology and the other on Meteorology, in which the contributions of many authors are co-ordinated into a coherent survey of the whole subject. Many of the recent papers on climatology and meteorology may indeed be regarded as marginal notes to one or other of Hann's books. But even in those monumental works discontinuities are still apparent in the classification. Geographical areas or meteorological elements are the dividing lines rather than the several constituent parts of the general circulation of the atmosphere. A similar comment may be made on the *Physics of the Air* by W. J. Humphreys—it is evident, on reading the book, that the author has selected for discussion certain features of atmospheric processes rather than endeavoured to meet the wishes of the reader who wants to understand, as Maxwell said, "the real go of things." In many meteorological works of the nineteenth century it is the life-history of the author that one realises rather than that of the earth's atmosphere. Nevertheless, Hann's successor F. M. Exner has added to the services which the Austrian Meteorological Institute has rendered to the science by his book on *Dynamische Meteorologie*, which in its second edition forms a coherent and almost continuous representation of the dynamics of the atmosphere.

The majority of the contributions to meteorological theory have been very ineffective. There are two anthologies of papers which may be cited as examples of the work which has been done. The first is in three volumes published by the Smithsonian Institution: Cleveland Abbé's translations of papers on dynamical meteorology[1]. The second is that compiled by Marcel Brillouin and entitled *Mémoires originaux sur la circulation générale de l'atmosphère*[2]. We shall add a brief one of our own containing indeed Brillouin's own paper on "Vents contigus et nuages," a development of Helmholtz' analysis, a paper by Coriolis on the relation of wind to pressure, a paper by James Thomson, and papers or books published in the United States by W. Ferrel, F. H. Bigelow, and F. J. B. Cordeiro. We shall quote the titles, arranged in chronological order, and indicate the anthologies from which they have been selected by the initials C.A., M.B., and S. respectively.

1686 *Halley*. An historical account of the trade-winds and monsoons by E. Halley. Phil. Trans. vol. xxvi, 1686–87, pp. 153–168. (M.B.)

1735 *Hadley*. Concerning the cause of the general trade-winds by Geo. Hadley. Phil. Trans. London, 1738, vol. xxix, 1735–36, pp. 58–62. (C.A. and M.B.)

1835 *Coriolis, G.* Sur la manière d'établir les différens principes de mécanique pour des systèmes de corps en les considérant comme des assemblages de molécules. Paris, Journ. École Polytechn. vol. xv (24ᵉ cahier), pp. 93–125. (S.)

1835 *Coriolis, G.* Sur les équations du mouvement relatif des systèmes de corps. Paris, Journ. École Polytechn. vol. xv, pp. 142–154. (S.)

1837 *Poisson*. On the motion of projectiles in the air, taking into consideration the rotation of the earth. (C.A.)

1843 *Tracy*. On the rotary action of storms. (C.A.)

1855 *Maury*. The physical geography of the sea and its meteorology. (M.B.)

1856–61 *Ferrel*. Essai sur les vents et les courants de l'océan. (M.B.)

—— Les mouvements des fluides et des solides par rapport à la surface de la terre. (M.B.)

—— Reply by Mr Ferrel to the Criticisms of Mr Hann, 1875. (C.A.)

1858 *Helmholtz*. On the integrals of the hydrodynamic equations that represent vortex-motions. (C.A.)

1859–62 *Braschmann and Erman*. The influence of the diurnal rotation of the earth on constrained horizontal motions, either uniform or variable. (C.A.)

1868 *Helmholtz*. On discontinuous motions in liquids. (C.A.)

1868 *Erman*. On the steady motions or the average condition of the earth's atmosphere. (C.A.)

1869 *Kirchhoff*. The theory of free liquid jets. (C.A.)

1871 *Colding*. Some remarks concerning the nature of currents of air. (C.A.)

—— On the whirlwind at St Thomas on the 21st of August, 1871. (Extract by Hann from Colding's memoir.) (C.A.)

1872 *Peslin*. Relation between barometric variations and the general atmospheric currents. (C.A.)

1873 *Helmholtz*. On a theorem relative to movements that are geometrically similar, together with an application to the problem of steering balloons. (C.A.)

1874 *Hagen*. The measurement of the resistances experienced by plane plates when they are moved through the air in a direction normal to their planes. (C.A.)

---

[1] 'The mechanics of the earth's atmosphere. A collection of translations.' 1st collection, *Smithsonian Report*, 1877; 2nd collection, *Smithsonian Misc. Coll.* vol. xxxiv, 1893; 3rd collection, *Smithsonian Misc. Coll.* vol. li, 1910. Washington, D.C.

[2] *Mémoires originaux sur la circulation générale de l'atmosphère*, Paris, 1900.

1874–75 *Hann*. On the diminution of aqueous vapour with increasing altitude in the atmosphere. (C.A.)
—— On the influence of rainfall upon the barometer. (C.A.)
—— Atmospheric pressure and rain (Additional Note). (C.A.)
—— The laws of the variation of temperature in ascending currents of air. (C.A.)
—— On the relation between the difference of pressure and the velocity of the wind according to the theories of Ferrel and Colding. (C.A.)
1875 *Sohncke*. The law of the variation of temperature in ascending moist currents of air. (C.A.)
1875 *Reye*. On rainfall and barometric minima. Additional Note by Reye, (C.A.)
1876–83 *Guldberg and Mohn*. Studies on the movements of the atmosphere. (C.A.)
1877 *Oberbeck*. On discontinuous motions in liquids. (C.A.)
1881 *Kerber*. The limit of the atmosphere of the earth. (C.A.)
1881 *Sprung*. On the paths of particles moving freely on the rotating surface of the earth and their significance in meteorology. (C.A.)
1882 *Oberbeck*. The movements of the atmosphere on the earth's surface. (C.A.)
—— On the Guldberg-Mohn theory of horizontal atmospheric currents. (C.A.)
1884 *Hertz*. A graphic method of determining the adiabatic changes in the condition of moist air. (C.A.)
1884–1906 *von Bezold*. On the thermodynamics of the atmosphere. Five communications. (C.A.)
—— Theoretical considerations relative to the results of the scientific balloon ascensions of the German Association at Berlin for the promotion of aeronautics. (C.A.)
—— On the reduction of the humidity data obtained in balloon ascensions. (C.A.)
—— On the changes of temperature in ascending and descending currents of air. (C.A.)
—— On the theory of cyclones. (C.A.)
—— On the representation of the distribution of atmospheric pressure by surfaces of equal pressure and by isobars. (C.A.)
—— The interchange of heat at the surface of the earth and in the atmosphere. (C.A.)
—— On climatological averages for complete small circles of latitude. (C.A.)
1886 *Siemens*. Conservation de l'énergie dans l'atmosphère de la terre. (M.B.)
1887 *Möller*. La circulation atmosphérique entre les hautes et les basses pressions. La répartition des pressions et la direction moyenne du vent. (M.B.)
1888 *Oberbeck*. On the phenomena of motion in the atmosphere. (Two papers.) (C.A. and M.B.)
1888–89 *Helmholtz*. On atmospheric motions. (Two papers.) (C.A. and M.B.)
1890 *Helmholtz*. The energy of the billows and the wind. (C.A.)
1890 *Rayleigh*. On the vibrations of an atmosphere. (C.A.)
1890 *Margules*. On the vibrations of an atmosphere periodically heated. (C.A.)
1890 *Ferrel*. Laplace's solution of the tidal equations. (C.A.)
1892 *Thomson, James*. On the grand currents of atmospheric circulation. Bakerian Lecture 1892. Phil. Trans. 183, A, pp. 653–684. (S.)
1893 *Pockels*. The theory of the movement of the air in stationary anticyclones with concentric circular isobars. (C.A.)
1898 *Brillouin*. Vents contigus et nuages. Ann. bur. cent. météor. 1896, part 1, pp. B. 45–150, Paris. (S.)
1898 *Bigelow, F. H.* Report on the international cloud observations, May 1, 1896 to July 1, 1897. U.S. Weather Bureau. (S.)
1900 *Neuhoff*. Adiabatic changes of condition of moist air and their determination by numerical and graphical methods. (C.A.)
1901 *Pockels*. The theory of the formation of precipitation on mountain slopes. (C.A.)
1901 *Margules*. The mechanical equivalent of any given distribution of atmospheric pressure and the maintenance of a given difference in pressure. (C.A.)

1904 *Gorodensky.* Researches relative to the influence of the diurnal rotation of the earth on atmospheric disturbances. (C.A.)

1904 *Margules.* On the energy of storms. (C.A.)

1908 *Gold.* The relation between wind-velocity at 1000 metres altitude and the surface pressure distribution. (C.A.)

1908 *Bauer.* The relation between "potential temperature" and "entropy." (C.A.)

1910 *Cordeiro, F. J. B.* The atmosphere, its characteristics and dynamics. New York. (S.)

1910 *Rayleigh.* Note as to the Application of the Principle of Dynamical Similarity. Report of the Advisory Committee for Aeronautics 1909–1910, p. 38, London, H.M.S.O. (S.)

1911 *Rayleigh.* The Principle of Dynamical Similarity. Report of the Advisory Committee for Aeronautics, 1910–11, p. 26, London, H.M.S.O. (S.)

1911–12 *Bjerknes, V.* and others. Dynamic meteorology and hydrography. (S.)

1921 *Bjerknes, V.* On the dynamics of the circular vortex with applications to the atmosphere and atmospheric vortex and wave-motions. Geofysiske Publikationer, vol. II, No. 4, Kristiania. (S.)

Those who are familiar with the practical applications of meteorology in forecasting, which is the ultimate touchstone of theory, will agree that many of these contributions, including those which express the most brilliant mathematical reasoning, are not in fact used, not because there is any unwillingness to use them, not even because they involve difficult mathematics, that could be got over, but because they are not adapted for practical use and cannot be so adapted even by accomplished mathematicians.

We hope to consider more fully the work of these contributions in Volume III. We think it will be found, on examining them, that the failure of practical application is a necessary consequence of assuming as the starting-point of the theory conditions which are not apparently the conditions from which the practical student is compelled to start. The conditions are assumed in order to make the theory of fluxions, or whatever other postulate may be adopted, workable. If we take, for example, Halley's theory of the trade-winds we find he assumes a wind to start from the North because there is convexion at the equator. Tracing back the trade-winds we ought to find them proceeding from a calm[1], but the most definite form of trade-wind comes from a complicated circulation that is as far from calm as the trade-winds themselves. The region of the equator to which they move is regarded as an exceptionally hot region; but the intertropical regions of the oceans as we know them are not regions of marked temperature-difference but rather of marked uniformity; if we wanted steep temperature-gradients we should not look for them at the equator.

Similarly Hadley introduced the conception of the rotation of the earth deviating the trade-wind from a North to a North East wind. Having achieved its object of making a North Easter its influence seems to have been supposed

---

[1] We should like here to point out that as a corollary of any theoretical representation of the atmospheric processes we are justified in considering that a process for producing to-day's conditions from yesterday's must be equally effective in deducing yesterday's conditions from to-day's, and that tracing backwards is quite as good a test of the effectiveness of a theory as tracing forward, and in many cases more than equally effective.

to cease; but no conception of dynamical processes can be satisfied by supposing that the rotation of the earth has no rotative influence on a North Easter.

So with most of the other contributions, the theory starts from assumed premisses which are not really represented on the maps when we have them. The proposition of Coriolis's expressing the relation of pressure to velocity in the atmosphere under balanced forces, which was rediscovered by Ferrel, has become a recognised part of meteorological science. A proposition of Cordeiro's, to the effect that a vortex column with vertical axis moving Eastward becomes directed to the North East by the influence of the rotation of the earth, might become so if meteorologists were agreed that a cyclonic depression is a quasi-permanent vortex with approximately vertical axis. Otherwise the results of many of these memoirs hardly come within the region of practical meteorology.

It is customary to regard Aristotle's work as being merely speculative; "hopelessly speculative" is a description which I have seen. But let us remember that Aristotle or Theophrastus had recognised the reversal of the North Easter, streaks of cloud from the South as a sign of coming storm, and had developed a classification of the winds which is more than a little suggestive of the polar front. His discussion of the formation of dew, rain, hoar-frost, snow or rainbows would do no discredit to a philosopher with fifteen hundred years' more experience at his command than had Aristotle, who was in fact very painfully limited as to his facts through having no instruments with which to observe. Making due allowance for their greater advantages in that respect all the philosophers whose papers we have referred to suffer from a similar disadvantage, their work is speculative because their facts, though much more abundant than Aristotle's, are not sufficient to give them a satisfactory basis of axioms, laws or principles from which to start. Like Aristotle's their reasoning is rigorous within the limits of its postulates, but the material to which they apply their reasoning is not the atmosphere that practical meteorologists have to deal with but a substitute, a creation of their own, made by omitting this or adding that. We must not be understood to be depreciating in any way the ability or distinction of the authors of these contributions. We may echo in this connexion a remark of Sir Gilbert Walker's in summarising an unsuccessful attempt at forecasting rainfall in Britain for a long period in advance, "in the present state of our knowledge an Isaac Newton would probably do no better."

It will be remembered that the science of the solar system was evolved by Newton from the laws of Kepler, and that there is a world of difference between the system of Copernicus with his circles, and the system of Kepler with his ellipses. A new world was created by the illumination of an inductive genius.

Looking back on the history of meteorological theory we may recognise three different schools of theorists. The first and most popular, which may be called the physical school, includes those who, following closely Aristotle's mode of thought, have propounded theories based largely upon the convexion of warm air without any quantitative expression of the meteorological elements involved—the principal names are Halley, Hadley, Maury, Ferrel, J. Thomson, Keller, Thom, Espy, Eliot, Dove; the second, or dynamical school, represented by Ferrel, Guldberg and Mohn, Helmholtz, Brillouin, Margules, Siemens, Oberbeck and Rayleigh, have based their conclusions upon the application of the general equations of motion to an idealised atmosphere; the third, or inductive school, represented by Buys Ballot, Buchan, De Tastes and Duclaux, Hildebrandsson, Teisserenc de Bort, Hann and Köppen, have endeavoured to form a working picture of the actual structure and general circulation of the atmosphere before formulating a thermal or dynamical theory. In sympathy with this school of thought we propose to devote Vol. II of this Manual with the title of "Comparative Meteorology" to setting out in as much detail as the limits of space permit the present state of our knowledge of the structure of the atmosphere, its general circulation, the sequence of its changes, and its local disturbances.

In this connexion we recall a personal incident of fifty years ago. Being interested for certain personal reasons in questions in experimental physics that were to be set in an examination in Natural Sciences in Cambridge in 1876, we acquired a copy of the papers set in the year 1874 when Clerk Maxwell happened to be the examiner. The first question was as follows: "Show how by observations of the motion of a body the resultant force acting upon it can be determined." The suggestion seemed at the time to be a very perverse one because all those who are familiar with the method of fluxions are aware that the universal practice is to assign the forces and deduce therefrom the motion, and to proceed in the opposite manner seems at first sight intellectually a childish mode of procedure. But Maxwell gave the answer himself in his little book on *Matter and Motion*; the full simplicity of his exposition of the science of the solar system on the principle of eliciting the forces from observation of the motion is only made perfectly clear by some notes which have recently been added to connect up the theory with Einstein's modification.

It seems more than possible that the true theory of meteorology will never be evolved by the iteration of marginal notes until they fill the page, and that the true course of progress is to accept Maxwell's hint to develop the representation of the motion to such a degree of perfection that the forces will be deduced from it, instead of supposing that we can specify the forces and that nothing but the method of fluxions is necessary to deduce the motion.

ADDENDA, 1932, AND NOTES

METEOROLOGY IN LONDON A CENTURY AGO (p. 117)

A Meteorological Society was instituted in London in 1823 and lasted till 1841. In 1839 it published a volume of transactions. Lord Robert Grosvenor, M.P., was president, Captain Sir John Ross, R.N., K.C.S., a vice-president, Robert Carr Woods, M.G.S.P., registrar, and W. H. White, M.B.S., a secretary; Sir John Herschel was a member of council.

A list of the principal Stations of the Society, seventy-one in number, includes two in London, twenty others in England (Dr Dalton, F.R.S., in charge at Manchester, John Ruskin at Oxford), two in Wales, two in Scotland, three in Ireland, one in the Channel Islands, ten in British Dominions overseas, namely four in Australia, one each in Newfoundland, Canada, the Cape of Good Hope (the Astronomer Royal with C. Piazzi Smyth, assistant), Gibraltar, Jamaica and Mauritius, three each in France (MM. Arago and Biot) and in Belgium (Quetelet), two each in Germany (Dove at Berlin and Kämtz at Halle), Portugal, Holland and Denmark (Oersted at Copenhagen), one each in China, Switzerland, Italy, Spain and Russia (Küpffer at St Petersburgh), one in Brazil and eleven in the United States (W. C. Redfield at New York, Loomis at Ohio, Silliman at New Haven, Conn.).

An appendix contains hourly observations for 36 hours of 1838, for the summer solstice at Alost and in lat. 53° 14′, long. 9° W, and for the vernal equinox at Copenhagen and Dundee, at London for both epochs. A second appendix gives a retrospect for 1837 at eight stations in Great Britain which includes a series of maxima, minima and range of pressure and temperature at London in the several months of July to December for periods of about 50 years and corresponding tables for shorter periods elsewhere.

The volume of transactions contains papers on observations in many parts of the world, including one by W. C. Redfield on the meteorology of New York and one by Lieut. George Grey on Tenerife, with notices of destructive gales and other phenomena.

The place of honour in the volume, "first wicket down," following some introductory remarks by a vice-president and some rather perfunctory directions for meteorological observations on land and sea by the Registrar, is given to a paper entitled "Remarks on the Present State of Meteorological Science" by John Ruskin, Esq. of Christ Church College, Oxford (sic), from which the following extracts are taken as indicating the ideas which the Society had in mind. Ruskin was then twenty years old:

"The comparison and estimation of the relative advantages of separate departments of science, is a task which is always partially executed, because it is never entered upon with an unbiassed mind; for, since it is only the accurate knowledge of a science which can enable us to perceive its beauty, or estimate its utility, the branch of knowledge with which we are most familiar, will always appear the most important. The endeavour, therefore, to judge of the relative *beauty* or *interest* of the sciences, is utterly hopeless. Let the astronomer boast of the magnificence of his speculations,—the mathematician of the immutability of his facts,—the chemist of the infinity of his combinations, and we will admit that they all have equal ground for their enthusiasm....

"We do not, therefore, advance any proud and unjustifiable claims to the superiority of that branch of science for the furtherance of which this Society has been formed, over all others; but we zealously come forward to deprecate the apathy with which it has long been regarded, to dissipate the prejudices which that apathy alone could have engendered, and to vindicate its claims to an honourable and equal position among the proud thrones of its sister sciences. We do not bring meteorology forward as a pursuit adapted for the occupation of tedious leisure, or the amusement of a careless hour. Such qualifications are no inducements to its pursuit by men of science and learning, and to these alone do we now address ourselves. Neither do we advance it on the ground of its interest, or beauty, though it is a science possessing both in no ordinary degree....

"For its interest, it is universal; unabated in every place, and in all time. He, whose kingdom is the heaven, can never meet with an uninteresting space,—can never exhaust the phenomena of an hour; he is in a realm of perpetual change,—of eternal motion,—of infinite mystery. Light and darkness, and cold and heat, are to him as friends of familiar countenance, but of infinite variety of conversation; and while the geologist yearns for the mountain, the botanist for the field, and the mathematician for the study, the meteorologist, like a spirit of a higher order than any, rejoices in the kingdoms of the air....

"There is one point, it must now be observed, in which the science of meteorology differs from all others. A Galileo, or a Newton, by the unassisted workings of his solitary mind, may discover the secrets of the heavens, and form a new system of astronomy. A Davy in his lonely meditations on the crags of Cornwall, or, in his solitary laboratory, might discover the most sublime mysteries of nature, and trace out the most intricate combinations of her elements. But the meteorologist is impotent if alone; his observations are useless; for they are made upon a point, while the speculations to be derived from them must be on space. It is of no avail that he changes his position, ignorant of what is passing behind him and before; he desires to estimate the movements of space, and can only observe the dancing of atoms; he would calculate the currents of the atmosphere of the world, while he only knows the direction of a breeze. It is perhaps for this reason that the cause of meteorology has hitherto been so slightly supported; no progress can be made by the enthusiasm of an individual; no effect can be produced by the most gigantic efforts of a solitary intellect, and the co-operation demanded was difficult to obtain, because it was necessary that the individuals should think, observe, and act simultaneously, though separated from each other, by distances, on the greatness of which depended the utility of the observations.

"The meteorological Society, therefore, has been formed, not for a city, nor for a kingdom, but for the world. It wishes to be the central point, the moving power, of a vast machine, and it feels that unless it can be this, it must be powerless; if it cannot do all it can do nothing. It desires to have at its command, at stated periods, perfect systems of methodical, and simultaneous observations; it wishes its influence and its power to be omnipresent over the globe, so that it may be able to know, at any given instant, the state of the atmosphere at every point on its surface. Let it not be supposed that this is a chimerical imagination—the vain dream of a few philosophical enthusiasts. It is co-operation which we now come forward to request, in full confidence that if our efforts are met with a zeal worthy of the cause, our associates will be astonished, *individually*, by the result of their labours in a body. Let none be discouraged, because they are alone, or far distant from their associates. What was formerly weakness, will now have become strength."

     *   *   *   *   *   *   *   *

The work at which the Meteorological Society of London aimed was mainly climatological. The number of co-operating "stations" in the United States is suggestive of the concurrent development of meteorological study in that country with the encouragement of the Smithsonian Institution. At the time, Redfield and Loomis were actively engaged upon the life-history of storms leading ultimately to the weather-map of which early details are indicated in a memoir by E. R. Miller in vol. LVIII of the *Monthly Weather Review* and other papers in the same journal.

In 1850 another Meteorological Society, of which James Glaisher, Superintendent of the meteorology of the Royal Observatory at Greenwich, was the prime mover, was founded in London with climatology as its main object. It is now the Royal Meteorological Society with His Majesty's patronage. In 1856 the Scottish Meteorological Society was founded with Alexander Buchan as its chief officer, and again with the climatology of Scotland as its object. These were supported by voluntary effort, with a small subvention from the State grant after 1876 in order to enable the country to take its part in the international organisation. In 1858 the British Rainfall Organization was established by J. G. Symons on a self-supporting basis, partly as a scientific enterprise expressing voluntary effort and partly as a valuable factor in the solution of the important problem of water-supply for the centres of population in the British Isles.

It is only within recent years that the British State authorities have recognised any general responsibility for climatological work. The original Meteorological Department of the Board of Trade was set up in 1854 under Admiral Robert FitzRoy, the captain of H.M.S. *Beagle* in which Charles Darwin made his famous voyage. Its purpose was to develop the study of the meteorology of the sea in the interest of H.M. Navy and the Mercantile Marine. And when the study of weather by synoptic charts based on telegraphic reports was recognised, with the encouragement of the British Association, the Department included that side of meteorology in its duties. After FitzRoy's death in 1865 the control of the enterprise was entrusted by Government to a committee of Fellows appointed by the Royal Society as described on p. 305, while climatology and rainfall still remained matters of private enterprise. It was only in 1919 that the whole duty of superintending the climatological work was included with that for marine meteorology and forecasting within the scope of a Government office. The Scottish Society was incorporated with the Royal Meteorological Society and the control of the British Rainfall Organization passed directly to the Meteorological Office. It remains for the Officers and Council of the Society to find means for maintaining and extending the scientific interest in the subject apart from any corporate responsibility for the compilation and publication of data.

     *   *   *   *   *   *   *   *

The story of the great effort of the Meteorological Committee of 1867 to 1877 under the guidance of Sir F. Galton to comprehend the sequences of events in weather is not without its importance for the science. We quote from *Nature*, December 5, 1931, p. 925:

"During twelve years a strenuous effort was made chiefly under Galton's guidance to use the records of the instruments at seven observatories like Kew to interpret the sequence of weather shown on the daily maps. Meanwhile forecasts were suspended. For twelve years every curve drawn by the instruments was reproduced with the associated maps for public information. The systems of closed isobars, now known as the cyclonic depression and the anticyclone, were the most striking features of the sea-level maps. The track and behaviour of every one of them during the twelve years was 'set in a notebook learned and conned by rote'. They were enthroned as nature's proctors for the guidance of weather, and under their guidance forecasting was restored in 1879 by direction of those who had been critical in 1865.

"The method of forecasting on the basis of cyclones and anticyclones was set out effectively by the Hon. Ralph Abercromby. The problem of the sequence of weather seemed to be solved. It was supposed that the progress of storms across the Atlantic could certainly be anticipated. Further investigation of the upper air by shell-bursts, by balloons, by the study of clouds under a new directing council, would clear up all the outstanding difficulties.

"Somehow the effort failed. The maps used were bounded on the west by the Irish coast. Cyclones and anticyclones suggested rules of behaviour but did not keep to them, and a magnificent effort was made to find out the how and why of the irregularities by maps of the Atlantic and adjacent continents for the days of the year of polar exploration 1882–3; but no one has ever been able to make out a rational account of the behaviour of cyclonic depressions. The Norwegian school has traced with marked success the influence of the movement of the different kinds of air of which they are composed, but the origin of the currents of which they are composed is not yet understood. Disappointed, the council turned its attention to harmonic analysis of hourly values as a method of discovering the latent causes of weather, and Galton transferred his interest to eugenics."

The conclusion which asks for attention is that the cyclonic depression, as a mild example of the tropical revolving storm, and its correlative the anticyclone are not the controlling agencies of the atmosphere; that there is something behind them or above them which should be sought for—something underlying, or more strictly speaking controlling from above, the suggestive picture of atmospheric integration.

"To summarise, our survey leads to the conclusion that meteorology has been mistaken in regarding the cyclonic depression and the anticyclone of a hypothetical sea-level map as the principal agents in control of the energy of weather, and asks for the consideration of the currents in the free air indicated by straight isobars as expressing the real estate of energy in the atmosphere, while the differences in the distribution of entropy and water-vapour associated therewith represent its potential estate."

The physical and dynamical volumes III and IV lead to the conclusion that the distribution of pressure at the so-called sea-level should be accounted for by a synthesis of the structure of the atmosphere which will give the observed pressures. The potential element of the atmosphere seems to be entropy obtained by the gain or loss of the energy of the sun's radiation or by the condensation or evaporation of water. Since the evaporation of water itself depends for the most part upon the energy received from radiation, the question arises whether the energy thus employed should be expressed as part of the entropy of the air which carries the water—that is however mainly a question of terminology.

ANCIENT TRADITIONS

Personification of the powers of the air: among the Eskimo (p. 7). The following appeal to the powers of nature comes from the shores of Alaska, through the Eskimo of Point Barrow:

"'Woman, Great Woman down there! Turn it aside, turn it aside from us, that evil!
"'Come, come spirit of the deep, one of thine earth-dwellers calls upon thee;
"'Prays thee to bite our enemies to death,
"'Come, Spirit of the Deep!'

"As soon as the two had sung it once, all present joined in a wailing, imploring chorus; they had no idea of what they were praying to, but they felt the power of the ancient words their fathers had sung. They had no food to give their children on the morrow; and they prayed the powers to make a truce for their hunting, to send them food for their children.

"And so great was the suggestive power of what had passed, in this wild place too near to the elemental forces, that we could almost see it all; the air alive with hurrying spirit forms, the race of the storm across the sky, hosts of the dead whirled past in the whirling snow; wild visions attended by that same rushing of mighty wings of which the wizards had spoken.

"So ended this battle with the storm; a contest between the spirit of man and the forces of nature. And those present could go home and sleep in peace, confident that the morrow would be fine.

"And in point of fact, so it proved."

(*Across Arctic America*, Narrative of the Fifth Thule Expedition, by Knud Rasmussen, G. P. Putnam's Sons, 1927, p. 277.)

In a subsequent section of the volume Rasmussen gives an angakoq's ideas of the power that they call Sila "which is not to be explained in simple words. A great spirit, supporting the world and the weather and all life on earth, a spirit so mighty that his utterance to mankind is not through common words, but by storm and snow and rain and the fury of the sea; all the forces of nature that men fear. But he has also another way of utterance, by sunlight, and calm of the sea, and little children innocently at play, themselves understanding nothing."

A line-squall in classical literature, 5th century B.C. (p. 72). Dr Crichton Mitchell contributes the following from Edinburgh:

"In the same year, the consul Valerius advanced with an army against the Aequi, but failing to draw the enemy into an engagement he commenced an attack on their camp. A terrible storm, sent down from heaven, of thunder and hail prevented him from continuing the attack. The surprise was heightened when after the retreat had been sounded, calm and bright weather returned. He felt that it would be an act of impiety to attack a second time a camp defended by some divine power. His warlike energies were turned to the devastation of the country." (Canon Roberts' translation of Livy's *History*, Book II, Ch. 62.)

Rig-Vedic meteorology (p. 11). Upon the publication of this first volume I received from Rao Saheb M. V. Unakar of the Meteorological Department, Poona, a typewritten copy of an interesting and substantial work on the meteorology of the modern Punjab and North-West frontier of India derived from the Sanskrit psalms, or hymns of praise, collected in the Rig-Veda which dates from 1000 B.C. "The importance of the regular recurrence of the seasons is indicated by their being personified and honoured as deities."

I am not aware that the work has been published. The subject is referred to in a communication to the *Meteorological Magazine*, vol. LXV, 1930, by Rao Saheb M. V. Unakar and Dr S. N. Sen.

The Book of Enoch. Mr C. J. P. Cave notes that in the Book of Enoch, a compilation of religious writings nominally attributed to the patriarch, which dates from the end of the second century B.C., there is a scheme of winds which can be resolved into a 12-part wind-rose. It associates *fruitfulness* or *prosperity* with rain from all four cardinal points, *snow* with the NW quadrant, *desolation* and *locusts* with the NE and SW quadrants, *cold* and *drought* from north of east. *Heat* and *drought* from south of east, *desolation, destruction* and *burning* with *drought* from south of west.

Palestine is suggested as the country of origin. The book was well known in the early Christian centuries, was lost about A.D. 800 and recovered in three Ethiopic manuscripts brought home from Abyssinia in 1773 by James Bruce, the famous traveller.

Survival of the prehistoric May Year (p. 52). "On most of the commons of England the Lord of the Manor has the right of pasture and the right of the hay from Candlemas (Feb. 2) to Lammas (Aug. 1) and the tenants have the right of common from August to Candlemas." (James Gow, *Selected Addresses*, Macmillan & Co. Ltd., 1924, p. 81.)

BIBLIOGRAPHIES

Cloud-atlases (pp. 212–13). Collections of cloud-forms have been received from: Commission Internationale pour l'Étude des Nuages (Atlas international provisoire), Paris, 1929; Barcelona 'Atlas internacional dels núvols i dels estats del cel,' Extracte de l'obra completa, 1930; A. McAdie, 'Clouds' and 'Cloud-Formations as hazards in aviation'; T. Okada, photographs by Watanabe, Isimaru and Yosinari; S. Fujiwhara, a cloud-atlas for Japan.

Palaeometeorology. In the period that has elapsed since the first issue of this volume a good deal of attention has been given by meteorologists in association with geologists and palaeontologists to what is comprised within the title "palaeometeorology." One of the arresting contributions is the late Dr Alfred Wegener's theory of the progressive drifting apart of the land-masses that now form separate continents.

The subject hardly comes within the range of modern meteorology, but for those who are interested we may cite: A. Wegener, *Die Entstehung der Kontinente und Ozeane*, Dritte Auflage, Braunschweig, 1922; 'Discussion on geological climates,' *Proc. Roy. Soc. London* CVI (B), 1930, pp. 299–317; G. C. Simpson, 'The climate during the Pleistocene period,' *Proc. Roy. Soc. Edin.* vol. L, pt iii, 1930, p. 262; R. DeCourcy Ward, 'The literature of climatology,' *Annals Assoc. Amer. Geog.* March 1931, p. 34.

NOTES

THIS volume was written for the purpose of indicating to the general reader, and explaining where it seemed necessary, and possible, the ideas and methods which the author of a paper on meteorology might expect his hearers to have in mind when he was addressing a scientific society. It involves a mental exercise of no small difficulty. On looking through the printed sheets I find that I have still something to explain. I have assumed without any attempt at explanation a knowledge of certain arithmetical, algebraical and trigonometrical processes, extending as far as the formulae for sine-curves and cosine-curves, with words of such recondite import as period, amplitude and phase. I have also employed the symbol Σ for the sum of a series of numerical values, and even introduced the signs of differentiation and integration, only once in the case of each, it is true, but there they are—and I confess that I am unable, for many sufficient reasons, to find an easy substitute for the use and practice of the elementary text-books of mathematics.

In the domain of experimental dynamics and physics, moreover, there are many ideas which I have assumed to be within the reader's common knowledge. Some of these which are of peculiar importance in meteorological practice will be recapitulated at the end of the volume which gives an account of the normal general circulation and its disturbances (Vol. II). One that occurs in this volume, the idea connoted by the words "ion" and "ionisation," I ought to notice here, because it has formed the connecting link between electricity and ordinary matter.

The name ion was selected by Faraday for the component parts into which a molecule of a chemical compound is resolved in a solution by the electrolytic action of an electric current. Of the component ions one is always positively charged and the other negatively. The positively charged ions consist of atoms of oxygen, chlorine, or some other corresponding element or radicle, and the electropositive ions consist of atoms of hydrogen, potassium, or some other metallic element or radicle. It is by the motion of the ions with their electric charges that the electric current makes its way through a solution, the positive ions going with the current, the negative against it. A gas may conduct electricity in the same way when it contains free ions; and these may be produced by radioactive agents, ultra-violet light, very hot bodies, the combustion of flame, and in other ways. The conduction of electricity through the atmosphere is now, therefore, attributed to the free ions which exist in it, and its capacity for conducting electricity is attributed to its ionisation. The ions in the atmosphere may be atoms of hydrogen or oxygen or they may be aggregates of these atoms with some other material.

Among the meteorological terms which are used without explanation I find the word "föhn," with which is associated certain cloud-forms which are called "lenticular." Föhn is the name given, on the northern side of the Alpine range, to the winds which have come from the South and crossed the ridge. As the air approaches the ridge it is cooled by the ascent, and water is condensed from it. It passes over the ridge and, by some process of eddy-convexion, replaces the much colder wind of the valleys on the Northern side by air which is warm, dry and disagreeable. Similar action is shown in the chinook winds of North America which come from the West over the Rocky Mountains and reach the plains on the East side as relatively warm winds which are very dry. Corresponding phenomena are to be observed in all parts of the world. The literature of föhn winds is very voluminous. They ought not to be confused with katabatic winds that come down the slopes of mountain-ranges, because the air on the slope is cooled by contact with the surface chilled by radiation; the physical processes are quite different.

One other assumption which is to be noticed on reading through the text is a supposed familiarity with a variety of languages. We need perhaps offer no apology for giving a biographical notice of a French writer in the French language, and it is in many ways appropriate that opinions should be expressed in the language in which they were originally given. But statements of scientific facts and laws bear translation easily; we offer for the reader's acceptance the translation of a few passages.

1. *Horace's experience of thunder and lightning in a clear sky* (p. 98). Translation by Sir R. Fanshawe, 1652:

I that have seldom worshipped heaven,
As to a mad sect too much given,
    My former ways am forced to balk,
    And after the old light to walk.
For cloud-dividing-lightning Jove
Through a clear firmament late drove
    His thundering horses and swift wheels.

2. *H. Leiter's conclusion as to the climate of North Africa* (pp. 89–90).

"The increase of temperature and diminution of rainfall in North Africa within historical times which is often asserted cannot be proved; indications of the opposite rather are discernible."

3. *The influence of Kämtz's handbook* (p. 141).

"There is hardly a single work anywhere on physical geography or climatology that is worth noting which has appeared in the course of the last forty years and has not been based more or less upon Kämtz's work."

4. *The original enunciation of Buys Ballot's law* (p. 303).

"Great barometric differences, within the limits of our country, are followed by stronger winds; and the wind is in general perpendicular, or nearly so, to the direction of the greatest barometric slope in such a way that a decrease of pressure from North to South is followed by an East wind and a decrease from South to North by a West wind."

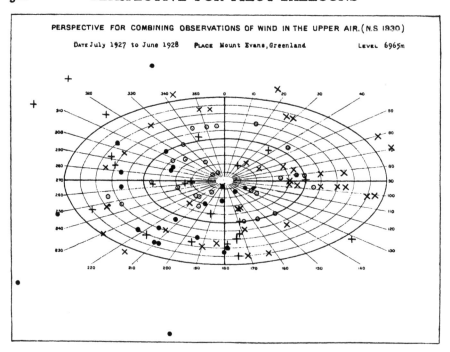

Fig. (for p. 271). Alternative for wind-roses or mean values. A diagram of the results of pilot-balloon observations for the level of 6965 m at Mount Evans, Greenland, tabulated in *Reports of the Greenland Expeditions of the University of Michigan*, Part 1, 1931, W. H. Hobbs, Director, S. P. Fergusson, Editor.

     ● June, July, August.   ⊙ November, December, January, February.

     × September, October.   + March, April, May.

Velocities are represented by the azimuth of the mark and its distance from the centre. A point on the outer ellipse represents a velocity of 20 metres per second.

INDEX